Future of Business and Finance

The Future of Business and Finance book series features professional works aimed at defining, analyzing, and charting the future trends in these fields. The focus is mainly on strategic directions, technological advances, challenges and solutions which may affect the way we do business tomorrow, including the future of sustainability and governance practices. Mainly written by practitioners, consultants and academic thinkers, the books are intended to spark and inform further discussions and developments.

Peter Wollmann • Doris Pemler
Mersida Ndrevataj
Editors

# The Sustainable Organization – Volume 2

## How Organizations Address the 17 UN SDGs Using the 3-P-Model

Springer

*Editors*
Peter Wollmann
Bonn, Germany

Mersida Ndrevataj
IUAV University of Venice
Venice, Italy

Doris Pemler
Ciudad Condal GmbH
Zürich, Switzerland

ISSN 2662-2467 ISSN 2662-2475 (electronic)
Future of Business and Finance
ISBN 978-3-032-23801-6 ISBN 978-3-032-23802-3 (eBook)
https://doi.org/10.1007/978-3-032-23802-3



This Springer imprint is published by the registered company Springer Nature Switzerland AG
The registered company address is: Gewerbestrasse 11, 6330 Cham, Switzerland

"The Truth is Rarely Pure and Never Simple"
—Oscar Wilde, *The Importance of Being Earnest*

# Contents

# About the Editors

**Peter Wollmann** has been a responsible manager, initiator, mentor, and facilitator in large, predominantly global transformations and strategic developments for almost 40 years in large organizations and since 2017 as independent consultant. He is also a member of an advisory board of an R&D oriented company. Peter holds a degree in mathematics and physics from the University of Bonn. He is the author and publisher of numerous books and articles on strategy, leadership, organization design and project and project portfolio management.

**Doris Pemler** is an entrepreneur, artist, and researcher based in Zurich. She received her M.A. in Egyptology, Sudan Archaeology and Archaeology of the Ancient Near East from Humboldt University zu Berlin and FU Berlin. Her research focuses on iconography, gender and intercultural communication in Ancient Egypt and Sudan. As a photographer, she contributed to several exhibitions in Germany and Switzerland. As CEO of Ciudad Condal GmbH, she leads an aspiring company focused on import and distribution of high-end sustainable products in Switzerland.

**Mersida Ndrevataj** is an urban scholar with a background in architecture, urban planning, and environmental psychology. She holds a PhD in Urban Planning and Public Policy from IUAV University of Venice, where her research focused on the implementation of public policies in cultural heritage management. She is currently a lecturer at IUAV University of Venice and collaborates with several COST Action research networks focused on urban studies. Her work brings together academic research and field-based inquiry by adopting audiovisual and sensory approaches to investigate urban spaces, everyday socio-spatial practices, and lived experiences.

# Part I

# Introduction

# The Framing of the Book: Why a Second Volume Is Needed and How It Reflects Current Developments of Sustainability and UN SDG Coverage Activities

1

Peter Wollmann, Doris Pemler, and Mersida Ndrevataj

**Abstract**

In this chapter, the general direction and framing of the book will be explained, its necessity after the publication of the first volume, its purpose and its extended focus stressed, and some first considerations on the current state of the broad-based Movement represented by the nearly infinite Activity Landscape to reach Sustainability and UN SDG Coverage placed. As a completion, facts on perception traps for judgment are introduced and the book's structure is explained. For this, the general linkage between the first and the second volume is described, which includes the continuation of using the same terminology, the same underlying thinking styles represented by the used joint "language" for the explorations and evaluations (the so-called Three-Pillar Model, see https://www.peterwollmann.com/three-pillars), the same categorization of the activities and actors of the Movement, the activity landscape, as in the previous book that makes a brief recap in this chapter necessary. In the context of the recap, the fundamental connection between the term "Sustainability" from a UN point of view and the UN SDGs, the Sustainable Development Goals (https://sdgs.un.org/goals) is touched again. For the first rough evaluation of the current state of the Movement's nearly infinite landscape of activities and projects to reach

P. Wollmann (✉)
Bonn, Germany
e-mail: pw@peterwollmann.com

D. Pemler
Ciudad Condal GmbH, Zürich, Switzerland
e-mail: doris@ciudad-condal.ch

M. Ndrevataj
IUAV University of Venice, Venice, Italy
e-mail: mndrevataj@iuav.it


P. Wollmann et al. (eds.), *The Sustainable Organization – Volume 2*, Future of Business and Finance, https://doi.org/10.1007/978-3-032-23802-3_1

sustainability for organizations, regions, countries, or for the world as a whole and to cover the UN SDGs, some significant use cases are discussed and a former high-ranking UN official interviewed to give at least some clues. In the further process, a particular focus is the critical discussion about the discrepancy between the public perception of the Sustainability Movement and its actual state—and some neurological reasons for the occurring perception biases (already briefly touched in the first book in Chapter 1). As a result, the cautious optimism regarding the Sustainability Movement's (or its Activity Landscape's) unbroken momentum is substantiated and the stage prepared for the numerous detailed use cases described in the following chapters. The book again will not at all strive for a detailed systematic and comprehensive overview but try to give a taste of the dimension of the Movement's Activity Landscape to reach Sustainability and UN SDG Coverage and the reached status. Additionally, an outlook on the upcoming Volume 3 as well as on further related book projects in the future is provided.

## 1.1 Introduction: Reasons and Background for a Second Volume

This book is the second volume of the book *The Sustainable Organization—How Organizations Address the 17 UN SDGs Using the 3-P-Model* published by SpringerNature in June 2025.[1] This second volume will be based on the same methodical setting as the first book, which includes the clarification and origin of key terms and concepts like Sustainability, UN SDGs, UN SDG Coverage, and the Three-Pillar Model[2] (details see below) as the leading thought model. As the previous book, the second volume understands the nearly infinite Landscape of Sustainability and UN SDG Coverage activities with all its acting players and affected stakeholders as a sort of an overwhelming Whole World Transformation Project Portfolio with millions of single activities—we therefore name it the Sustainability and UN SDG Coverage Movement.

A comprehensive and detailed overview of this complex—partly top-down driven,[3] partly bottom-up driven[4]—movement is impossible in the scope of one or two or even more books; therefore, we continue following our concept and goal from the first book to provide for the reader solid meta-level ideas (of parts) of the Movement's activity landscape, its complexity, and its former and current

[1] Wollmann, P., Pemler, D., Ndrevataj, M.: The Sustainable Organization—How Organizations Address the 17 UN SDGs Using the 3-P-Model; Cham 2025, SpringerNature (https://link.springer.com/book/10.1007/978-3-031-89549-4)

[2] See: https://www.peterwollmann.com/three-pillars)

[3] For example, as an impact of the UN SDG planning and reporting system.

[4] For example, also many grassroot activities (https://en.wikipedia.org/wiki/Grassroots), often according to the grassroot organizing model of Marshall Ganz (https://en.wikipedia.org/wiki/Marshall_Ganz)

momentum by describing some few fundamental high-level considerations and some interesting and insightful activities of key players, including their related individual case studies—without claiming in any way to be exhaustive. We are fully aware that the Movement's situation is contradictory and confusing,[5] not only in terms of public perception[6] (and in media) but also in terms of content. We also know that supposedly good and bad news occur in a mix that is difficult to assess, and that it is demanding to form a uniform, consistent opinion about where the Movement actually stands and what its chances of success are—especially since people tend to overestimate bad news in their perception (more on this later in part 1.2.4). The aim is rather to create a basic, flexible understanding on a meta-level, more from top-down, and to underpin this with some case studies (see 1.2.2) and therefore to generate curiosity and openness for further research and evidence.

In the first book, we documented in the first chapter some key messages for our readers, based on our evaluation, knowledge, and insights as of 2024[7]—and it might be interesting in the second volume to challenge these messages and to check if they are still valid:

- There is a broad movement of activities and actors in place.
- There is a lot more progress already achieved than perceived in the public perception.
- In general, there is a perception bias in the judgment of large transformational movements as people tend to expect linear developments (also as a consequence of neurological imprints).
- The Movement is to a very large degree well rooted in the middle of society and irreversible.
- The Movement will force organizations' (and private persons') journeys through unknown territories, which requires the mindset of a so-called Traveling Organization (more details see below).
- The coordination and cooperation across activities, projects, and transformations is a Herculean task and can only be handled in an explorative and agile way.

Especially the second, third, and fourth statements will be regarded and checked on his validity in the book, starting with this chapter regarding (political) developments in 2025 (first comments, see Sects. 1.2.2–1.2.5). We assume that the other statements are less controversial. We will run an overall resumé in the last chapter (Conclusions and Takeaways).

[5] As also the situation of the world as a total: in an obviously very confusing state, perhaps at a turning point with many contradictory and conflicting developments occurring and where a "new clear pattern" is not (yet) foreseeable (a typical Terra Incognita Situation).

[6] It has nevertheless to be stated that there is a systematical underestimation of movements in place, see, for example, the discussion of the so-called 3.5% rule for the success of Movements that non-violently protest against an authoritarian government (see: https://en.wikipedia.org/wiki/3.5%25_rule), which shows positive and negative use cases. As there are examples supporting the hypothesis, the statement that the impact of the Movement might be underestimated can be regarded as proven.

[7] Cited from this chapter, https://link.springer.com/book/10.1007/978-3-031-89549-4

Another reason for a second volume is that we experienced after the publication of the first volume a significant increase of our networks by new players in this field, offering many so-far unknown use cases from different fields and parts of the world, broadening our perspective and focus tremendously.

Overall, we might sum up the following key reasons for a second book:

- Overwhelming mass of new material, new exciting use cases and considerations arising after the publication of the first book, complementing the key messages of it.
- Significant number of new authors joining the author community and contributing new insightful use cases and insights.
- Urgent need to check the validity of our statements, outcomes, and insights from the first book.
- New insights on the current intensive controversial discussion on the state and future development of the momentum of the Sustainability and UN SDG Coverage Activity Landscape—is there a serious setback or is the development irreversible and unstoppable long term?
- Some new additional fundamental considerations on the difficulties to evaluate and judge historical (political, societal, economic) disruptions and connected developments/transformations and their underlying neurological dispositions.

Thus, we will in our second book try to find a convincing (but always preliminary) answer to the question about the situation of the Sustainability and UN SDG Coverage Activity Landscape from above: is there a serious setback or is the development unstoppable long term? To anticipate: despite all the adversity, the newly gained pieces of the puzzle unexpectedly show—contrary to public perception in the media—a clear and unbroken momentum, a growing gap between what is officially happening at the political level and what is actually taking place in the private and public sectors.

This is plausible also on second view as it seems impossible for such a complex, diverse system to be controlled by just a few individual players—even if simplified linear thinking suggests otherwise (more on thinking and perception traps see below under 1.2.4)—a lot of control also comes from the bottom up, from the sheer multitude of activities. We will elaborate on this in more detail later on, but it might already at this point be useful to refer to a quote from Hannah Arendt: "No one actually possesses power; it arises between people when they act together, and it disappears as soon as they disperse again."[8] This means that the necessary momentum can also be generated bottom-up—and not just top-down in a hierarchical manner—at least if the approach is based on an ethic of responsibility rather than an ethic of conviction (with its danger of slipping into ideology).[9] We will come back

[8] Hannah Arendt: Vita activa oder Vom tätigen Leben, München/Zürich, 2019, Piper Verlag, S. 252 (translated).

[9] Hannah Arendt generally advocated a radically different concept of power, one that is understood not vertically but horizontally and cooperatively. In his book *Der Sinn von Politik ist Freiheit—Warum Hannah Arendt uns Zuversicht in schwieriger Zeit gibt* (*The Meaning of Politics is*

to these statements and considerations in the last chapter on "Conclusions and Takeaways" when reflecting the outcome of our book in detail.

## 1.2 The Sustainability and UN SDG Coverage Movement's Activity Landscape and Its Momentum

### 1.2.1 Start with an Anecdote

It is well known that there is a widespread impression that we are living in unprecedented historic times—meaning times of great uncertainty and upheaval for us all. Historic upheavals and the developments they trigger with often long-term effects are frequently difficult to assess, particularly with regard to their further course and consequences. It lies in human nature that most people find this very unsettling (we will discuss the underlying neurological dispositions in detail in the following sections). But first, we would like to introduce an anecdote as food for thought to change the thinking perspective.

The anecdote is about a conversation between Henry Kissinger and Zhou Enlai in 1972 in the context of the diplomatic rapprochement between the People's Republic of China and the USA. According to the anecdote, Henry Kissinger asked Zhou Enlai about his opinion on the impact of the French Revolution. "Too early to say," Zhou Enlai replied. The first spontaneous interpretation of this statement was that "given that the French Revolution of 1789 had occurred nearly 200 years before, Zhou Enlai was expressing the long view of history in a very witty and Oscar Wildean way. News of this quote flew quickly around the chattering classes in the west, and it was soon used as evidence that the Chinese (especially Chinese intellectuals and leaders) took the long view of things, that they were a patient civilization, and that, when they thought about the future, it was hundreds of years distant".[10]

Even though it seems that there was a misunderstanding—and Zhou Enlai did not refer to 1789 French Revolution but to the 1968 student uprising in Paris leading to a sort of country shut down for a few weeks in this summer[11]—the anecdote animates us to think more in terms of long-term historical periods: It is also not yet possible to make a definitive assessment of the final social, cultural, and political results of the 1968 movement because it is highly nonlinear and occurs in dialectically driven waves, divided into an innumerable number of individual sub-currents, some of which run counter to each other, without always following clear patterns. If you like, you can compare the 1968 movement with today's Sustainability and UN

---

*Freedom—Why Hannah Arendt Gives us Confidence in Difficult Times*), Winfried Kretschmann, the first Green Party prime minister of a German Federal State, explains in Chap. 3 why this particular model of thinking is of paramount importance in climate policy (2025, Patmosverlag, Ostfildern).

[10] https://professorbuzzkill.com/2017/08/07/qnq-26-zhou-enlai/

[11] Ibid and https://www.oxfordreference.com/display/10.1093/acref/9780191826719.001.0001/q-oro-ed4-00018657

SDG Coverage Movement; it is of comparable dimension and complexity. Looking at historical examples and the current Movement, it means that it could take decades to understand the real momentum of those in between all contractionary perceptions—and therefore the Sustainability and UN SDG Coverage Movement needs ongoing observation, analysis, and reflection—and the collection of fundamental considerations and practical use cases.

### 1.2.2 Some Unexpected or Often Missed Perceptions of Developments

We would like to present some cases of developments that we did not expect and that look completely controversial to the public perception of a backlash for Sustainability and UN SDG Coverage. We begin our exploration with a few initial unsystematic and random observations and then examine them systematically in order to arrive at a plausible overall picture.

#### 1.2.2.1 Use Cases 1 and 2

The first case is based on an article from the leading German magazine *Der Spiegel*. Christian Stöcker[12] wrote April 6, 2025 under the headline "The overlooked revolution—A column by Christian Stöcker"[13] about the new "Global Energy Review 2025" from IEA (International Energy Agency) on worldwide electricity generation and demand. The demand increased in 2024 by 4.3% for different reasons (one is the demand to run AI), and the generation development looks like this:

> In 2024, 80% of the growth in global electricity generation was provided by renewable sources and nuclear power. Together, they contributed 40% of total generation for the first time, with renewables alone supplying 32%. New renewable installations hit record levels for the 22nd consecutive year, with around 700 GW of total renewable capacity added in 2024, nearly 80% of which was solar PV. Generation from solar PV and wind increased by a record 670 TWh, while generation from natural gas rose by 170 TWh and coal by 90 TWh. In the European Union, the share of generation provided by solar PV and wind surpassed the combined share of coal and gas for the first time. In the United States, solar PV and wind's share rose to 16%, overtaking that of coal. In China, solar PV and wind reached nearly 20% of total generation.[14]

This is, according to *Der Spiegel*, aligned with other sources. According to the International Renewable Energy Agency (IRENA), 585 GW of renewable peak capacity was added. Important: Capacity is not the same as the amount of electricity generated. But even in terms of the amount generated, electricity from solar, wind, and water accounts for "almost three-quarters of total growth," according to the

[12] Professor at HAW University of Applied Sciences Hamburg: Hochschule für Angewandte Wissenschaften Hamburg

[13] https://www.spiegel.de/wissenschaft/mensch/energiemarkt-strom-aus-sonne-und-wind-die-uebersehene-revolution-a-483e4a52-b3ee-492d-9f76-41e573ebcb42

[14] https://www.iea.org/reports/global-energy-review-2025/key-findings

IEA. According to the World Nuclear Industry Status Report by consultant and antinuclear activist Mycle Schneider, 8.2 GW gigawatts of nuclear power capacity were added to the grid during the same period.

Overall, the generation capacity of renewable sources was extended significantly faster than coal and gas and also nuclear sources, probably as a consequence of the unbeatable low price. This finding contradicts the current public discussion that transports another perception. But in the end, energy generation follows economic principles like the sustainable (lower) price.

On October 12, 2025, the same author summarized under the headline "These facts will change your worldview,"[15] an overview of enormous changes that are currently taking place around the world. We have selected 5 of his 10 fact statements:[16]

1. Around the planet, nothing is being expanded as massively in terms of electricity as renewable energies, that is, electricity from the sun and wind, as well as battery storage. Investment in so-called ElectroTech is about twice as high as in fossil fuels. 92.5% of newly built electricity generation capacity worldwide in 2024 was renewable. In 2023, the figure was 86%, and in 2022, it was already 83%.
2. In the first half of 2025, China added twice as much renewable energy capacity as the rest of the world combined.
3. Since at least 2024, China has been earning tens of billions of dollars more from exports of ElectroTech than the USA, the world's largest fossil fuel exporter, which earns from exports of oil and gas. The gap is growing. Between 70% (wind) and 90% (solar, batteries) of global production capacity for ElectroTech is located in China.
4. In China and India, $CO_2$ emissions from electricity generation are now falling, which is sensational news. This is due to renewable energies.
5. The market for combustion engines has been shrinking since 2017. Just as only renewables and storage are growing in electricity generation, only electric motors are growing in mobility and at an accelerating rate. This is called disruption: a superior, easier-to-manufacture, and ultimately cheaper product is displacing the old, expensive competition. This is how the market economy works.

The interesting outcome of this—reduced and focused—overview is that a systematic momentum has clearly been triggered, which is already having a demonstrable impact on national economies and was triggered by price-driven disruption, among other factors. The mentioned facts in the five statements above are publicly available, so it is worth asking why public perception, which often focuses only on narrowly defined areas, is so different. It seems as if there are various barriers to forming an overall picture (and basing this on pure economic logic).

The impression gained from reading the two analytical articles in the magazine can be easily confirmed many times over when accessing the official information

[15] https://www.spiegel.de/wissenschaft/mensch/energiewende-diese-fakten-werden-ihr-weltbild-veraendern-a-37b25092-46d0-403e-8f49-d0afea2d9758

[16] Cited from ibid., translated into English by DeepL.

platforms of the IEA and IRENA, with their collection of data and facts verified multiple times. Based on its official database, IEA claims, for example, on its website: "As temperatures around the world continue to break records, the case for swiftly transforming the global energy system has never been stronger. The IEA's Net Zero Roadmap lays out a pathway for the world's energy sector to reach net zero emissions by mid-century, limiting global warming to 1.5 °C, as called for by the Paris Agreement, and avoiding the worst effects of climate change. The rapid growth of some clean energy technologies in recent years—including electric cars, solar PV and batteries—has kept the door to 1.5 °C open."[17]

In this context, IEA underlines the crucial role of innovation and delivers diverse promising examples, for example, on the fast development of innovation helping secure future battery markets and mineral supplies[18] or on progress in the field of carbon dioxide removal (CDR),[19] only to mention two analytical IEA reports out of numerous examples. The suspicion arises that the diversity of the innovation landscape, its broad progress, and the successes already achieved may not yet have been reflected in public perception, which tends to be based on selective reporting (to prevent reading times over more than 5–10 min).

There might be another perception bias as the key momentum moved from the Western part to the east: "The global energy innovation landscape is changing. The United States, Japan and Europe led energy technology innovation over the past century. But today's innovation landscape has shifted. China became the largest single country for energy patenting in 2021, overtaking Japan and the United States. More than 95% of China's energy patenting in 2022, the latest year for which data are available, was related to low-emissions technologies."[20]

Or to summarize for a first preliminary conclusion: there is a lot more going to reach the "IEA's Net Zero Roadmap" in terms of number of projects and innovation with a lot of potential and additionally the innovation focus has moved to China—and we rarely find the respective results in the available main media in the West. A recommendation might be to access the source data and analysis by visiting the websites of, for example, IEA, IRENA, etc., on a regular basis to form your own opinion based on the original sources. In order not to exceed the scope of the introductory chapter, we will leave it at these notes and indications connected with cases 1 and 2.

#### 1.2.2.2 Use Case 3

A third case is also based on an article on October 4, 2025, in the leading German magazine *Der Spiegel*, this time under the headline "Ecological

---

[17] Cited from https://www.iea.org/topics/climate-change

[18] https://www.iea.org/commentaries/how-can-innovation-help-secure-future-battery-markets-and-mineral-supplies

[19] IEA Report on Cross-Cutting Technologies & Infrastructure, see: https://www.iea.org/reports/cross-cutting-technologies-infrastructure

[20] Cited from IEA: https://www.iea.org/topics/energy-innovation

Restructuring—Trump's fight against windmills will fail in the long run. A column by Ullrich Fichtner."[21] We quote parts from this:

> Research into sustainability is currently making unprecedented progress. It may seem insignificant compared to the spectacle put on by today's populists, but the crucial question is who will have the greater staying power. Next Tuesday,[22] Pollutec,[23] a trade fair for environmental technology, energy and waste issues, water and recycling management, will begin in Lyon. In these times ... this sounds almost like a joke. Pollutec? ... Can you imagine anything more boring? Well, I think it couldn't be more exciting...the researchers, start-ups, inventors, and entrepreneurs gathered in Lyon are moving steadfastly toward a better world. Better in the sense of: cleaner, healthier, more sustainable. Better for rivers, forests, and oceans, for the soil, mountains, air, and climate, and therefore also more livable for humanity.
>
> I understand that at first glance—and compared to the Daily Show of world news—it may seem small when 1500 exhibitors set up their stands in Lyon, sit down in conferences and explain their projects. But Pollutec, this one event, is itself only a placeholder for the millions of players involved in the ecological and economic transformation currently taking place in the world's economies. ...
>
> Assuming that there are around 30,000 notable trade fairs worldwide, which is a plausible estimate. If we further assume that 10 to 15% of these are dedicated to CleanTec[24] topics, then we arrive at around 3000 to 4500 exhibitions per year that focus on environmental and sustainability issues.
>
> If each of these 3000 to 4500 trade fairs had only 1000 exhibitors and visitors, that would already amount to three to four and a half million people. It is likely that the actual number is much higher. So, we are talking about a whole host of experts in ecological transformation and a mass of people who are actively interested in it.
>
> I'm harping on about this a little because it reflects a pattern that has recently emerged concerning our distorted perception of the reality that surrounds us. As we know, modern media behavior is dominated by quick headlines and bad news. Spectacular scandals are appealing, while protracted reform successes are not. The former immediately seem immensely important, while the latter seem insignificant and "boring." However, they only become apparent in retrospect, sometimes with great force.

Ullrich Fichtner's well-founded hypothesis is that Pollutec in Lyon will turn out to be much more important than all the futile attempts by politicians "to stop the march of progress" that they are in contrary compared to researchers and scholars, the "false giants of the present," terribly effective for a while, but in the long run only a minor disruption to the course of the world.[25] In the following, Ullrich Fichtner refers to so-far unknown but important companies or organizations like "Plastic Vortex,"[26] which might actually be able to capture 100% of all plastic waste from

---

[21] https://www.spiegel.de/ausland/klimapolitik-donald-trumps-kampf-gegen-windmuehlen-wird-langfristig-scheitern-a-db1631f7-bb55-4a4b-9d79-2e48ef6b2766 (translated into English by DeepL)

[22] October 7, 2025.

[23] https://www.pollutec.com/en-gb.html#/

[24] Abbreviation for "Clean technologies"

[25] quoted in essence from his article ibid.

[26] https://www.plasticvortex.co

rivers and streams before it reaches the oceans; or AC Biode,[27] with its "Plastalyst," which found a CO2-free, energy-saving, endlessly scalable method for chemically breaking down plastics that are difficult to recycle; or Cactile,[28] a company whose "Cereus" process turns roofs, facades, and garden walls into rainwater reservoirs and water tanks, without sealing any additional surfaces and powered solely by natural gravity, only to show some exciting companies of Ullrich Fichtner's list. We highly recommend checking the documented links and visit the respective websites to collect more detailed pieces of information and to form your own opinion based on the original sources.

The special takeaways from this case are, on the one hand, the fast development of innovation and game-changing new products and procedures (btw: this time not connected with AI) and, on the other hand, his critics on modern media creating wrong perceptions. It shows again that the world is larger and more diverse than expected, and that one needs the whole picture to solidly judge the situation.

#### 1.2.2.3 Use Case 4

At almost the same time when the US President Trump held his speech at the UN headquarters—where he called climate change a "hoax"—a large number of green entrepreneurs met nearby in Manhattan for the Climate Week NYC as *Der Spiegel* reported on September 26, 2025, under the headline "The Mutineers of Manhattan,"[29] stressing that "according to the organizer Climate Group, over a thousand such events have been publicly announced for this week. Behind the facades, countless confidential talks are likely to take place in which deals are initiated. Trump or no Trump: hardly anyone here in the center of US high finance wants to forego green business. Because it is running at full speed. Almost €1 trillion has been invested worldwide in climate-friendly technologies, EU Commission President Ursula von der Leyen announced in New York—more than twice as much as in the development of oil, natural gas, and coal. This is mainly for economic reasons: '90% of new renewable energy projects are cheaper than fossil fuels.'"

The special takeaway is that the Sustainability and UN SDG Coverage Movement has to a large part not mostly ethical but economic reasons that makes it irreversible and unstoppable—contrary to public perception. It might be a problem that according to public perception the Sustainability Movement and its Activity Landscape have mostly ethical roots. Of course, there are ethical reasons to preserve the world for future generations and keep it livable. But the economic reasons—often based on the laws of natural sciences, especially physics—are logically an at least equally important driver for the Sustainability Movement.

[27] https://acbiode.com

[28] https://www.cactile.fr

[29] https://www.spiegel.de/wissenschaft/mensch/investoren-gegen-donald-trump-die-aufstaendischen-von-manhattan-a-7472cc21-2641-4efb-ad86-32a3188dbd52

#### 1.2.2.4 Summary of 1.2.2

The four cases—which are a small selection from similar cases—show at least that there is some evidence that the media-driven public perception might be wrong in many cases. We selected these cases to encourage our readers to always question your own perceptions and consider other perspectives as well.

### 1.2.3 Insights from a Former UN Official

We decided to additionally run an interview with Ovais Sarmad, who served the United Nations Climate Change secretariat (UNFCCC)—COP23 to COP 27 (2017 to 2023). Prior to UNFCCC, he served as the director of Cabinet & chief of staff to the Director General at the International Organization for Migration (IOM). At IOM, he worked in several management and policy development capacities over a period of 27 years.

Currently, he serves as the vice chair of GHG-Protocol Steering Committee and in a number of roles in diverse national and international organizations. His other areas of focus include engaging with and supporting activities and organizations around the world (a) that promise to innovate and accelerate sustainable development and growth without damaging the environment; (b) contribute to public policy discourse and development; (c) empower youth through quality education at all levels; and (d) advise, mentor, and coach individuals and organizations in multilateralism, diplomacy, and geopolitical issues on climate change and migration and organizational change and development.

Ovais is a role model for active former UN official; after stepping down from the UN, he has developed a worldwide network in all sectors and—as a participant in many international conferences and summits—a remarkable overview of "what's really going on around Sustainability and UN SDG Coverage" and "how key organizational principles, procedures and processes have to be changed to adapt the new developments of the framework preconditions in the last decades." Ovais is very actively and transparently present on diverse platforms like LinkedIn.[30]

[30] See https://www.linkedin.com/in/ovais-sarmad/

We asked Ovais some questions to collect his insights and consideration from the last 2.5 years after leaving UN for the book:

**Question**: Ovais, after leaving UNFCCC you have been very actively involved outside the official UN formal framework on issues relating to sustainability in a much broader sense than just climate change. What are your key insights so far?

**Answer:** I left UNFCCC and the UN system in May 2023 after a professional career spanning over 40 years, 33 years of which was with the UN organizations and gained a new outside-in-perspective which is different compared to my previous inside-out-perspective. The current UN system was established after World War II, that means 80 years ago, based on the situation, demands and insights in the forties of the last century. It has been inevitable that the UN system would become over time "old legacy" in terms of its organizational principles, working styles, bureaucracy, etc. Especially the UN multilateralism concept could not follow the change of the world in the last eight decades. It is thus crucial to achieve a fundamental reform and significant reset for all parts of the UN and its ecosystem and to redefine its multilateralism concept and practice in this context. Here, I would give the perspective in relation to United Nations Climate Change (UNFCCC) organization and its ecosystem where I was for the last 6 years of my career using a heading of: "Beyond United Nations Multilateralism to a Multilateral Approach that is Practical-Pragmatic-Purposeful Inclusive and Democratic to deal with Global Issues."

I especially see today main objectives for Sustainability and UN SDG Coverage as follows:

- **A democratic initiative** in terms of reducing bureaucratic boundaries and rules to get ground-level engagement and demands of society and individuals faster and more easily translated into action, helping sustainability, preserving resources etc. In this context it's time for a "Remake-Reimagine-Reform-Reset" of the UN Climate Change process, its COPs, the COP SBs, the negotiations, the climate weeks and all the ancillaries in order to make it leaner and more focused.
- **A better prioritization** of the vast landscape of potential activities to 3–5 focus points (high level of effectiveness and efficiency instead of the hundreds of complex agenda items). This would reduce complexity and allow more focused momentum. Especially picking up the most innovative and transformative initiatives and working out national–regional–global implementation plans for them—in the spirit of the implementation of The Paris Agreement—would help as those initiatives would have the potential that everyone can agree upon in principle and take them away and translate them into their respective transition plans. This would especially allow building more focus on the power of technological innovation—such a plan can and should make climate action a reality at scale and pace that is necessary rather than a political issue of rich versus poor or developed versus developing economies "fight."
- **A better understanding and coverage of public perception** issues of the UN work (for the public, the UN work is often not very visible and understood). An external perspective of the UN system, its working and achievements, and

especially the current form of UN multilateralism needs to be created, preferred by people from outside the UN system.

There are different groups of academics and practitioners (including former UN officials) in place which try to schedule realization path ideas and proposals for such a transformation.

**Question:** Do you experience a critical setback for the numerous activities beyond the narrower formal and political (UN) sphere or has the movement still a strong momentum? What are the key examples?

**Answer:** I would really like to make a difference between those parts of the Sustainability and UN SDG Coverage Activity Landscape which were/are originally initiated and monitored in the formal UNFCCC and COP processes—referred to as Party negotiations—and those which were independently initiated and monitored elsewhere to contribute, referred to as Non-party stakeholder Climate Action. For the first category a constructive critical look at the performance from outside the system is necessary to identify the demands for refinements and redirections—and there are enough events and opportunities in place to present, discuss and regard the critical perspectives from outside. The second category also contains surprisingly innovative and transformative stand-alone activities and many of them have a very strong momentum which I have been able to verify personally in a large number of cases over the past 2 years. It is of high importance to have a comprehensive and not a selected view of the overall landscape.

For me, a particularly important key question is how to achieve genuine representation (especially in the first category). Around 70,000 participants, in-person and online, were involved in COP30 in Belém, including many government representatives, of course, but also interest groups, NGOs, etc. But how well is civil society and general public represented or involved in the discussion and decision-making process? What platforms for participation are conceivable? Could models such as those used in the context of the change in the Irish constitution (citizen participation via a constitutional convention)[31] also be applied on a global scale to Sustainability and UN SDG Coverage measures? There are many comparable models thinkable. It is in any necessary to offer more co-creation opportunities for the civil societies.

**Question:** What is your perspective on the mutual impact of the development of the political sphere and the development of the movement beyond it mentioned above?

**Answer:** The old dividing lines that used to separate the world—the distinctions between rich and poor, highly developed and less developed countries; first, second and third world, North and South—are, in my opinion, outdated. They are old patterns that are no longer valid (like the old multilateralism concept based on these patterns). During the last Covid pandemic, among other things, we noticed that all parts of the world were affected in their own way, and it is no different with climate

[31] See https://en.wikipedia.org/wiki/Constitutional_Convention_(Ireland)

change. Covid also highlighted how interdependent, inter-connected and inter-related the global society, economics, safety, security is.

We need another view today and should concentrate on the following (instead of dividing):

- Creation and retainment of **credibility, transparency, and integrity**—this is the new "currency" of action to get things done.
- **Co-creation**—the different players must be really and seriously involved, not only UN and its ecosystem, governments, NGOs,etc., but also representatives of the civil societies of every part of the world (see my perspective on the previous question and the Irish example). Co-creation creates co-ownership, stewardship, empowerment—and it does not foster competition but collaboration.
- **Re-imagination of the future**—especially regarding the fast and partly unbelievable technical progress providing new solutions, results, and deliveries in a reasonable amount of time.
- **Focus and prioritization**—concentrate on the most promising new technologies (like reduction of methane, use of biomass management for clean energy production, green hydrogen production).
- **Embrace change**—we learnt in the pandemic what we can change fast, why not transfer the positive experience to the Sustainability and UN SDG Coverage challenge. We need for this an ethics of responsibility (not of conviction) with the focus on purposeful, pragmatic, practical proceeding and credible delivery of results with active involvement of the civil societies. This will guarantee the needed momentum for the Sustainability Movement. Especially vivid civil societies are a success factor—and through co-creation they are strengthened.
- **Accept the multipolar world and learn to manage it**. The key challenges are to understand how the new multipolar world works (especially in terms of interconnectedness and interdependence) in general and in terms of Sustainability, what the impact on modern multilateral concepts is and how progress can be sustainably achieved under these preconditions. In terms of the results of COP30 does this means they are not great and unreservedly satisfactory—but better than commonly expected and they overall strengthen the movement.[32] We will have to calibrate expectation and interpret results more long-term and compared with what is realistically possible.

**Question:** Which will be the key fields that need the most attention in the future?
**Answer:** I already partly mentioned these fields:

- Focus on "Technological Advancement and Artificial Intelligence (AI) Advancement"

[32] The German magazine Spiegel evaluated the results like a football play with the result 2:1, not an overwhelming success but a winner (see: https://www.spiegel.de/wissenschaft/klimakonferenz-cop30-in-belem-das-taugen-die-ergebnisse-die-analyse-a-05965fb4-dc1a-448a-adc8-e802c52811ed). See also for example: https://www.linkedin.com/feed/update/urn:li:activity:7398272641377386496/

- Focus on "Sustainability of Environment and Resources"—never take these for granted
- Focus on "Reimaging an Innovative, Transformative Future"

### 1.2.4 Coping with Perception Biases in General

In general, the human brain prefers clear, simple relationships under stable external conditions that do not change significantly over the long term. This directly results in a preference for easily recognizable patterns, transparent cone-to-one correlation contexts of causes and consequences, linear (in contrast to exponential) developments, and complete clarity at the outset of projects or transformations, rather than clarity that is laboriously achieved over time. Complex situations and (economic, political, or social) systems with multiple causal relationships, dialectical developments, and even the familiar S-curves of the life cycles of products and services are difficult to understand.

Connected with these neurological patterns of thinking, we find, among others, the following thinking and perception traps:

- There is always a pattern we know in what we observe (from the past, which means we have in any case some applicable precedence).
- The development of change is most probably linear (that is the assumption of an illusionary "standard" of predominantly linear change processes).
- One must know and fix everything in detail before one start (project, transformation, journey; it is the so-called "Waterfall-Model"), no start into uncertainty (too risky).
- Today, we are in the most demanding challenging time ever (the human brain likes to smooth out experiences in retrospect [ex-post] and cuts off extreme peaks in favor of a more easily imaginable linearity—but it also likes to continue the development of the past linearly into the future and so refuses to learn from past experiences about large transformations).
- The past was—considered ex-post—a lot easier to manage (problem: one understands ex-post developments always easier than in between).
- The future can be derived linearly from the present, using the existing assumptions and paradigms, namely the assumption of an illusionary "standard" of predominantly linear change processes (the alternative option cannot be imagined).
- If a large movement like the UN SDG Coverage has not reached after 50% of the time 50% of the goals, it has failed (which neglects the S-Curve-alike development, for example, of innovation).
- Momentum is nearly always created top-down through leaders with power, never bottom-up (the hierarchical pattern often even used in our times, the idea of an army-alike character of movements, see controversial statements from Hannah Arendt mentioned above).
- Rate every negative, critical message higher than a positive one—attention for always the worst case happening (the weird delight in failure and ruin).

These perception traps lead to a significant lack of ability to realistically assess and evaluate the real status, realization or failure probability, and obstacles of large long-term transformations and change processes with overarching societal, political, economic, environmental, and organizational impacts—especially when their lifecycle exceeds decades or even generations. Thus, a solid knowledge of these perception traps and an efficient coping practice to identify and avoid or reduce them is necessary. As even fundamental decisions are often made based on not challenged perception, a large degree of discipline and self-refection is demanded. One should never spontaneously believe what sounds plausible without challenging, also using different perspectives.

### 1.2.5 Summary of Sect. 1.2

Perhaps, to keep it brief, we should limit ourselves to two statements:

- One should always assume that what one thinks one recognizes may be quite different, especially in the case one does have the full picture and only concentrated on smaller scopes.
- There is—maybe surprisingly—some evidence that the Sustainability Movement (or Sustainability and UN SDG Coverage Activity Landscape) has an unbroken momentum, only slightly irritated by some political developments but nevertheless over the full period irreversible (we will give more evidence in the more than 30 chapters of the book).

We will refer to these statements in the last chapter "Conclusions" again, after appreciating the outcome and the most important takeaways of all chapters.

## 1.3 Useful Recaps from Volume 1[33]

As already mentioned, there is a strong general linkage between the first and the second volume, which includes the continuation of using the same terminology, the same underlying thinking styles represented by the used joint "language" for the explorations and evaluations, the same categorization of the activities and actors of the Movement's activity landscape as in the previous book that makes a brief recap in this chapter necessary.

### 1.3.1 The Term "Sustainability"

In the previous book, *The Sustainable Organization Volume 1*,[34] we clarified the term Sustainability in the context we use it also here like this:

[33] Partly quotation from Volume 1.

[34] https://link.springer.com/book/10.1007/978-3-031-89549-4

"'Sustainability'—contextualized in terms of its scope (the world as a whole, regions, single organizations, etc.)—is in place as well in literature as in the daily linguistic usage, and often the terms ESG[35] and CSR[36] are subsumed under the umbrella term 'Sustainability'. In some countries—like Germany for example—the term of 'Sustainability' often has a strong ethical and moralistic dimension and triggers a fundamental dispute between representatives of ethics of conviction and ethics of responsibility whereas other countries have a mainly pragmatic approach."[37]

As already stressed in Volume 1, it is unnecessary to go too deeply into the discussion of the precise definitions, because the case studies included as well in Volume 1 as also in this book each present their own definitions tailored to the specific case,[38] so we handle the use of the term in a pragmatic way. UN takes the perspective "Sustainability for the World" and respectively formulates a meta-level target answering the question how we can sustain the world in a human, social, economic, and environmental perspective not only to prevent negative but to generate positive impacts for future generations. Coming from this meta-level target, the UN SDGs (Sustainable Development Goals; see below)[39] are the concrete goals for the world's transformation journey to sustainability recently. The defined SDGs were recently completed by a groundbreaking "Pact for the Future" designed to turbo-charge implementation of the UN SDGs and adopted in the General Assembly in New York on September 22, 2024.[40] On the level of the single organization, Sustainability and its facets have consequently to be concretely defined in detail by each individual player in both the public[41] and the private sectors, which includes addressing Sustainability not only within the organizations themselves but also in terms of their impact on the environment and the world as a whole.

[35] ESG = Environmental, Social and Governance.

[36] CSR = Corporate Social Responsibility.

[37] See this chapter of https://link.springer.com/book/10.1007/978-3-031-89549-4

[38] For detailed definitions see: Helmold, M.; Treu, J.; Fritz, J.; Hummel, F.: "ESG, CSR und SDG als langfristiger Wettbewerbs-vorteil—Nachhaltigkeit durch innovative Konzepte, Methoden und Tools"; Wiesbaden. Springer Fachmedien GmbH, SpringerGabler © 2024.

[39] https://sdgs.un.org/goals and https://www.globalcompact.de/en/our-work/sustainable-development-goals

[40] See: https://www.un.org/en/unis-nairobi/press-releaseunited-nations-adopts-ground-breaking-pact-future-transform-global#:~:text=Photo%2FLoey%20Felipe-,PRESS%20RELEASE%3AUnited%20Nations%20adopts%20ground%2Dbreaking%20Pact%20for%20the,a%20Declaration%20on%20Future%20Generations. "The Pact for the Future, the Global Digital Compact, and the Declaration on Future Generations open the door to new opportunities and untapped possibilities," said the Secretary-General during his remarks at the opening of the Summit of the Future. The areas touched are "Peace and Security," "Sustainable Development, Climate and Financing for Development," "Digital Cooperation," "Youth and Future Generations," "Human Rights and Gender," and the "Summit Process."

[41] The public sector includes the governmental and also the so-called "Civil Society Organizations (CSOs)": Non-State, not-for-profit, voluntary entities formed by people in the social sphere that are separate from the State and the market.

**Fig. 1.1** UN SDG (source: https://www.un.org/sustainabledevelopment/), used with friendly permission of UN. The content of this publication has not been approved by the United Nations and does not reflect the views

### 1.3.2 The UN SDG

As already touched above, the more general term of Sustainability is closely linked to the 17 UN SDGs (see Fig. 1.1), which form so to say the master plan for the development of the world by 2030 to make it a significantly better place in many perspectives. The history of the ambitious underlying transformation program was described in more detail in Chapters 1 and 3 of the previous book[42] in which we also stressed that the landscape of sustainability and UN SDG Coverage activities with all its acting players and affected stakeholders is nearly infinite.

When you consider that each SDG has been broken down into specific targets, which are being pursued in every country on earth by a large number of institutions and organizations with tailored measures involving many stakeholders, you can imagine that the number of acting players and affected stakeholders is nearly infinite. So, the reader might get an idea of the great challenge to generate a half-way solid overview of the status of the landscape of activities and projects to reach sustainability for organizations, regions, countries, or for the world as a whole and to cover the UN SDGs.

[42] Wollmann, P.; Pemler, D.; Ndrevataj, M. (Eds.): The Sustainable Organization—How Organizations Address the 17 UN SDGs Using the 3-P-Model. Cham Springer Nature, © 2025; (https://link.springer.com/book/10.1007/978-3-031-89549-4)

### 1.3.3 The Common Professional Language or Thinking Style: The Three-Pillar Model

For our endeavor—to continue creating a basic, flexible understanding of the landscape of sustainability and UN SDG Coverage activities on a meta-level, to underpin this with some case studies, and, therefore, to generate curiosity and openness for further research and evidence—it is important to introduce a "common professional language"—or perhaps better the applied "analytic thinking style or thinking model." In our case, this "common professional language" is the so-called Three-Pillar (Mindset) Model, based on three key building blocks (pillars). It was developed between 2019 and 2021 for the so-called VUCA[43] world by a global author community of more than 60 professionals from the public and private sectors[44] and published at Springer Nature in up to now five books (eBook and print).[45] The first three books were already translated into German (by AI) and very successfully published in July 2023 and April and June 2024.[46] The model has since then been frequently applied—and its applicability and usefulness proven—in various global use cases (so far > 80) in the private and public sectors. The goal of the Three-Pillar Model is to help understand and navigate the complex world of today's corporate territory in all sectors.

(1) The First Pillar: **Sustainable Purpose**
Sustainable Purpose is the raison d'être of an organization having a guiding objective means bringing orientation and certainty to the people for their joint endeavor and success. This is especially important in times of crisis and transformation. It is also the key source for the narrative of transformation.[47]

---

[43] VUCA = Volatile, Uncertain, Complex, Ambiguous.

[44] The very diverse global author community consists of practitioners, leaders, and academics from all continents and more than 20 countries. They represent knowledge stemming from their roles within more than 20 large global public institutions (including UN Agencies) and global enterprises and more than 30 different local public organizations or national companies.

[45] Wollmann, P.; Pemler, D.; Ndrevataj, M. (Eds.): The Sustainable Organization—How Organizations Address the 17 UN SDGs Using the 3-P-Model. Cham Springer Nature, © 2025

Wollmann, P.; Stricker, L.; Pemler, D. (Eds.): Reading an Unknown Organization—Leaders in New Roles Thriving in Challenging Environments with the 3-P-Model Mindset. Cham Springer Nature, © 2024.

Wollmann, P.: Püringer, R. (Eds.): Transforming Public and Private Sector Organizations—Implementing Sustainable Purpose, Travelling Organization and Connectivity for Resilience. Cham: Springer Nature, © 2022Wollmann, P.; Kühn, F.; Kempf, M.: Püringer, R. (Eds.): Organization and Leadership in Disruptive Times—Design and Implementation of the 3-P-Model. Cham: Springer Nature, © 2021.

Wollmann, P.; Kühn, F.; Kempf, M. (Eds.): Three Pillars of Organization and Leadership in Disruptive Times—Navigating Your Company Successfully through the 21st Century Business World. Cham: Springer Nature, © 2020.

[46] https://link.springer.com/book/10.1007/978-3-031-23988-5, https://link.springer.com/book/10.1007/978-3-031-55279-3, https://link.springer.com/book/10.1007/978-3-031-57163-3

[47] Everybody in the organization, but also its key stakeholders in the entire ecosystem and organization environment, must know what the organization stands for and what value and societal contribution it creates.

(2) The Second Pillar: **Traveling Organization**

Traveling Organization means for an organization accepting to be in a permanent state of flux and transformation while interacting with the external requirements of its journey with rapid learning and adaptivity. Ideally, this is the mindset of loving the unknown and being well able to manage it.[48]

(3) The Third Pillar: **Connectivity**

Connectivity is all about the ability to interconnect on the journey with all needed parties and resources inside and outside the silos, thereby creating high efficacy and consistency and during this process building also shared learning networks.[49]

Exemplifying the narrative of the Three-Pillar Model to get a deep understanding normally starts with the Traveling Organization. In one's imagination, this term is able to evoke the image of a very agile, flexible team embarking on an unforeseeable journey to explore unknown territories with the underlying expectation that the framing fundamental understanding of the unknown territories might change daily as a consequence of new circumstances, hurdles, and encounters.

The success of the journey of a Traveling Organization especially through unknown territories demands a clear lighthouse, a convincing Sustainable Purpose for the journey as orientation and direction. And without Connectivity, survival and success on a journey through unknown territories would not have been possible; the Traveling Organization has to connect with everybody in the team and with all what's important along the chosen path.[50]

### 1.3.4 Categories Within the Movement's Activity Landscape of Sustainability and UN SDG Coverage

As already stressed above and worked out in Volume 1, we regard the landscape of activities, projects, transformations, actors, and affected stakeholders somehow as the world's change and project portfolio on all (environmental, ecological, societal, ethical, organizational, etc.) sustainability and UN SDG Coverage topics ("to make the world a better place"), a list with maybe some million entries. This means that even an overview is only possible on a very advanced meta-level or based on sufficiently exemplary and paradigmatic use cases.

---

[48] One has an intuitive idea of a Travelling Organization from the metaphor of an expedition to Africa in former times. The participants of the Travelling Organization must be curious, open to learn, courageous, keen to experiment, and they deal well with uncertainty, stress, unforeseen incidents in a totally new world. This mindset of the Travelling Organization is the core of a Learning Organization. It is impossible in and for the Travelling Organization not to learn. Learning is a question of survival.

[49] This means that the term organization has to be thought of within a network perspective, including interfaces to adjacent organizations and across organizational borders. In many cases, it might be impossible to learn effectively without collaborating closely with people from outside your silo.

[50] More details in Chap. 2 or at https://www.peterwollmann.com/three-pillars

In Volume 1, we fixed the following categories[51] of the activities, and we will also use them in Volume 2 as we have use cases for each category:

- **In the public sector (1): International and national institutional actors**
  These include, for example, UN Agencies like UNFCCC and UNESCO, governmental organizations on the respective topics in the countries or regions or on EU level (partly with significant regulatory tasks), academies, universities, scientific research institutions, think tanks, and consulting governments.
- **In the public sector (2): All sorts of NGOs with respective purposes (a special stake in sustainability topics and the UN SDG Coverage)**
  There are numerous global, regional, and local foundations and NGOs working in fields around environmental topics, on inclusion, promotion of women and gender equity, on reduction of poverty, on education, on health topics, etc. The cooperation between NGOs and the other categories of players is getting continuously closer.
- **Larger well-established private sector organizations worldwide**
  In the private sector, all enterprises that decided to contribute to sustainability topics named Corporate Social Responsibility (CSR)[52] or Environmental, Social, Governance (ESG)[53] and that kept the engagement in spite of political pressure in some parts of the world are players of interest. Enterprises often advertise their own sustainability effort in their communication[54] reflecting their stakeholders' expectations.
- **Private or public sector start-ups with a special stake in sustainability topics and the UN SDG Coverage**
  Often, those start-ups are specialized in one important aspect of sustainability and UN SDGs, for example, in the context of a new way to create "green energy" or to implement a circular economy at a special place or to produce clean water or to develop new LifeScience/MedTech products. In many cases, such types of start-ups are supported by established enterprises or governmental programs (see, e.g., new legislation in Germany to support R&D[55] projects). How different venture capital is granted for start-ups in the USA and Europe regarding the start-up purposes connected with sustainability and UN SDGs was touched in the previous book.[56]

---

[51] See Volume 1.

[52] See, for example, https://en.wikipedia.org/wiki/Corporate_social_responsibility#:~:text=Corporate%20social%20responsibility%20(CSR)%20or,volunteering%20through%20pro%20bono%20programs%2C

[53] https://en.wikipedia.org/wiki/Environmental,_social,_and_governance

[54] For example, 10% Profit Pledge for Sacred Forest (for the indigenous people of the Arhuaco in Colombia) by Rituals Cosmetics or Mindful Travelling Campaign by Austrian Airways.

[55] Research & Development.

[56] See Chapter 18 of Volume 1.

## 1.4 A Brief Overview of the Book's Structure

**Part I** of the book ("**Introduction**") contains two chapters. The introductory chapters enlighten the reasons and challenges for this book, its framing, and the chosen proceedings and structure. The first chapter also contains a recap of the terminology and other fundamentals, already defined and used in the first book, which will be used again in the second book. The second chapter builds on this and describes additionally the well-tailored update of the "User Manual" of the Three-Pillar Model on Sustainability and UN SDG Coverage Following Current Developments, which means how the model—or better the thinking style and respective language—has to be best tailored and applied for the topics of the book.

**Part II** ("**Fundamentals**") tackles some fundamental considerations like in the third chapter on crucial institutional, cooperation, and communication preconditions for achieving Sustainability and UN SDG Coverage of the UN SDG. The fourth chapter outlines a key area for Sustainability and UN SDG Coverage—and one key focus of the book—the task to advancing sustainable development in urban and rural areas.

**Part III** ("**Global Institutions to Foster Sustainable Cooperation and Improve Communication**") has a focus on global institutions and their crucial roles to foster sustainable cooperation and improve communication to achieve Sustainability and UN SDG Coverage (Chaps. 5–9), all authors coming from a close UN context. These considerations on roles, concepts, and cooperation demands start with a unique use case for the sustainable development of a whole country (Qatar) from the perspective of the Multilateral Diplomacy Division of UNITAR,[57] followed by a chapter about the importance of SDG 16[58] linked with the role of international law and strong institutions in post-conflict settings and by a chapter developing a concept for multilateral state-to-state dispute resolution mechanism within the WIPO[59] framework to protect fundamental right and enhance access to justice[60] in line with SDG16. Part III is finished by a description of the process from global goals to ground-level impact: translating climate negotiation into local and regional action illustrated in detail with an amazing specific example in the United Kingdom, and last but not least a consideration on the urgent requirement of better communication and education to cover the purpose of sustainability and UN SDGs. Thus, Part III gives some rare insights from the world of international institutions and cooperation.

**Part IV** ("**Advancing Sustainable Development in Urban and Rural Areas**") is a core of the book—as already mentioned—and touches cases how to advance sustainable development in urban and rural areas (Chaps. 10–16). The concepts and case studies cover sustainability and UN SDG Coverage in agriculture, the

[57] UN Institute for Training and Research.

[58] Peace, Justice, and Strong Institutions.

[59] World Intellectual Property Organization.

[60] IP right is crucial in a world needing innovation and solid legal preconditions for R&D projects and their funding.

development of a rural areas in Germany and in India, a concept on innovative tourism in mountain areas in Northern Italy, a study on horizontal networks for the implementation of proximity welfare centers, insights from sustainability proceedings in World Heritage Policies and Management in Italy, and the contribution of co-working spaces to Sustainability and UN SDG Coverage in three different countries.

**Part V** (**"Conclusions and Takeaways"**) covers the conclusions and stresses selected takeaways for the reader.

Thus, the second book also offers a significant number of use cases from different fields and thus offers every reader the opportunity to select the chapter that interests them. We want to take our readers on a journey through partly unknown territory extending perspectives and even adding new unexpected key insights and conclusions to those from the first book.

## 1.5 Volume 3

While creating and compiling this book, we received far more ideas and material than could be included in Volume 2; for this reason, we have simultaneously written a Volume 3 focusing on especially "Academic Institutions to Enable Sustainable Practice and UN SDG Coverage" and "Cross-Sectoral Approaches to Sustainability from Industry to Fashion." As things stand, the third volume will have the title *The Sustainable Organization Volume 3—Academic, Cross-Sectoral, and Industrial Pathways to Achieving the UN SDGs* and the following contents:

***Part I: Introduction and Fundamentals***

**Chapter 1: The Framing of the Book: Why a Third Volume is Needed, What Its Focus Is, and How It Reflects Current Developments of Sustainability and UN SDG Coverage Activities**

Peter Wollmann, Doris Pemler, Mersida Ndrevataj

**Chapter 2: A Brief Update on the Three-Pillar Model on Sustainability and UN SDG Coverage**

Peter Wollmann

**Chapter 3: The End of Meroe: A Surprisingly Modern Lesson in Climate History**

Doris Pemler

***Part II: Academic Institutions to Enable Sustainable Practice and UN SDG Coverage***

**Chapter 4: Sustainable Data Management: Leveraging the Data Value Chain for UN SDG Achievement Through the Three-Pillar Model**

Sebastian Fischer, Claudia Lemke, Benjamin Rausch

**Chapter 5: Transformative Education and the Earth Charter Enabling Sustainable Development Through Transformative Learning in Higher Education**

Maike Land, Claudia Lemke, Jacobus Bracker, Sam Crowell

**Chapter 6: Educating Purpose-Driven Innovators for Circularity: Insights from Global Industry–Academia Collaboration**

Albena Björck, George Toye, Cedric Gysel

**Chapter 7: Improving Corporate Environmental Responsibility: The Impact of Perceived Stakeholder Pressure, Benefits, and Barriers in Switzerland and Germany**

Carlo Pugnetti, Nadine Berchtold, Stefan Hunziker

**Chapter 8 From Metrics to Meaning: Fostering Sustainability by Reclaiming Leadership as a Conscious and Archetypal Way of Being**

Jacqueline Hofste

*Part III: Cross-Sectoral Approaches to Sustainability from Industry to Fashion*

**Chapter 9. Case study from Swiss Post: Sustainability Matters—How to Transform an Organization**

Vivien Büchler-Bezzola and Christian Plüss

**Chapter 10: Case Studies on Organizational Resilience and Sustainability Through Decision-Oriented Governance Systems**

Christoph Schade

**Chapter 11: Remote Monitoring Advancing Sustainability in the MedTech Industry**

Tom Günther

**Chapter 12: Driving the Circular Economy: How Multicycle Business Models Turn Circularity into Competitive Advantage**

Martin Hewing, Phillip Brandts

**Chapter 13: Collaborative Sourcing: A Three-Pillar-Based Approach to Meeting the 17 SDGs (UN Strategic Development Goals)**

Jürgen Scherer

**Chapter 14: Fashioning the Future: How Luxury Brands Champion Sustainability—An Interview with Laura Tinti**

Mersida Ndrevataj and Laura Tinti

*Part IV: Conclusions and Takeaways*

**Chapter 15: Desirable Impact of Our Sustainability Books on the Consulting Industry and for Academic Education and Research**

Ovais Sarmad, Claudia Lemke, Peter Wollmann

**Chapter 16: Toward a Reconstructive Foundation for Sustainability: A Philosophical Deepening of Sustainability Beyond Perception and Performance**

Ashkan Tashvir

**Chapter 17: Conclusions**

Peter Wollmann, Doris Pemler, Mersida Ndrevataj

## 1.6 An Outlook

During the creation of this book, many valuable ideas for future contributions were placed by existing and new interested authors, especially also on equality, diversity, and inclusion (EDI) perspectives—with EDI regarded from a Three-Pillar perspective not (mainly) as an ethical task but focusing on performance reasons, or economic and efficiency systemic logic. Such a book could offer a refreshing take on this topic and support an objectification in an atmosphere of a kind of culture war. The realization is planned for 2027.

And at the end of the decade—which means in 2029/2030—we consider a Sustainable Organization retrospective, a potential fifth volume of *The Sustainable Organization* with a focus on a check how the 2025/2026 first, second, and third book volumes' discussions, projections, analytical outlooks, etc., for the Sustainability Movement have turned out 5 years later, something like a "check-point," a "knowledge update," and "a revaluation" also as a learning point for historical developments, which is necessary for the epistemic evolution of ideas.

**Peter Wollmann** has been a responsible manager, initiator, mentor, and facilitator in large, predominantly global transformations and strategic developments for almost 40 years in large organizations and since 2017 as an independent consultant. He is also a member of an advisory board of an R&D-oriented company. Peter holds a degree in mathematics and physics from the University of Bonn. He is the author and publisher of numerous books and articles on strategy, leadership, organization design, and project and project portfolio management.

**Doris Pemler** is an entrepreneur, artist, and researcher based in Zurich. She received her M.A. in Egyptology, Sudan Archaeology and Archaeology of the Ancient Near East, from Humboldt University zu Berlin and FU Berlin. Her research focuses on iconography, gender, and intercultural communication in Ancient Egypt and Sudan. As a photographer, she contributed to several exhibitions in Germany and Switzerland. As CEO of Ciudad Condal GmbH, she leads an aspiring company focused on the import and distribution of high-end sustainable products in Switzerland.

**Mersida Ndrevataj** is an urban scholar with a background in architecture, urban planning, and environmental psychology. She holds a PhD in Urban Planning and Public Policy from IUAV University of Venice, where her research focused on the implementation of public policies in cultural heritage management. She is currently a lecturer at the IUAV University of Venice and collaborates with several COST Action research networks focused on urban studies. Her work brings together academic research and field-based inquiry by adopting audiovisual and sensory approaches to investigate urban spaces, everyday socio-spatial practices, and lived experiences.

# An Updated "User Manual" for the Three-Pillar Model on Sustainability and UN SDG Coverage

# 2

Peter Wollmann

**Abstract**

This chapter is both a recap of the Three-Pillar Model introduced in a series of books before and a tailored reference and adoption, a "user manual" for the readers of the current book. The user manual reflects the current developments "in the world of Sustainability and UN SDG Coverage (United Nations Strategic Development Goals, see https://sdgs.un.org/goals) and the respective ecosystems." The underlying question is how the model can be used and applied to understand, evaluate, judge, and manage the nearly infinite activity landscape striving for Sustainability and UN SDG Coverage in general and on the level of a concrete organization. To achieve this, we especially focus on the necessary capabilities described and demanded by the Three-Pillar Model.

**Please note**: The Three-Pillar Model, which serves as the common professional language and thinking framework for our exploration of activities aimed at achieving sustainability and covering the UN Sustainable Development Goals (SDGs), was already briefly described in Chap. 1 (Sect. 1.3.3) to provide context for the book's purpose, general direction, and framing. It is repeated here to complete and round out the recap of the Three-Pillar Model.

P. Wollmann (✉)
Bonn, Germany
e-mail: pw@peterwollmann.com


P. Wollmann et al. (eds.), *The Sustainable Organization – Volume 2*, Future of Business and Finance, https://doi.org/10.1007/978-3-032-23802-3_2

## 2.1 Introduction

The Three-Pillar (Mindset and Thinking Style) Model was developed between 2019 and 2021 by a global author community of at this time already more than 60 professionals from the public and private sector[1] to develop a new option how to cope with the so-called VUCA[2] world. The outcome was published at Springer Nature in up to now five books (eBook and print).[3] The first three books were already translated by Springer into German (by AI) and very successfully published in July 2023 and April and June 2024.[4]

The Three-Pillar Model cannot be mechanically applied and is not intended to be used as a simple problem-solving recipe set. Instead, it functions more like a meta-level thinking style or a mindset model for the user to tailor application exactly to their unique situation. Therefore, it demands some effort in neutral, unbiased thinking[5] to determine what exactly the Three-Pillar Model can contribute to a specific situation, how it can be effectively and efficiently applied, what concrete outcome might be reached, and which detailed measures are requested for long-term success. To help with this process, there is a set of questions at the end of the chapter as a backup[6] to animate the discussion and thinking processes needed in each individual situation. In this book, the application of the model is closely connected with the organizations' demands to achieve long-term sustainability and to cover (some of) the 17 UN SDGs[7] and reflects the latest developments in the surrounding ecosystem.

The great advantage of the Three-Pillar Model mindset is that by using it and getting used to it, formal, sophisticated, and precisely described operational proceedings are no longer necessary. Instead, engaging in the discussion of the individual questions simply starts every person's and organizations' unique journey

---

[1] The very diverse global author community consists of practitioners, leaders, and academics from all continents and more than 20 countries. They represent knowledge stemming from their roles within more than 20 large global public institutions (including UN Agencies) and global enterprises and more than 30 different local public organizations or national companies in different industries and sectors.

[2] VUCA = Volatile, Uncertain, Comply, Ambiguous.

[3] Wollmann, P.; Pemler, D.; Ndrevataj, M. (Eds.): The Sustainable Organization—How Organizations Address the 17 UN SDGs Using the 3-P-Model. Cham Springer Nature, © 2025. Wollmann, P.; Stricker, L.; Pemler, D. (Eds.): Reading an Unknown Organization—Leaders in New Roles Thriving in Challenging Environments with the 3-P-Model Mindset. Cham Springer Nature, © 2024.Wollmann, P.: Püringer, R. (Eds.): Transforming Public and Private Sector Organizations—Implementing Sustainable Purpose, Travelling Organization and Connectivity for Resilience. Cham: Springer Nature, © 2022.Wollmann, P.; Kühn, F.; Kempf, M.: Püringer, R. (Eds.): Organization and Leadership in Disruptive Times—Design and Implementation of the 3-P-Model. Cham: Springer Nature, © 2021.Wollmann, P.; Kühn, F.; Kempf, M. (Eds.): Three Pillars of Organization and Leadership in Disruptive Times—Navigating Your Company Successfully through the 21st Century Business World. Cham: Springer Nature, © 2020.

[4] See https://link.springer.com/book/10.1007/978-3-031-23988-5, https://link.springer.com/book/10.1007/978-3-031-55279-3, https://link.springer.com/book/10.1007/978-3-031-57163-3

[5] Like Hannah Arendt's "thinking without a banister" (https://medium.com/quote-of-the-week/thinking-without-a-banister-2b68a5120625).

[6] As back-up, content taken from the Volume 1, the previous book (see first entry at footnote 3)

[7] https://sdgs.un.org/goals and https://www.globalcompact.de/en/our-work/sustainable-development-goals

and has the potential to lead to meaningful results. Therefore, the demanding challenge is not to cover sophisticated methodical steps but to achieve unbiased thinking and discernment—as the gap between public perception and real situation in topics around Sustainability and UN SDG Coverage and the respective Activity Landscape is huge and the public debate and the formation of personal and organizational opinions are correspondingly difficult. This topic has already been discussed in some detail in the first chapter and will also be tackled in Chaps. 3 and 9.

## 2.2 The Three-Pillar Model: An Overview Including Current Application Challenges[8]

The Three-Pillar Model is based on three key building blocks (pillars), defining the Three-Pillar (Mindset or Thinking) Model for organizational design and leadership for the VUCA world. The author community's starting point in 2019 was to establish guiding principles for a world seemingly devoid of them. The model has since been frequently applied—and its applicability and usefulness proven—in various global use cases (so far > 80) in the private and public sectors. The goal of the Three-Pillar Model is to support understanding and navigating the complexities of today's corporate and institution territory in all sectors.

(1) The First Pillar: **Sustainable Purpose**

Sustainable Purpose is the *raison d'être* of an organization. Having a clear guiding objective provides orientation and certainty to people, fostering a sense of shared effort and success. This is especially important in times of crisis and transformation, as it becomes the key source for the narrative of transformation.[9]

It is obvious and well-known that sustainability, along with the alignment to (some of) the UN SDGs, forms a compelling sustainable purpose for an organization. Sustainability is understood here in two dimensions: sustainability of the organization itself and its contribution to sustainability on a global level. It is clear that achieving sustainability of organizations and their ecosystems as well as the coverage of at least some of the UN SDGs demand significant transformations.

The ***special challenge with Sustainable Purpose*** linked to a general idea of sustainability and the coverage of selected UN SDGs in detail lies in the perception bias, the gap between the public perception of status, success,

[8] Partly taken from Volume 1.

[9] Everybody in the organization, but also its key stakeholders in the entire ecosystem and organization environment, must know what the organization stands for and what value and societal contribution it creates.

reachability of sustainability targets including the judgment of concrete parts of the activity landscape, on the one hand, and the fact-based reality (of the comprehensive big picture), on the other hand. Here, one often finds doubt, discouragement, even despair up to doomsday mood. The perception bias was already tackled in Chap. 1—and the whole book's intention is to show at least puzzle pieces of another reality. It is therefore important "to clear the brain" with discipline, try to get a comprehensive, fact-based picture, and to start "to think without banister" (that means open-minded and without prejudices—the situation might be different from what is intuitively expected or from selective perception from media usually consumed. To foster this think style, positive use cases are crucial—they show what is possible and give an idea, what an organization can strive for). In any case, leadership is particularly important in this regard.

(2) The Second Pillar: **Traveling Organization**

Traveling Organization refers to an organization accepting to be in a permanent state of flux and transformation while interacting with the external requirements of its journey by means of rapid learning and adaptivity. Ideally, this is the mindset of loving the unknown and being well able to manage it.[10]

It is widely recognized that the transformational journeys of organizations to reach sustainability and align with (some of) the UN SDGs are explorative in nature and often have to pass through unknown territories.

The ***special challenge with a Traveling Organization Mindset and Capabilities*** is today, on the one hand, that the north star of a Sustainable Purpose in the context with Sustainability and UN SDG Coverage is challenged by a pessimistic view and a public perception bias already described above and in Chap. 1—which has an impact on motivation and momentum. Going on an explorative journey through the unknown and being confronted with change is always demanding but is even more challenging under these preconditions. On the other hand, there is one more challenging factor: In contrast to the research expeditions of the late nineteenth century through unknown territories in Africa or Asia (mentioned below and in the previous books as a metaphor for Traveling Organizations), which traversed these unknown territory but under relatively stable or linear, albeit unknown external conditions, the journeys today undertaken by organizations are through also uncharted territory, unknown territory but in which external conditions are quickly and unexpectedly constantly

[10] One has an intuitive idea of a Traveling Organization from the metaphor of an expedition to Africa in former times. The participants of the Traveling Organization must be curious, open to learn, courageous, keen to experiment, and they deal well with uncertainty, stress, unforeseen incidents in a totally new world. This mindset of the Traveling Organization is the core of a Learning Organization. It is impossible in and for the Traveling Organization not to learn. Learning is a question of survival.

changing in a nonlinear manner. This adds another dimension that must be taken into account: fast nonlinear changes of the preconditions/contexts. The resulting challenges for leadership are obviously tremendous—but manageable as this book shows in diverse chapters.

(3) The Third Pillar: **Connectivity**

Connectivity focuses on the ability to interconnect during the journey with all necessary parties and resources inside and outside the silos, thus creating high effectiveness and consistency and also building shared learning networks[11] during this process.

It is obvious and well-known that an enormous degree of connectivity is needed for transformational journeys of organizations to reach sustainability and align with (some of) the UN SDGs; this is, among others, a consequence of the high number of stakeholders and affected and/or involved persons.

The ***special challenge with Connectivity*** in the context of activities on Sustainability in general and the coverage of selected UN SDGs in detail is the increasingly strong fragmentation among stakeholders in terms of opinions, perceptions, and ethics (ethics of conviction versus ethics of responsibility), all of which are sometimes strongly linked to a certain emotionality. This makes it difficult to agree on a common starting point for cooperation or even neutral exchange. High, often unattainable expectations are met with defeatism, but also with total rejection. The situation often requires the utmost diplomatic skills and special integrative leadership qualities.

To exemplify the narrative of the Three-Pillar Model in general, one should perhaps start with the Traveling Organization. In our imagination, this term can evoke the image of a very agile, flexible team embarking on an unforeseeable journey to explore unknown territories. In former times, this meant, as already mentioned above, for example, a scientific expedition to Africa or Australia to explore territories previously unknown to us. It is intuitively clear—and supported by various travel reports and even video footage—what kind of team dynamics, personal attributes, and collective capabilities are essential for the successful execution of such an endeavor.

Imagining such an endeavor to succeed, it strongly needs a lighthouse, its own Sustainable Purpose to operate as orientation and direction. In the idea of an explorative expedition, the Sustainable Purpose becomes immediately clear: it is scientific curiosity and the urge to clarify and thus build up lasting knowledge. These sorts of Sustainable Purpose were usually not controversial for the team in any way. Why this might be different in current contexts around Sustainability and UN SDG Coverage was mentioned above. Nevertheless, thinking of a Sustainable Purpose is crucial, even though it might take more effort today than in former times.

[11] This means that organization has to be thought of within a network perspective, including interfaces to adjacent organizations and across organizational borders. In many cases, it might be impossible to learn effectively without collaborating closely with people from outside your silo.

A Traveling Organization does always need its own accepted Sustainable Purpose; leadership has to make this happen and to prepare the stage for it, among others, by the mentioned "thinking without banister." And during its journey into or through the unknown, the Traveling Organization also needs the capability to connect within the team but even more so with its surroundings. On a scientific Africa expedition in the nineteenth century,[12] multilayered Connectivity was an essential skill set—one had to connect between the team members, with indigenous people and their specific cultural traits, with animals, flora and fauna, geography, and many other unknowns. Without Connectivity, survival and success would not have been possible.[13]

Applied to the current situation, connectivity must also be established when this is extremely difficult for the reasons mentioned above. The overall success of the nearly infinite Activity Landscape for Sustainability and UN SDG Coverage is dependent on the degree of connectivity realization as the overall movement needs bottom support to be reached through a reasonable degree of participation.[14] This concept is based on the idea that power (as for the activity landscape movement) arises through participation of many from the bottom up. Force—the top-down concept, often confused with power, does not create power according to Hannah Arendt.[15] This is an important statement that has to be linked with the 3-P Model as it stresses the options for action and opportunities to exert influence (a Traveling Organization should always be aware of its scope for action, it is not a passive object but a proactive subject of a journey—and the mindset of the mentioned explorers at the end of the nineteenth century was exactly like this).

[12] See also: Wollmann, P.; Püringer, R.: "About Travelling in the Unknown in the 19th Century and Today. A Pattern for Leadership and Management in a 3-P Model Context" in: Wollmann, P.; Kühn, F.; Kempf, M.: Püringer, R. (Eds.): Organization and Leadership in Disruptive Times—Design and Implementation of the 3-P Model. Cham: Springer Nature, 2021. More under: https://www.springer.com/gp/book/9783030630331

[13] More details see: https://www.peterwollmann.com/three-pillars

[14] As a background for Hannah Arendt's concepts see for example: Lederman, S.: Arendt and the Current Participatory Moment, in Lederman, S.: Hannah Arendt and Participatory Democracy—A People's Utopia. Cham. Palgrave Macmillan. 2019. https://link.springer.com/book/10.1007/978-3-030-11692-7

And for a use case see: Ahmad, N.A.; Ismail, K.; Ramli, F.; Ismail, M.: Does Participation in Decision-Making Empower Local People? Understanding Empowerment Through Hannah Arendt's Theory of Power? In: Hamdan, K.R.; Buallay, A.: Artificial Intelligence (AI) and Customer Social Responsibility (CSR), Cham: Springer 2024. https://link.springer.com/chapter/10.1007/978-3-031-50939-1_2

[15] See: https://grattoncourses.wordpress.com/wp-content/uploads/2019/12/hannah-arendt-on-violence-harcourt-brace-jovanovich-1969.pdf, retrieved on June 13, 2023.

## 2.3 A Deeper Dive into the Three-Pillar Model Application in Unknown Territories[16]

Living in a VUCA world without many certainties, we increasingly find ourselves navigating through unknown territories. In such situations, we lack precedents to guide us, and our highly developed, sophisticated methods and tools, along with familiar rules and regularities, often become inapplicable—or even entirely obsolete (e.g., if stability or slow linear developments turn to become exponential). Furthermore, the carefully collected and analyzed historical data we rely on to predict the future cease to be relevant.

Realizing this has two important consequences. On the one hand, it means that we need another set of thinking models to give orientation more on a meta-level, if it is not possible on a concrete and detailed level. The Three-Pillar Model turns out to be extremely helpful in focusing on meta-capabilities to give a general orientation (see also previous section): how to formulate an exciting and authentic lighthouse direction (Sustainable Purpose) even in difficult environments, how to embark on agile and flexible journeys as an organization (Traveling Organization), without fearing the unknown territories, and how to effectively connect with all participants needed for the journey (Connectivity). The exact way to nail down and tailor these capabilities in detail has to be developed and decided in the given situation—which is possible with both the meta-capabilities and the right mindset in place. This means that the Three-Pillar Model can be well applied to the Unknown even though or—better because—it demands a lot of effort for the specific tailoring. A crucially needed base is to be able to mind thinking traps like perception biases and to understand that the future can be consciously reshaped by oneself and thus take a different course than predicted by linear thinking[17]—or formulated in another way, that in general future forecasts do only predict what is most likely to happen if people do not intervene and if nothing unexpected occurs—and as already stressed in Chap. 1, this a relevant source for perception biases.

On the other hand, we have a tremendous need to learn about journeys through unknown territories[18] under unclear preconditions even on a detailed level, for example, with a significant lack of authenticated, reliable facts and knowledge, making necessary a certain level of healthy suspicion to prevent unpleasant surprises. This sort of learning also needs the "thinking without banister," one might extend this to "thinking and learning without banister," without thinking traps and perception biases, always challenging the own perception.

---

[16] Partly reused from Volume 1.

[17] Quote by Hannah Arendt: "Predictions of the future are never anything but projections of present automatic processes and procedures, that is, of occurrences that are likely to come to pass if men do not act and if nothing unexpected happens; every action, for better or worse, and every accident necessarily destroys the whole pattern in whose frame the prediction moves and where it finds its evidence." From Hannah Arendt "On Violence", p.13, Houghton Mifflin Harcourt 1970.

[18] Such unknown territories also might occur in the context of needed innovation to fix issues in the context of Sustainability and UN SDG Coverage.

In unknown territories, this specifically means for each organization to learn quickly—it is a matter of survival. Conventional learning approaches are often insufficient in such contexts, making explorative, experimental, and iterative learning with an open mind essential. This must be coupled with intensive, ongoing exchange, including cross-organizational collaboration and mutual peer challenging. Such circumstances demand a new open learning mindset that abandons reliance on rigid specifications and prioritizes adaptability over simply applying lessons from past experiences.

Another valuable reference in uncertain times is Rebecca Solnit and her famous book *A Field Guide to Getting Lost*,[19] in which she works out how important it is to get lost to be able to perceive something important in a completely different and unbiased way. In unknown territories, it is important not to use well-known traditional associations to prevent misperceptions. Or even more directly formulated: The capability to cope with the unknown is based on the ability to get lost. Getting lost means—according to Rebecca Solnit—"that the world has become larger than your knowledge of it,"[20] and you are aware of this.

To paraphrase, a certain openness regarding the Unknown is necessary to be able to gain new knowledge. Rebecca Solnit completes her book with a Greek philosophical perspective: "How will you go about finding that thing the nature of which is totally unknown to you?"[21] which is a fundamental question for each journey through unknown territories. One of her recommendations is to find and use some unusual associations be it from religion, philosophy, literature, arts, or science to introduce radically new perspectives and connected narratives that might help you to cope and understand.

In this context, we have to consider three key statements:

- The future[22] is always—despite all efforts to make it predictable and controllable—an "unknown territory." It is a perception bias that one involuntarily tries to continue the present statistically and linearly into the future.[23] This leads to a clash to understand the character of historical developments and the options for shaping the future.[24]
- The best ways to reach sustainability—both for an organization within its ecosystem and for the world as a whole—remain only partly investigated, fixed, and explored concerning their systematic interactions. This, in turn, means that unknown territories will appear on the journey to reach the sustainability goals.

[19] Solnit, R.; A Field Guide to Getting Lost, New York, Viking Penguin, 2005.

[20] Solnit, R.; A Field Guide to Getting Lost, New York, Viking Penguin, 2005, page 22.

[21] Alleged Pre-Socratic philosopher Meno, cited in Solnit, R.; A Field Guide to Getting Lost, New York, Viking Penguin, 2005, page 4.

[22] Inclusive future innovation or the long-term impact of current innovation, respectively.

[23] It is partly a neurological imprint as our brains love special sorts of structure and patterns and have problem to understand light signals for disruption.

[24] See footnote 17.

- Even though there is a system of precise measures and quantification for the UN SDGs for a country and regional level in place, together with some reporting activities, the impact on the concrete organization is the same as can be seen regarding the topic of sustainability: there are so many points to be clarified that the journey will inevitably touch unknown territories.

The exploration of the best ways to secure the organization's future by reaching sustainability and coverage of (some of) the UN SDG will therefore also be a confrontation with a mostly so-far unknown world. In a context as such, one must first determine which parts of the existing knowledge and experience, if any, are applicable. In any case, it makes sense to approach the ambitions connected to sustainability and the UN SDGs from a meta-level, using the Three-Pillar Model (defining the sustainable purpose connected with the organization's and the world's sustainability and with the UN SDGs, developing some level of journey mindset for the respective transformation journey and capabilities and the ability to connect with all stakeholders and involved parties) combined with the insights of Rebecca Solnit to provide a fresh and unbiased view, always considering to "Leave the Door Open for the Unknown."[25]

In general, it should be noted that unbiased thinking and perception are of utmost importance, especially in unfamiliar territory. The concepts and suggestions mentioned by Hannah Arendt (Thinking Without Railings) and Rebecca Solnit (the ability to get lost and gain a new fresh, unbiased view) provide valuable insights in this regard.

## 2.4 Backup: Helpful Questions for the Application of the Three-Pillar Model[26]

Even though there is an intuitive approach to the three pillars of the Three-Pillar Model, and it quickly generates an image in one mind's eye, there is also the need to get specific and detailed in order to reach a successful application and a fruitful discussion. In this respect, the following questions might be helpful.

***Sustainable Purpose*** **Initial Remarks**: It is important to divide the (Sustainable) Purpose of a Traveling Organization into two parts:

- The "noble" Sustainable Purpose, this might be linked with the UN Sustainable Development Goals (UN SDGs), appealing to the value system of society and people (long-term purpose).
- The "basic" (or rather simple and operational) Sustainable Purpose is to cover a special customer need (situational, temporary, and more often changing or rather more often to be newly interpreted purpose).

[25] Solnit, R.; A Field Guide to Getting Lost, New York, Viking Penguin, 2005, page 4.

[26] Updated version from Volume 1.

Additionally, it is crucial to always regard sustainability in two respects: the sustainability of an organization and its long-term survival through continuous development. And, on the other hand, the organization's contribution to the sustainability of its environment or the world's sustainability as a whole. Both aspects should be reflected in the formulated sustainable purpose of the organization.

**Potential Questions:**

- Does the Sustainable Purpose aim to "make the world a better place," while also being financially (or reputationally) beneficial for the organization? Is the Sustainable Purpose convincing?
- Does the formulated Sustainable Purpose have the power and convincing force to cope with current doubts and biased public perceptions?
- Does the Sustainable Purpose credibly and comprehensively show the organization's social responsibility?
- Is it to be expected that the Sustainable Purpose is a key and highly prioritized factor in difficult decision-making processes, or does it always "lose" against profit interests?
- Does the Sustainable Purpose only concentrate on rankings ("we want to be number 1 in ….") and market positions ("we want to grow our market share to …") or on sense-making future organizational capabilities?
- Is the Sustainable Purpose formulated in a very abstract way ("we want to always satisfy our clients' demands …")?
- What importance does the Sustainable Purpose of an organization have for you as a person (differentiated, if necessary, between established organizations and start-ups)? How do you evaluate and judge the Sustainable Purpose of a new organization?
- Which sorts of purposes do attract people in your opinion? How do they have to be formulated? How do you evaluate and judge them for a new organization?
- How do you evaluate if the "official" Sustainable Purpose is meant seriously (which also means that it has a high priority) and to what extent does it attract and convince the people in the organization and its stakeholders?
- Do you have best (and worst) practice examples as your benchmarks? And how do you evaluate your Sustainable Purpose in the political current environment?
- What do you do to improve a Sustainable Purpose and its perception in your organization?

***Traveling Organization*** **Initial Remark**: The organizational capabilities of a Traveling Organization are closely linked to the readiness of an organization to transform, its maturity, and its political will to run a transformation journey with high priority. In our days, the readiness to transform is a key (strategic) attribute of an organization, a potential game-changer, and competitive advantage.[27] Thus,

[27] See also: Schmidt, H.: "Psychological Capabilities Required for Continuous Transformations—The Self on Permanent Journeys with a Travelling Organization" in: Wollmann, P.: Püringer, R. (Eds.): Transforming Public and Private Sector Organizations—Implementing Sustainable Purpose, Travelling Organization and Connectivity for Resilience. Cham: Springer Nature, © 2022.

questioning the existence and maturity degree of a Traveling Organization is crucial. Therefore, a "setting check" of your organization to analyze its ability for journeys also through unknown territory and/or its "readiness (and sustainable ability) to transform" should be mandatory, but kept simple (in application), easily understandable—and certainly built on a process of analytical discussion in a diverse team that includes different expertise, experiences, cultures, geographies, and genders. These capabilities should be especially stressed in the context of striving for sustainability and UN SDG Coverage.

**Potential Questions:**

- How common is it to start urgent change projects exploratively, which means with a detailed "big picture" and with targets developed during the journey?
- How accepted and easy is it to change the project plan during the journey if needed? How are the decision processes for such changes?
- How high is the trust level in the organization—and the culture of making mistakes (e.g., in basing on assumptions that turn on the journey out to be wrong)?
- How many people in the organization can cope with uncertainty and ambiguity? If you needed several people for a really difficult and demanding project through unknown territory very quickly, would you get a critical mass following?
- Are the colleagues and staff members mostly only able to act if they have a detailed plan (waterfall culture)?
- Is the organization's culture more based on curiosity or more solid-state-oriented?
- What importance does the character of an organization as a Traveling Organization have for you as a person (differentiated, if necessary, between established organizations and start-ups)? What are your first (hard or soft) indicators?
- How do you evaluate in more detail if the organization has the character and mindset/would be able to develop the character and mindset of a Traveling Organization? What are your criteria and checkpoints for evidence?
- Does the organization have a high diversity in thinking styles, leadership styles, experiences, careers? Are there regular exchanges on thinking traps, perception biases, animations through thinker like Hannah Arendt or Rebecca Solnit?
- Is there an adequate learning concept in place (especially in unknown territories)?
- How do you check if there is a critical mass in the organization ready to cope with unknown territory?
- What are your expectations for the organization's HR to foster a Traveling Organization mindset? Do you have best (and worst) practice examples as your benchmarks?
- What do you do to improve your organization's readiness to become a Traveling Organization?
- To what degree is your organization ready and capable of running journeys to reach sustainability and UN SDG Coverage?

***Connectivity*** **Initial Remark**: The organizational capability for Connectivity is based, among other things, on the cooperation and communication culture of the organization and on the personality of key players and their capabilities for Connectivity. In this, a special type of personality is needed, as described in a previous Three-Pillar book.[28]

**Potential Questions**

- How strongly is cooperation supported across units on all levels (board, senior management level, management level, staff)?
- How intensive are discussions about who exactly is responsible for something vs. how can an issue be fixed in the best and fastest way?
- How is the communication culture: absolutely open inside and outside the units or mostly restricted to the respective unit?
- If you have a quick inquiry: are you allowed to approach the responsible person without asking his/her boss before for permission?
- How trustful and vivid are the informal cooperation and exchange channels?
- How are the capabilities to connect and communicate in difficult environments?
- Which importance does Connectivity have within an organization, its ecosystem, and its societal environment for you as a person (differentiated, if necessary, between established organizations and start-ups)? What are your first (hard or soft) indicators for Connectivity?
- How do you evaluate if Connectivity is really in place (e.g., cross-unit cooperation) and not only a not-applied official behavior rule?
- How do you check if there is enough Connectivity in your peer network and on the level of your board?
- Which sort of benefit system is necessary to foster Connectivity, especially cross-silo cooperation? How do you check if it is in place? Which priority does this have to have in your opinion?
- Do you have best (and worst) practice examples as your benchmarks?
- What do you do to improve Connectivity and its perception?
- Additional: How do you analyze/realize the difference between official and unofficial org charts, the official and unofficial key influential persons, and their networks? What do the differences mean in your perspective? And what is their impact on Connectivity but also the Sustainable Purpose and a Traveling organization? And how does your organization cover connectivity in the context of sustainability and UN SDGs?

[28] Schmidt, H.: "Personality and Connectivity" in: Wollmann, P. et al.: Three Pillars of Organization and Leadership in Disruptive Times, Springer Nature, 2020.

**Peter Wollmann** has been a responsible manager, initiator, mentor, and facilitator in large, predominantly global transformations and strategic developments for almost 40 years in large organizations and since 2017 as an independent consultant. He is also a member of an advisory board of an R&D-oriented company. Peter holds a degree in mathematics and physics from the University of Bonn. He is the author and publisher of numerous books and articles on strategy, leadership, organization design, and project and project portfolio management.

# Part II

# Fundamentals

# Considerations on Crucial Institutional, Cooperation, and Communication Preconditions for Achieving Sustainability and Coverage of the UN SDG

3

Peter Wollmann

**Abstract**

In this chapter, the need for suitable, adaptive institutions is discussed, which are guaranteeing or at least fostering rule-based cooperation and supporting solid public knowledge and trusting perception of the requested institutional fundament for effective and efficient cooperation and communication to achieve Sustainability and UN Sustainable Development Goals (UN SDG, https://sdgs.un.org/goals) Coverage. In the discussion, different perspectives are taken. Firstly, the (external) view of a mathematician and project and project portfolio leadership expert (the author); secondly, the view from Rule of Law from a multilateral UN perspective (taken from a Dag Hammarskjöld Lecture given at Uppsala University on November 6, 2002); and thirdly, the Global Governance view of system theory (according to Helmut Willke). All perspectives are mainly based on the progress driving principles of gaining effectiveness and efficiency as system logic. It is obvious that the broad-based Movement as we might regard the nearly infinite Activity and Project

*This chapter is dedicated to Helmut Willke, a brilliant sociologist and systems theorist and good friend who passed away far too early. His reflections on systems theory, state theory, global governance, and global regime building in disruptive times were groundbreaking, and continuing them today with his expertise would be highly appreciated (for more, see https://en.wikipedia.org/wiki/Helmut_Willke).*

P. Wollmann (✉)
Bonn, Germany
e-mail: pw@peterwollmann.com





P. Wollmann et al. (eds.), *The Sustainable Organization – Volume 2*, Future of Business and Finance, https://doi.org/10.1007/978-3-032-23802-3_3

Landscape to reach Sustainability and UN Sustainable Development Goals Coverage with billions of entries, which was reintroduced in Chap. 1, needs some institutional and rule-based fundaments to reach the needed effectiveness and efficiency. Otherwise, every rule of engagement between actors would need to be bilaterally negotiated from scratch between each single activities and projects. In the chapter, it will be also discussed that the intense attack on international institutions and a rules-based world has evidently and clearly led to a broad-based countermovement, which will prevail in the long term—if only for reasons of effectiveness and efficiency. However, this process requires patience and active engagement, on the one hand, while, on the other hand, institutions and rules in a rapidly changing world must also be able to adapt more flexibly to new developments and external conditions in the future to keep or again obtain legitimacy. As a suitable meta-level mindset and thinking model for such challenging situations that need an understanding of nonlinear, potentially dialectic developments and that base on exploration of unknown territories and respective evaluation, the so-called Three-Pillar Model (see https://www.peterwollmann.com/three-pillars), could be well used as already proved in the previous book. The 3-P-Model and especially its metaphor of a Traveling Organization might also act as a very useful part of the urgently needed communication for the Movement actors but also for the general public.

## 3.1 Introduction

It is not easy to make a meaningful contribution on "Crucial Institutional, Cooperation and Communication Preconditions for Achieving Sustainability and Coverage of the UN SDG" to the current book, given the dimension of this topic and the somewhat contentious and fragmented debate on the state of the world (turning point, epochal change, paradigm shift, etc.) and thus also on the framework conditions for (trustworthy, effective, and efficient) international institutions that are necessary to achieve Sustainability and cover the UN SDGs, especially to make a contribution whose expiration date is not just a few weeks away.

When we—the editors and the author community—began planning a second volume in 2024, even before the manuscript for the first volume[1,2] of *The Sustainable Organization* was submitted (the reasons for a second volume are explained in

[1]Often referred to as VUCA or BANI world (VUCA means **V**olatile, **U**ncertain, **C**omplex and **A**mbiguous, BANI stands for **B**rittle, **A**nxious, **N**on-linear und **I**ncomprehensible).

[2]See: https://link.springer.com/book/10.1007/978-3-031-89549-4

detail in Chap. 1), one of my particular convictions was already that we have to touch the topic that the broad-based Movement to achieve Sustainability and cover the UN SDGs requires strong but flexible institutions, and that these must be well communicated and understood by the public—and this special convictions has not changed, in contrary. Whereas the requirements are unchanged—as I will try to show in this chapter—the questions how realistic the fulfillment of these requirements is and how it might look like are perhaps more open than 1 or 2 years ago. The importance of the topic under discussion is reflected in the whole part "Global Institutions to Foster Sustainable Cooperation and Improve Communication" of the book, with its Chaps. 5–9.

When assessing the current situation, it is important to also consider the usual pitfalls of thinking and perception. One of these is that narratives—especially those of elites—have strayed from the facts in crucial areas, as Hans Rosling[3] explains in his famous book from 2018 *Factfulness—Ten Reasons We're Wrong About the World—and Why Things Are Better Than You Think*.[4] The extensive survey of people in various (rich) countries about global conditions and trends revealed the alarming development hat in almost all cases, a clear majority of respondents to multiple-choice questions consistently and incorrectly chose the most negative conditions and trends, reflecting both a lack of factual knowledge and the cognitive bias of a "catastrophe thinking"—and, of course, inevitably also implicitly implying a false perception of the failure of global institutions. Yet, international institutions and their rule-based work have achieved nothing short of sensational successes in many fields over the past decades.[5] It is an interesting experience to try the respective tests and quiz from Rosling's Gapminder Foundation[6] with the mission "… to fight devastating ignorance with a fact-based worldview everyone can understand."

Neurologically caused thinking traps (assumption of linearity, tendency toward traditional explanatory patterns, etc.) and mindset-related thinking traps (the combination of the desire for everything to stay as it is with the desire for catastrophes) are naturally obstacles to correct assessments and thus also to the ability to transform. It should also be mentioned that linear thinking is often accompanied by an illusion of extensive controllability of even highly complex systems, meaning that the limits of direct influence, on the one hand, and positive results even from

[3] See: https://en.wikipedia.org/wiki/Hans_Rosling

[4] https://en.wikipedia.org/wiki/Factfulness

[5] For example, did you know that inn 2024 More than 60% of child refugees attend primary school? 85% of people asked were wrong. And that life expectancy in the poorest countries increased from 44 years in 1970 to 60 years now as an impact of a significantly reduced infant mortality?

[6] https://www.gapminder.org founded in 2005 by Hans Rosling with Ola Rosling and Anna Rosling Rönnlund.

developments that meander greatly in between, on the other hand, are difficult to recognize.

In addition to personal skills of constant self-reflection, a certain "insurance" against thinking traps is sufficient diversity in teams in the form of different thinking styles, skills, experiences, and influences, as well as fair treatment and productive use of these in team discussions. Or, in other words, the application of professional knowledge management procedure that will be briefly touched again in Sect. 3.3.4.

## 3.2 The UN Sustainable Development Goals (SDG) at a Glance

The chapter examines the institutional, interactional, and communicative (global) preconditions to achieve sustainability. The more general term of Sustainability is closely linked to the 17 UN SDGs (see Fig. 3.1), which form so to say the master plan for the development of the world by 2030 to make it a significantly better place in many perspectives. The history of the ambitious underlying transformation program was described in more detail in Chaps. 1 and 3 of the previous

**Fig. 3.1** UN SDG (source: https://www.un.org/sustainabledevelopment/), used with friendly permission of UN. The content of this publication has not been approved by the United Nations and does not reflect the views

book[7] in which we also stressed that the landscape of sustainability and UN SDG Coverage activities with all its acting players and affected stakeholders is nearly infinite.

When you consider that each SDG has been broken down into numerous specific (sub-)targets and complemented by additional activities, which are being pursued in every country or region on earth by a large number of official and private institutions and organizations with tailored measures involving many stakeholders, you can imagine that the number of acting players and affected stakeholders is nearly infinite. So, the reader might get an idea of the great challenge to generate a half-way solid overview of the status of the landscape of activities and projects to reach sustainability for organizations, regions, countries, or for the world as a whole and to cover the UN SDGs.

## 3.3 Different Perspectives on (Institutional) Preconditions to Achieve Sustainability

### 3.3.1 A Starting Note

I would like to start with a note on mathematics: since many projects related to or shaping the future must navigate uncharted territory, where traditional paradigms, methods, tools, data, and procedures may no longer apply. Thus, they must proceed in an exploratory manner rather than according to so-called waterfall models. The future cannot be calculated in detail and without doubt, no matter how much computing power is available, but scenarios might be able to describe potential futures. Institutions with knowledge hubs and rule-based procedures are helpful and necessary for this purpose, enabling well-founded decisions to be made even in terra incognita. It is important to bear in mind the limitations of the mathematics and logic that are often used: In 1930, Kurt Gödel formulated his Gödel's incompleteness theorems,[8] which state that in systems such as arithmetic, not all statements can be proven or disproven, which also means that the terms "true" and "provable" are not the same in mathematics. Put simply, Gödel's theorem states that any sufficiently powerful formal system is either contradictory or incomplete. In my view, this can be perfectly linked to project and project portfolio management and to the Sustainability Movement. For me, Gödel's Theorems underscore the need for flexible exploration and interaction in order to build the knowledge we need and to continuously develop it further—nothing is final in science and everything in a rule-based scientific process supported by institutions (like they exist in mathematics) as this proceeding gives insights and knowledge the necessary legitimacy—and legiti-

[7] Wollmann, P.; Pemler, D.; Ndrevataj, M. (Eds.): The Sustainable Organization—How Organizations Address the 17 UN SDGs Using the 3-P-Model. Cham Springer Nature, © 2025; (https://link.springer.com/book/10.1007/978-3-031-89549-4).

[8] See https://en.wikipedia.org/wiki/Gödel%27s_incompleteness_theorems

macy ultimately means effectiveness and efficiency in further application, thus it is a strategic success factor at the end and driver for further development (see also 3.3.4). It is at the end a mindset and thinking style to act accordingly on the basis of these insights.

### 3.3.2 The Project or Project Portfolio Management Perspective: Effectiveness and Efficiency in the Focus

The urgent requirement of strong but flexible institutions can be first easily deduced logically and operationally from the principles of effective and efficient project and project portfolio management (at the end mathematical or project management thinking), regarding the fact that the broad-based Sustainability Movement is a large, nearly infinite landscape of activities, projects, transformations, actors, and affected stakeholders somehow presenting the world's change and project portfolio on all (environmental, ecological, societal, ethical, organizational, etc.) Sustainability and UN SDG Coverage topics ("to make the world a better place"), it is a list with maybe some billion entries.[9] It is also obvious that there are many content-wise or resources-demand-wise interfaces between the single activities that require at least a minimum of cooperation and alignment for effectiveness and efficiency reasons.[10] Regarding the vast number of different activities and projects, it is not an option to base only on bilaterally negotiated single commitments from each individual to each individual, it is crucial to have a certain fundament of multilaterally discussed and fixed basic principles, rules, and procedures in order to reduce complexity and increase effectiveness and efficiency. The development of the concepts (including methods and tools) of project and project portfolio management shows that a certain set of "norms" and "trusted procedures" is crucial for success. However, the challenge is to keep the set flexible but always legitimized and prevent it to become too

[9] See Chap. 1 where we stressed—as in Volume 1—that even a very rough Movement overview is only possible on a very advanced meta-level or based on selected sufficiently exemplary and paradigmatic use cases that cover larger parts of the landscape without any claiming of completeness. Understanding the landscape in principle does not mean understanding it in detail, on the contrary. And central steering in detail is an impossible task.

[10] This is also supported by the 1984 written book *The Evolution of Cooperation* by the political scientist Robert Axelrod that expands upon a paper of the same name written by Axelrod and evolutionary biologist W.D. Hamilton (details see https://en.wikipedia.org/wiki/The_Evolution_of_Cooperation). In summary, the book shows that success is correlated with the following characteristics:

- Be nice: cooperate, never be the first to defect.
- Be provocable: return defection for defection, cooperation for cooperation.
- Don't be envious: focus on maximizing your own 'score', as opposed to ensuring your score is higher than your "partner's".
- Don't be too clever: or, don't try to be tricky. Clarity is essential for others to cooperate with you.

bureaucratic (see trends to agile project management concepts—but even agile projects are rule-based).[11]

Thus, the elimination of rule-based action and connected (trustworthy) institutions—as partly under discussion—would result in an immense increase in complexity, rendering many actions completely inefficient and ineffective. This cannot be sustained for long because it requires a great deal of resources, energy, time, and money to be spent on "the same clarifications/agreements over and over again," making the entire system inefficient and ineffective. Even those players who expect to benefit from a departure from institutions and rule-based systems will not achieve this in the long term, because the effort required to manage complexity tends to increase exponentially. This is why there is now a broad countermovement against the dismantling of international institutions and rule-based action—both for fundamental and partly ethical reasons, but above all for reasons of pure expediency. And—as mentioned above—on closer inspection, as in Rosling's book *Factfulness*, it is easy to see how successful the work of many international institutions has been in recent years, despite many misguided development (which is inevitable for such a complex set of tasks).[12] It is a severe issue that the perception in public—and even among experts—is different, more negative that requires more and better communication. These insights are so important because the analysis of the factors contributing to the success or failure of large projects or transformations has shown that cultural issues, which require intensive communication and persuasion, are of paramount importance.[13]

One final note on mathematics: since many projects related to or shaping the future must navigate uncharted territory, where traditional paradigms, methods, tools, data, and procedures may no longer apply. Thus, they must proceed in an exploratory manner rather than according to so-called waterfall models. The future cannot be calculated in detail and without doubt, no matter how much computing power is available, but scenarios might be able to describe potential futures. Institutions with knowledge hubs and rule-based procedures are helpful and necessary for this purpose, enabling well-founded decisions to be made even in terra incognita. It is important to bear in mind the limitations of the mathematics and logic that are often used: In 1930, Kurt Gödel formulated his Gödel's incompleteness theorems,[14] which state that in systems such as arithmetic, not all statements can be proven or disproven, which also means that the terms "true" and "provable" are not the same in mathematics. Put simply, Gödel's theorem states that any sufficiently powerful formal system is either contradictory or incomplete. In my view, this can be perfectly linked to project and project portfolio management and to the

[11] See, for example, Scrum Framework: https://en.wikipedia.org/wiki/Scrum_(project_management).

[12] And now in parallel the challenging demand of the UN80 Initiative https://www.un.org/un80-initiative/en

[13] See, for example, https://www.pmi.org/learning/library/systematic-biases-culture-project-failures-5578

[14] See https://en.wikipedia.org/wiki/Gödel%27s_incompleteness_theorems

Sustainability Movement. For me, Gödel's theorems underscore the need for flexible exploration and interaction in order to build the knowledge we need and to continuously develop it further—nothing is final in science and everything in a rule-based scientific process supported by institutions (like they exist in mathematics) as this proceeding gives insights and knowledge the necessary legitimacy—and legitimacy ultimately means effectiveness and efficiency in further application, thus it is a strategic success factor at the end and driver for further development (see also 3.3.4). It is at the end a mindset and thinking style to act accordingly on the basis of these insights.

### 3.3.3 The Rule of Law at Home and Abroad Perspective: The Multilateral Legal View Requiring Consistency

Another approach to demonstrating the necessity of rule-based action and corresponding international institutions was laid out by Lakhdar Brahimi,[15] Special Representative of the UN Secretary-General for Afghanistan, in his Dag Hammarskjöld Lecture given at Uppsala University on November 6, 2002,[16] under the headline "The Rule of Law at Home and Abroad," stressing that in our days national rule-based acting and strong institutions need an equivalent on an international level in order to resolve the increasing number of issues and cases that need to be resolved between different countries. In his plaidoyer, Lakhdar also covers ethical as well as effectiveness and efficiency aspects, which are more often well connected than assumed (maybe one reason to resist the temptation to argue too much from a purely ethical perspective as an ethicist of conviction and to focus more on the results of actions as an ethicist of responsibility).

He starts with a country perspective stating that "bringing peace, security and stability back to a country coming out of conflict requires helping them build or rebuild their institutions and structures, thereby permitting a culture of peace, tolerance and justice to take root again."[17]

Lakhdar, who refers to "The Rule of Law" as a key component of the "Enlightenment Project" of the French, British, and other European thinkers of the eighteenth century and incorporated in the American Constitution and enshrined in the Universal Declaration of Human Rights of the United Nations in 1948, counteracts the impression that the "Rule of Law" principles are refers to Gandhi, Nehru, Nelson Mandela, liberation movements, and intellectuals in the Third World to underline its universality. And regarding international rule-based acting, Lakhdar underlines that "the concept of international law applicable horizontally between States is not a new one. It arose in both the West and the East out of the self-interest of tribal and feudal chiefs, kings and emperors, who all felt the need for some rules

[15] Author of the important Brahimi Report in 2000: https://peacekeeping.un.org/en/brahimi-report-0

[16] http://www.daghammarskjold.se/wp-content/uploads/2014/08/Lakhdar.pdf

[17] Ibid, page 9.

on warfare, the treatment of prisoners of war and the status of visiting diplomats and plenipo-tentiaries, and later, on issues of trade, treaties and conventions between States"—similar considerations as described above focusing on "practicality—including for reducing complexity." This development accelerated in the twentieth century, especially as a consequence of the two World Wars, also triggered by modern travel and communication, and encouraged by "the dreams of the visionary idealist" that "converged with the needs of the pragmatic statesman and businessman into the creation of the League of Nations after World War I and the United Nations after World War II." In this context, it became obvious that international law needs further development to tackle the limitations of the Rule of Law, which applied at the national level. This urgent need is evident in the case of issues that transcend national borders, such as combating climate change, dealing with a global pandemic, or achieving many of the UN Sustainable Development Goals (SDGs).

In summary, Lakhdar's considerations describe, on the one hand, a historical movement that has existed for centuries and, on the other hand, effective and efficient solutions for intergovernmental and supranational issues that inevitably need to be addressed in an increasingly interconnected global world. This means strong, legitimized institutions and rule-based procedures, well communicated.

### 3.3.4 Global Governance Perspectives from System Theory: The Complex Global Knowledge Societies

In this paragraph, I will touch below on some—connected—system-theoretical considerations from the famous sociologist Helmut Willke[18] on Government and Governance theoretically emphasizing this. It is interesting with which system theory-driven considerations Helmut Willke[19] approaches the topic of global institutions and rule-based cooperation, for example, as presented in his book from 2014 "Regieren—Politische Steuerung komplexer Gesellschaften" (translated: Governing—Political Governance of Complex Societies").[20] In it, Willke uses "governance" as a "comprehensive term to describe the problem areas in management, leadership, transformation, and strategic orientation of actors and institutions within a particular society."

The governance form of a system describes its principles, concepts, methods, and instruments that coordinate the system or with which it coordinates itself. The expansion of "government" to "governance" means that actors in civil society now become part of the political process and thus actively contribute to political decision-making".[21] This principle has—obviously—already established itself

[18] https://en.wikipedia.org/wiki/Helmut_Willke

[19] Professorships in Political Theory and Global Governance at Bielefeld University and Zeppelin University Friedrichshafen.

[20] Willke, H.: "Regieren—Politische Steuerung komplexer Gesellschaften"; 2014, Wiesbaden, Springer VS

[21] Ibid page 73, translated from German using DeepL.

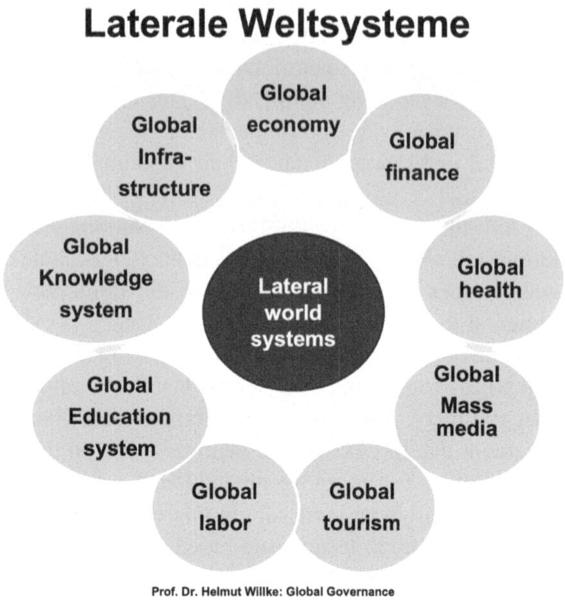


**Fig. 3.2** Lateral world systems according to Helmut Willke, taken from a slide deck of a lecture for Zurich University in 2015. (See slide 9 from https://www.business.uzh.ch/dam/jcr:ffffffff-ac4e-3bfb-ffff-fffff652f726/10_GLO_MNU_2015.pdf, retrieved on February 1, 2026)

empirically, as can be easily seen from the parties involved in international issues, such as the COPs on climate change, but also in many other contexts.

In lessons on Global Governance at Zurich University in 2015,[22] Willke introduced the so-called lateral world systems as a consequence of the emergence of globalized knowledge societies (see Fig. 3.2).

As societal systems on a national level, the exemplarily mentioned lateral world systems also have their own logic, procedures, specific languages, complexity, etc., which means that they do not simply fit to each other or are able to communicate and interact with each other without frictions. It is crucial to build a structural linkage based on purpose-driven, learning-oriented, agile interaction that will need the support of flexible institutions and basic rules, based on respect for the self-organization and self-management capabilities of social functional systems like in an international context the lateral world systems. It is the establishment and continuous optimization of hinges between self-referential systems,[23] which can be understood as one of the drivers to create and document knowledge. The focus reveals a different perspective and a different understanding of politics: context-oriented coordination of and interaction between largely self-governing knowledge-based systems instead of activist, detailed top-down control of content. Since Willke

[22] https://www.business.uzh.ch/dam/jcr:ffffffff-ac4e-3bfb-ffff-fffff652f726/10_GLO_MNU_2015.pdf

[23] See: https://www.degruyterbrill.com/document/doi/10.1524/9783486845327.218/pdf?licenseType=restricted&srsltid=AfmBOorvIxB8_NtuNZ275NdvAWB_oC0XJBY9L39PS6gPBDx-GyxulFsp8

fundamentally understood systems not as fixed, but as changing through learning he intensively worked on the field of knowledge management[24] and the establishment of global knowledge societies.

The dynamics of the global or globalized knowledge society inevitably lead, according to Willke,[25] to

- Global networking
- Digitalization
- Establishment of expertise as a core resource
- Global competition
- Institutional dynamics

The final outcome of this evolutionary, versatile, and incremental (step-by-step) process is evidently still open. Willke identified two serious contradictions for the future development of (global) governance in the global knowledge society: "... on the one hand, there is the conflict between the traditional sovereignty of nation states and the increasingly dense network of international, transnational, and global contexts and multi-level systems; on the other hand, there is the conflict between the ignorance of the majority and the democratic deficit of expert and specialist communities. Both contradictions are direct consequences of the ongoing transformation of nation-state-organized industrial societies into globalized knowledge societies. Historically speaking, this transformation is still in its infancy and therefore many of its consequences and lines of development are not yet foreseeable. However, it is already clear that this transformation of society will be as profound in the medium term as the transformation of Europe's agrarian societies into industrial societies was 200 years ago. It is not surprising that, under these conditions, governance is becoming more demanding and complicated. This also means that innovative concepts, models, and instruments of governance are becoming more important. The challenge for governance is to find ways to manage the contradictions between the demands of democracy and the demands of expertise."[26]

As a consequence, Willke calls—basing on the analysis of his book on *Climate Crisis and Social Theory*[27]—for a (new) form of governance in lateral world systems and sees opportunities for overcoming the climate catastrophe only in a network of internationally active nongovernmental organizations, international organizations, and governments that understand their own limitations. He accurately

[24] See: Willke, H.: "Einführung in das systemische Wissensmanagement (Carl-Auer Compact)", 2018, Carl-Auer Verlag, Heidelberg.

[25] https://www.business.uzh.ch/dam/jcr:ffffffff-ac4e-3bfb-ffff-fffff652f726/10_GLO_MNU_2015.pdf slide 10

[26] Willke, H.: "Regieren—Politische Steuerung komplexer Gesellschaften"; 2014, Wiesbaden, Springer VS, page 135/136, translated from German using DeepL.

[27] Willke, H.: "Klimakrise und Gesellschaftstheorie—Zu den Herausforderungen und Chancen globaler Umweltpolitik"' 2023, Campus Verlag Heidelberg.

observed how business and science, politics and law were moving far too slowly toward understanding the necessity of networking.[28]

In another context,[29] he discusses how to give rise to new forms of legitimacy through procedures to these (expertise-based) organizations and institutions: "The competent judgment of expert communities, reached in accordance with Merton's rules of scientific discourse (Merton 1973), cannot claim formal democratic legitimacy, but it can claim a derived form of legitimacy, which I call legitimacy through expertise (Willke 2007 [Smart Governance. Governing the global knowledge society. Frankfurt a. M.: Campus], chap. 3.3.). This derivative of legitimacy derives its persuasive power not from mere arithmetic (majority rule), but from the better argument."[30] Factors that contribute to democratic legitimacy are primarily participation, representation, transparency, accountability, rule of law, and governance competence.[31]

To summarize: Willke, who was a great optimist, did not believe in activist politics, as he called it, but rather in various self-referential systems with their own logics, which can be coordinated appropriately through rule-based networking, interaction, and communication at their context-related interfaces. The institutions involved are expertise- or knowledge-based and must legitimize themselves through the appropriate procedures. The corresponding processes in the context of a global knowledge society are exploratory, incremental, volatile, and ambiguous. He believed in systemic intelligence that at the end results on balance positive results (s. Rosling's book *Factfulness*, mentioned above) and that the end systemic and economic intelligence (effectiveness and efficiency) will prevail.

Willke would therefore have liked a comment of Prof. Henrik Müller[32] in the (political) magazine *Der Spiegel* on February 1, 2026: "Its position as the leading international reserve, investment, and financing currency is based not only on the sheer size of the American financial markets, which allow international investors and central banks to conduct large-volume transactions at any time. It also needs to be anchored in a reliable constitutional state and predictable institutions. This combination gives the US unique advantages."[33] The comment touches on a system of

---

[28] quoted in essence from "In memoriam Helmut Willke" by Dirk Baecker in SOZIOLOGIE Jahrgang 53 • Heft 2 • 2024, see: https://publikationen.soziologie.de/index.php/soziologie/de/article/view/1808

[29] Willke, H.: "Regieren—Politische Steuerung komplexer Gesellschaften"; 2014, Wiesbaden, Springer VS.

[30] Ibid page 124, translated from German using DeepL. On page 127 Willke gives an overview of potential Forms of this kind are legitimacy through procedure (according to Luhmann 1969), through self-organization and self-control (according to Black 2009; Buchanan and Keohane 2006; Hechter 2009) and in particular, legitimacy through expertise (according to Sinclair 2005; Willke 2006, p. 65 ff.).

[31] Ibid.

[32] https://de.wikipedia.org/wiki/Henrik_Müller

[33] https://www.spiegel.de/wirtschaft/unternehmen/donald-trump-risiko-fuer-den-dollar-drohender-wertverlust-und-schuldenlast-a-a9d9a2d8-1a41-4c38-8c83-f579cb4db8c3 translated from German using DeepL

special interest of Willke, the global finance system and its logic, and the institutional and the rule-based preconditions for success in it.

A similar interesting discussion is on the role of trust in systems, societies, and organizations—also a rule-based precondition for success.[34] Trust, that is, building on predictability, reliability, fairness, transparency, etc., is a driver for effective and efficient systems that have competitive advantages.

### 3.3.5 General Consensus and Consequences

All three perspectives advocate—for reasons of effectiveness and efficiency—for global institutions and rule-based action; both are indispensable. Due to the complexity of the situation, which affects various vertical levels (from local to regional and national to global), excellent communication and opportunities for participation are required. All three perspectives see a high need for (incremental) development (including expertise), legitimation (of institutions), coordination (between institutions and other players), etc., while at the same time requiring a high degree of flexibility and rapid adaptability to new external circumstances.

The challenge is to define and establish the requested institutional cornerstone, including its interactions in an agile, flexible, in any case not too rigid, bureaucratic way to keep the needed flexibility for adaption of new situations and for running the tasks that mainly or at least often will have project character (e.g., see the support of the Sustainability Movement or similar cases). We know how difficult this challenge is already on a significantly smaller scale, for example, in a global company, namely to define the project portfolio concept in such a way that, on the one hand, it enables a certain degree of operational efficiency and controllability or predictability and guarantees the best results across all projects, but, on the other hand, does not stifle innovation, independence, and pragmatic solutions.

At this point, it makes sense to refer to Ovais Sarmad's[35] fully compatible considerations from the upcoming contribution to Volume 3 of *The Sustainable Organization*. Ovais calls in the context of Sustainability and UN SDG Coverage, among others, for an initiative to reducing bureaucratic boundaries and rules of the existing organizations to get ground-level engagement and demands of society and individuals faster and more easily translated into action (institutional adaptation of changed situation and regaining legitimacy), a better prioritization of the vast landscape of potential activities using criteria of high level of effectiveness and efficiency (complexity reduction), and a better understanding and coverage of public perception issues of the UN work for the public, as the UN work is often not very visible and understood (communication).

[34] See https://symposium.org/how-the-trust-economy-could-benefit-us-all/

[35] Ovais Sarmad served the United Nations Climate Change secretariat (UNFCCC)—COP23 to COP 27 (2017 to 2023). Prior to UNFCCC, he worked as the director of Cabinet and chief of staff to the Director General at the IOM.

## 3.4 The Three-Pillar Model

The nearly infinite landscape of sustainability and UN SDG Coverage activities can be best understood and handled applying the meta-level mindset or "analytic thinking style or thinking" of the Three-Pillar Model, based on the three key building blocks (pillars) "Sustainable Purpose," "Traveling Organization," and "Connectivity." It was developed between 2019 and 2021 for the so-called VUCA[36] world by a global author community of more than 60 professionals from public and private sector[37] and published at Springer Nature in up to now five books (eBook and print).[38] The first three books were already translated into German (by AI) and very successfully published in July 2023 and April and June 2024.[39] The model has since then been frequently applied—and its applicability and usefulness proven—in various global use cases (so far > 80) in the private and public sectors.

(1) The First Pillar: **Sustainable Purpose**

Sustainable Purpose is the raison d'être of an organization, having a guiding objective means bringing orientation and certainty to the people for their joint endeavor and success. This is especially important in times of crisis and transformation. It is also the key source for the narrative of transformation.[40]

The (global) institutions frequently mentioned and described above usually do not have any problem with presenting an animating, convincing Sustainable Purpose. The challenge is the translation of the defined Sustainable Purpose into the organizational design in detail, including processes, rules, and procedures, as the—often legally demanded—requirements are complex and the number of the organizations' sponsors high that leads to sophisticated contracts

---

[36] VUCA = Volatile, Uncertain, Complex, Ambiguous.

[37] The very diverse global author community consists of practitioners, leaders, and academics from all continents and more than 20 countries. They represent knowledge stemming from their roles within more than 20 large global public institutions (including UN Agencies) and global enterprises and more than 30 different local public organizations or national companies.

[38] Wollmann, P.; Pemler, D.; Ndrevataj, M. (Eds.): The Sustainable Organization—How Organizations Address the 17 UN SDGs Using the 3-P-Model. Cham Springer Nature, © 2025. Wollmann, P.; Stricker, L.; Pemler, D. (Eds.): Reading an Unknown Organization—Leaders in New Roles Thriving in Challenging Environments with the 3-P-Model Mindset. Cham Springer Nature, © 2024.Wollmann, P.: Püringer, R. (Eds.): Transforming Public and Private Sector Organizations—Implementing Sustainable Purpose, Travelling Organization and Connectivity for Resilience. Cham: Springer Nature, © 2022.Wollmann, P.; Kühn, F.; Kempf, M.: Püringer, R. (Eds.): Organization and Leadership in Disruptive Times—Design and Implementation of the 3-P-Model. Cham: Springer Nature, © 2021.Wollmann, P.; Kühn, F.; Kempf, M. (Eds.): Three Pillars of Organization and Leadership in Disruptive Times—Navigating Your Company Successfully through the 21st Century Business World. Cham: Springer Nature, © 2020.

[39] See https://link.springer.com/book/10.1007/978-3-031-23988-5, https://link.springer.com/book/10.1007/978-3-031-55279-3, https://link.springer.com/book/10.1007/978-3-031-57163-3

[40] Everybody in the organization, but also its key stakeholders in the entire ecosystem and organization environment, must know what the organization stands for and what value and societal contribution it creates.

and easily to a certain level of bureaucracy organizational rigidity and inflexibility that is a contradiction to the second pillar.

It might make sense to stress that the Sustainable Purpose of the (global) institutions under discussion is usually connected to expertise- or knowledge-driven content.

(2) The Second Pillar: **Traveling Organization**

Traveling Organization means for an organization accepting to be in a permanent state of flux and transformation while interacting with the external requirements of its journey with rapid learning and adaptivity. Ideally, this is the mindset of loving the unknown and being well able to manage it.[41]

It is obvious that the environment of the (global) institutions under discussion is constantly and rapidly changing and that they therefore need a high level of flexibility and adaptability—especially when they enter on their journey unknown territory without precedence. In this case, they have to act in an explorative, experimental way—but this contradicts their "constitution" described above. This dilemma requires creative solutions.

(3) The Third Pillar: **Connectivity**

Connectivity is all about the ability to interconnect on the journey with all needed parties and resources inside and outside the silos, thereby creating high efficacy and consistency and during this process building also shared learning networks.[42]

The global institutions' need of building networks, interconnecting, context-oriented coordination and interaction is obvious, especially to build circles of high expertise and to negotiate projects, actions, and procedures at common interfaces in a fair and transparent way.

Exemplifying the narrative of the Three-Pillar Model to get a deep general understanding of the concept normally starts with the Traveling Organization. In one's imagination, this term is able to evoke the image of a very agile, flexible team embarking on an unforeseeable journey to explore unknown territories with the underlying expectation that the framing fundamental understanding of the unknown territories might change daily as a consequence of new circumstances, hurdles, and encounters.

The success of the journey of a Traveling Organization demands, on the one hand, the respective organizational mindset and capability for journeys of this type and, on the other hand, a clear lighthouse, a convincing Sustainable Purpose for the journey as orientation and direction. And without Connectivity, survival and success

[41] One has an intuitive idea of a Traveling Organization from the metaphor of an expedition to Africa in former times. The participants of the Traveling Organization must be curious, open to learn, courageous, keen to experiment, and they deal well with uncertainty, stress, unforeseen incidents in a totally new world. This mindset of the Traveling Organization is the core of a Learning Organization. It is impossible in and for the Traveling Organization not to learn. Learning is a question of survival.

[42] This means that the term organization has to be thought of within a network perspective, including interfaces to adjacent organizations and across organizational borders. In many cases, it might be impossible to learn effectively without collaborating closely with people from outside your silo.

on a journey through unknown territories would not have been possible, the Traveling Organization has to connect with everybody in the team and with all what's important along the chosen path.[43]

## 3.5 Conclusion

In the long run, the intrinsic logic of (global and national) systems striving for optimized effectiveness and efficiency demands from different perspectives respective suitable global knowledge-based institutions with high legitimacy, rule-based interaction, cooperation, and procedures and a reasonable degree of public support for the concept to achieve Sustainability and UN SDG Coverage and the respective Movement, which demands adequate public communication. However, reaching Sustainability and UN SDG Coverage including the institutions' interactions is clearly a highly complex, explorative, and incremental step-by-step creation and optimization task with many variables to be taken into consideration. Expectations of politicians, experts, and the public have to be respectively calibrated. In addition, the (interim and final) results of the transformation journey will for sure change over time under changed conditions, this has to be taken into account—and it is not a sign for bald planning or failure but for the success of an explorative, flexible proceeding.

Applying the Three-Pillar Model to this demanding situation, a double dilemma becomes transparent: on the one hand, a communicative and perception challenge as such complex journeys—for which there are no comparable precedents in the personal environment of the general public and even of most experts—are difficult to understand in their entirety and therefore difficult to assess in terms of their value potential long-term outcome, their opportunities, etc. On the other hand, respectively changing institutional directions and agreements between a large number of different parties according to the journey's outcomes are generally "somewhat more cumbersome" when it comes to the necessary adjustments to adapt changed conditions. Or in other words: rules and institutions need a solid base but have nevertheless to be flexible and remain open for adaptation. Changes of rules and institutions must be negotiated fairly, which takes time, yet results are needed fast as the situations change rapidly. And when changes are finally fixed, the results are not easily adaptable in detail.

These dilemmas need proper solutions. Some first ideas on this can be found in Chaps. 5–9 of this book.

A last remark on perception and communication: the mentioned dilemmas are naturally even more explosive and difficult to resolve in times of general uncertainty. And in the concrete case of Sustainability and UN SDG Coverage, all of this is further enhanced by massive perception and analysis challenges, already mentioned in Chap. 1: the discrepancy between the public perception of the Sustainability Movement and its actual state—and also some neurological reasons for the

[43] More details in Chap. 2 or at https://www.peterwollmann.com/three-pillars.

occurring perception biases that cause perception developing in most critical ways, influenced by the headlines of the rough short-term political discussion. But taking a more long-term perspective and cope in the right way with perception biases and thinking traps,[44] cautious optimism regarding the Sustainability Movement's (or its Activity Landscape's) unbroken momentum is substantiated.

**Disclaimer** The considerations in the chapter were not made by someone from the UN system, the diplomatic service, or from a lawyer specializing in this field, or an expert in systems theory, international relations, etc., but rather from an external, mathematical, and project-oriented perspective, albeit based on experience as lead editor with the global Three-Pillar Author Community and its eight published books, as well as numerous interactions and exchanges on related top. That is the reason why the considerations focus on practicality and effectiveness and efficiency that will prevail in the end over more "ideological" views.

**Peter Wollmann** has been a responsible manager, initiator, mentor, and facilitator in large, predominantly global transformations and strategic developments for almost 40 years in large organizations and since 2017 as independent consultant. He is also a member of an advisory board of an R&D-oriented company. Peter holds a degree in mathematics and physics from the University of Bonn. He is the author and publisher of numerous books and articles on strategy, leadership, organization design, and project and project portfolio management.

[44] See in detail in Chap. 1.

# A Special Book Focus: Advancing Sustainable Development in Urban and Rural Areas

4

Mersida Ndrevataj

**Abstract**

For a long time, urban centers and rural areas have been conceptualized as separate socio-spatial domains, assumed to face distinct challenges requiring differentiated planning and policy responses. This longstanding division constitutes a structural obstacle to meaningful progress, as contemporary global challenges—including climate change, socioeconomic inequality, and resource scarcity, just to mention a few—operate across relational urban–rural systems that exceed fixed geographical boundaries. These processes reveal explicitly the interdependence of territorial scales and call into question spatial governance models grounded in spatial fragmentation. Drawing on insights from contributions presented in the second volume of *The Sustainable Organization* (particularly Chaps. 10–16), this chapter engages the urban–rural divide as a critical site of theoretical and practical reconsideration. It finally sustains that integrated approaches to spatial planning can better operationalize the UN Sustainable Development Goals by reconfiguring governance beyond sectoral and spatial separations, positioning urban–rural linkages as a foundational framework for advancing sustainable development.

M. Ndrevataj (✉)
IUAV University of Venice, Venice, Italy
e-mail: mndrevataj@iuav.it





P. Wollmann et al. (eds.), *The Sustainable Organization – Volume 2*, Future of Business and Finance, https://doi.org/10.1007/978-3-032-23802-3_4

## 4.1 The Legacy of the Urban–Rural Conceptual Divide

In 1996, the United Nations Population Fund (UNFPA) stated in its *State of World Population* report: "The growth of cities will be the single largest influence on development in the twenty-first century."[1] A demographic milestone shift has indeed characterized the mainstream discourse of spatial planning in recent decades.

In the *United Nations World Urbanization Prospects* (2007), it is noted that the world was approaching a historic turning point between the urban and rural areas in 2008 (see Fig. 4.1), when more than half of the global human population would be living in urban areas, as a consequence of rapid urbanization rhythms under an increasing world population number passing from 6.2 billion to 9.2 billion by 2050.[2]

Today (in 2025), 45% of the 8.2 billion people who populate the planet live in urban areas, double the 20% of the 2.5 billion back in 1950. According to the future projections, most of this growth is expected to occur in Africa and Asia, predominantly in small- and medium-sized cities.[3]

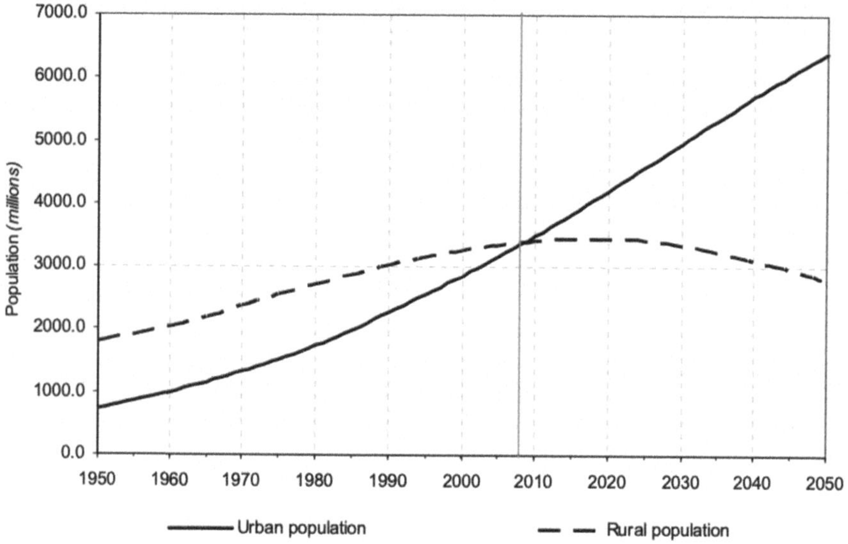


**Fig. 4.1** UN estimates and projection of the global urban and rural population. (Source: United Nations, *World Urbanization Prospects: The 2007 Revision*. See page 4 of: https://www.un.org/development/desa/pd/sites/www.un.org.development.desa.pd/files/unpd_egm_200801_presentation_heilig.pdf)

[1] UNFPA (1996). *The state of world population 1996: changing places: population, development and the urban future*. New York: United Nations.

[2] United Nations (2008). *World Urbanization Prospects: The 2007 Revision*. New York: United Nations.

[3] United Nations (2025). *World Urbanization Prospects 2025: Summary of Results*. New York: United Nations.

While the statistical growth of the urban population is not particularly disputed, nor is the recognition of urbanization as a process that affects all nations around the world, the trajectories of territorial development on a global scale are manifold and difficult to reduce to a simple urban–rural dichotomy.[4] This distinction retains its analytical usefulness only if the two categories are considered in relation to each other, rather than as antagonistic or hierarchically ordered terms.

Living in an increasingly urbanized world should not imply focusing our attention almost exclusively on urban areas. On the contrary, it requires recognizing the dense interconnections between urban and rural areas: rural areas continue to provide a fundamental base of material, ecological, and energy resources, as well as a space where many of the externalities produced by urbanization and globalization accumulate. From this perspective, urban and rural areas do not represent separate entities, but components of a single interdependent territorial ecosystem.

Yet the demographic turn toward urbanism has been accompanied by a parallel and often neglected transformation: the restructuring of rural territories in response to urban growth. The relationship between urban expansion and rural change is neither linear nor unidirectional. Rural areas are not simply left behind as cities grow; they are actively reconfigured by urbanization—in social and spatial terms. Understanding this relational dynamic requires moving beyond the statistical opposition of "urban" and "rural" toward a more integrated, systems-level account of territorial development.

> Far from being an arbitrary clustering of people and buildings, the city is the nucleus of a wider zone of activity from which it draws its resources and over which it exerts its influence. The city and its hinterland represent two phases of the same mechanism which may be analyzed from various points of view.[5] Robert Park and Ernest Burgess (1925:182)

In the critical urban theory tradition, we can find some essential conceptual tools for this reframing. Manuel Castells (1972)[6] challenged the naturalization of "the urban" as a self-evident demographic analytical category, arguing that cities are socially produced through the contradictions of capitalist accumulation. Henri Lefebvre (1970)[7] extended this insight with his hypothesis of "complete urbanization": the progressive permeation of all territories—rural and urban alike—by the logic of capitalist space production. More recently, Neil Brenner and Christian Schmid have translated Lefebvre's framework into an operational research program

[4] In 2020, the United Nations Statistical Commission endorsed the *Degree of Urbanization methodology*—a global standard for measuring settlement patterns, distinguishing cities, towns and suburbs, and rural areas, using population density on a 1 $km^2$ grid, thereby moving beyond a simple urban–rural dichotomy and enabling more consistent definition and identification of settlement types across different countries and over time. Nevertheless, a linear narrative of the "urban age" continues to dominate mainstream discourse in sustainable development.

[5] Park, R. E., & Burgess, E. W. (1925). The City. University of Chicago Press.

[6] Manuel Castells, *The Urban Question: A Marxist Approach* (London: Edward Arnold, 1977; orig. Paris: Maspero, 1972).

[7] Henri Lefebvre, *The Urban Revolution*, trans. Robert Bononno (Minneapolis: University of Minnesota Press, 2003; orig. Paris: Gallimard, 1970).

through the concept of "extended urbanization" and the notion of "operational landscapes"—the vast, often invisible hinterland territories of resource extraction, agricultural production, and logistics that structurally support metropolitan life.[8] Taken together, these theoretical lineages invite a fundamental rethinking of the urban–rural relationship: not as a fixed spatial divide but as a dynamic, contested, and historically produced system of interdependence.

It is against this theoretical and empirical backdrop that the present chapter engages the urban–rural divide as both a conceptual problem and a governance challenge. Drawing on contributions from this second volume of *The Sustainable Organization* (Chaps. 10–16 particularly), it examines how integrated territorial approaches—approaches that transcend the inherited separation between urban and rural policy domains—can operationalize the ambitions of the United Nations 2030 Agenda for Sustainable Development (Fig. 4.2), and in particular SDG 11 (Sustainable Cities and Communities) and more specifically its Target 11.8 (Strong National and Regional Development Planning) calling for better urban–rural linkages, and Target 11.6 (Reduce the Environmental Impact of Cities), which addresses the reduction of the adverse environmental impact of cities.

**Fig. 4.2** UN SDG (source: https://www.un.org/sustainabledevelopment/), used with friendly permission of UN. The content of this publication has not been approved by the United Nations and does not reflect the views

[8] Brenner, N. (Ed.). (2014). *Implosions/Explosions: Towards a Study of Planetary Urbanization*. Jovis; Brenner, N., & Schmid, C. (2015). Towards a new epistemology of the urban? *City*, 19(2–3), 151–182; Brenner, N., & Katsikis, N. (2020). Operational landscapes: Hinterlands of the Capitalocene. *Architectural Design*, 90(1), 22–31.

## 4.2 The Urban Development Landscape and Discourse in the Twenty-First Century

> Ne pas essayer trop vite de trouver une définition de la ville; c'est beaucoup trop gros, on a toutes les chances de se tromper.[9] George Perec (1974:119)

The narrative of an "urban world" is not only inadequate for understanding the divide between urban and rural areas, it is equally reductive when applied to urban territories themselves. The convergence of data toward a predominantly urban world obscures the heterogeneity of urbanization processes across different areas of the world, flattening multiple and divergent urban realities into a single, unnuanced portrait.[10] The distinction between Global North and South thus introduces a second level of complexity that the dominant discourse on world urbanization tends to ignore, while still producing the same a dichotomous effect on the "nature of the city": a geopolitical hierarchy between Northern and Southern cities that is not merely descriptive but carries analytical and political implications—positioning Northern cities as the norm and Southern cities as deviations from it.[11] Seen in this light, the urban–rural dichotomy appears less as an ontological separation between distinct spheres and more as a classificatory effect that obscures the material, ecological, and political interdependencies underpinning urbanization.[12] The question is therefore not only one of transcending the urban-rural divide, but of surfacing the internal dichotomies within what is defined as urban itself: formal and informal, global and local, center and periphery.[13]

It is within this complex and contested urban landscape that certain defining contemporary phenomena emerge with particular force. While the urban areas remain the primary site of social, economic, and political power, with their many facets across the Global North and South, they are home to some contemporary phenomenon, two of which are particularly dominant: globalization[14] and the right to the city.[15] The two are distinct but closely related—the former describes the struc-

[9] English translation: "Don't be too hasty in trying to define the city; it is much too big and there is every likelihood that you will get it wrong." Source: Georges Perec, *Espèces d'espaces* (Paris: Galilée, 1974).

[10] See Robinson, J. (2006). *Ordinary Cities: Between Modernity and Development*. Routledge.; Roy, A., & Ong, A. (Eds.). (2011). *Worlding Cities: Asian Experiments and the Art of Being Global*. Wiley-Blackwell.

[11] See Roy, Ananya (2009). *The 21st-Century Metropolis: New Geographies of Theory*. Regional Studies, 43(6), 819–830.

[12] See Brenner, N., & Schmid, C. (2014). *The "Urban Age" in Question*. International Journal of Urban and Regional. Research, 38(3), 731–755.

[13] See Watson, V. (2009). Seeing from the South: Refocusing urban planning on the globe's central urban issues. *Urban Studies*, 46(11), 2259–2275.

[14] See Saskia Sassen, *The Global City: New York, London, Tokyo* (1991).

[15] The right to the city was originally formulated by Henri Lefebvre in *Le Droit à la ville* (Paris: Anthropos, 1968) and has since been adopted as a guiding principle by UN-Habitat and formally incorporated into the New Urban Agenda. For a contemporary theoretical elaboration, see David Harvey, *Rebel Cities: From the Right to the City to the Urban Revolution* (London: Verso, 2012).

tural forces that reshape urban economies, spatial forms, and power relations in unequal terms across world regions; the latter articulates the political claim, expressed from conditions of marginalization and informality, to collectively participate in and benefit from the city machine. Yet, in the Anthropocene era—in which human activity has become the dominant force shaping Earth's biophysical systems—a third distinctive condition emerges as central, both to how we define and how we govern the city. While recognizing the ecological impacts of human activity, Jason Moore (2016)[16] introduces the concept of Capitalocene,[17] arguing that the ecological crisis is the direct material consequence of capitalist ways of organizing nature, space, and society and not from humanity as an undifferentiated species. This extends the political economy critique tradition toward a full account of the ecological costs of capitalist urbanization, as highlighted also by Brenner and Katsikis (2020), who frame the rural areas as operational landscapes of the Capitalocene.[18]

All the above observations bring us to the concluding question of this section: what is today's Urban Agenda for Sustainable Development? The answer, this chapter argues, cannot be singular. If the theoretical tradition from Castells (1977) to Moore (2016) has taught us anything, it is that the city is not a universal object, urbanization is not a uniform process, and the urban–rural relationship is not a fixed spatial fact but a historically produced, politically contested, and ecologically embedded system of interdependencies.

At the institutional level, the 2030 Agenda and SDG 11 provide the most authoritative contemporary framework,[19] reinforced by the New Urban Agenda (Habitat III, Quito, 2016),[20] which calls for integrated territorial approaches that transcend the inherited separation between urban and rural policy domains. Yet, their transformative potential remains conditional on the critical corrections that urban theory provides: without them, institutional frameworks risk reproducing, under a sustainability vocabulary, the same spatial hierarchies they set out to dismantle.

What emerges is, in Brenner's terms, an *urban constellation* (2014)[21]—not a singular agenda but a plural, polycentric, and irreducibly tensioned field of urban conditions and claims: between growth and redistribution, between formal governance and informal practice, between urban ambition and rural resilience, between the metabolic demands of cities and the ecological limits of the planet. Any serious

---

[16] See Moore, J.W. (Ed.). (2016). *Anthropocene or Capitalocene? Nature, History, and the Crisis of Capitalism*. PM Press.

[17] See: https://en.wikipedia.org/wiki/Capitalocene.

[18] See Brenner, N., & Katsikis, N. (2020). Operational landscapes: Hinterlands of the Capitalocene. *Architectural Design*, 90(1), 22–31.

[19] United Nations (2015). *Transforming Our World: The 2030 Agenda for Sustainable Development*. A/RES/70/1. United Nations General Assembly.

[20] UN-Habitat (2016). *Urbanization and Development: Emerging Futures. World Cities Report 2016*. UN-Habitat, Nairobi.

[21] Brenner, N. (Ed.). (2014). *Implosions/Explosions: Towards a Study of Planetary Urbanization*. Jovis.

engagement with sustainable development must hold these tensions together rather than dissolve them artificially.

## 4.3 The Rural Development Landscape and Discourse in the Twenty-First Century

Compared to urban development, rural development as a distinct scholarly field emerged later in the academic and policy discourse—and has been cast primarily as a passive recipient of modernity rather than as a territory with its own development logic and future.[22] Moreover, the reflection on rural areas has developed in a primarily dichotomous relationship with the urban, with the rural consistently positioned as the other of the urban. Critical planning theory, however, has contributed to challenge this dichotomy and argued for a more integrated understanding of urban and rural as distinct but functionally, environmentally, economically, and institutionally linked spatial entities rather than opposed categories.

Today, the discourse on rural development is shaped by several intersecting phenomena that both reflect also the complex urban–rural relationship. Rural territories—including mountain areas—are more and more emerging as sites of experimentation for new models of tourism, such as slow tourism, that explicitly position themselves against the overtourism associated with urban areas, emphasizing the quality of the travel experience and a deeper engagement with local communities and their culture. Its counterpart, *undertourism*—a tourism management strategy of promotion of lesser-known or under-visited destinations, redistributing visitor flows toward territories that can benefit economically from tourism without suffering its negative effects[23]—is framed as a rural development tool. Yet, while framed as sustainable tourism models, they are driven by an urban demand for escape and therefore carry the risk of becoming a future consumption landscape rather than a genuine and self-determining economic territory.

A second defining discourse is shaped by demographic dynamics. If the dominant narrative in urban studies is one of overpopulation and densification, in the Global North's rural studies it is one of depopulation and abandonment; a narrative that does not apply to the Global South, where rural territories have different issues. In response to depopulation, and accelerated by the COVID-19 pandemic, a counter-movement has emerged: urban residents migrating toward rural and mountain areas—as digital nomads, second-home buyers, or permanent settlers, in search of a slower and more rooted way of life.[24] This movement is seen by some as a revitalization opportunity for territories in decline and by others as a threat. The arrival of new residents tends to drive up housing prices, reshape local services, and displace

[22] See Pelucha, M., & Kasabov, E. (2019). Rural studies and rural development policy. Taylor & Francis; Brown, D.L., & Shucksmith, M. (2024). *Rethinking Rural Studies*. Edward Elgar.

[23] Peltier, D. (2019). Travel megatrend: Undertourism is the new overtourism. *Skift*, February 5.

[24] Halfacree, K. (2012). Heterolocal identities? Counter-urbanisation, second homes, and rural consumption in the era of mobilities. *Population, Space and Place*, 18(2), 209–224.

longstanding communities—generating what has been termed *rural gentrification*[25]: the same market logic documented in urban contexts, transposed into rural settings.

> Despite the dominance of urbanism in the modern world we still lack a sociological definition of the city which would take adequate account of the fact that while the city is the characteristic locus of urbanism, the urban mode of life is not confined to cities.[26] Louis Wirth (1938:1)

Wirth's observation, drawn from his essay "Urbanism as a Way of Life" (1938), retains a striking relevance—and not only in the direction its author intended. If the urban mode of life is not confined to cities, the inverse risk is equally real: that rural territories become progressively shaped by urban assumptions, values, and development models imported by those who migrate toward them. It is precisely this risk that we can term an *urban way of rural development*. Considering that the power of the urban narrative does not stop at the city boundary, it can shape how rural territories are perceived, planned, and inhabited, often in ways that subordinate locally rooted development logics. The recent declining attractiveness of urban settings—driven by housing unaffordability, congestion, environmental degradation, and the spatial freedoms unlocked by remote work—has pushed growing numbers of people toward rural areas, perceived as natural realms that permit a slower and more sustainable way of life. This movement can generate new forms of rural habitation and underlie a risk, behind its apparent opportunity: that the endogenous ways of life of rural territories are displaced rather than strengthened by this influx.

The 2030 Agenda does not include a dedicated rural development goal, addressing rural territories only indirectly through a series of sectoral goals—food, water, biodiversity, poverty—Instead, the OECD's rural policy paradigm (2006, 2020), advocates for a territorial approach that valorizes rural assets and capacities, promotes rural innovation, and explicitly rejects the image of rural territories as mere operational or compensation landscapes serving urban demand.[27] Central to this framework is the recognition that rural territories are not homogeneous: the OECD distinguishes between rural areas within functional urban regions, rural areas close to cities, and remote rural areas—each requiring differentiated policy responses rather than a single development model. Thus, highlighting the internal divide, and recalling for a place-based approach, in addition to integrated and cross-sectoral development.

---

[25] Phillips, M. (1993). Rural gentrification and the processes of class colonisation. *Journal of Rural Studies*, 9(2); Sherman, J. (2023). "Please Don't Take This": Rural gentrification, symbolic capital, and housing insecurity. *Social Problems*, 70(2), 491–510.

[26] Wirth, L. (1938). Urbanism as a way of life. *American Journal of Sociology*, 44(1), 1–24.

[27] See OECD (2006). *The New Rural Paradigm: Policies and Governance*. OECD Rural Policy Reviews. OECD Publishing; OECD (2020). *Rural Well-Being: Geography of Opportunities*. OECD Rural Studies. OECD Publishing.

## 4.4 Toward Urban–Rural Linkages for an Integrated Sustainable Development Approach

> Tradition has broken down. ...Rural influences neutralise the town. Urban influence neutralise the country. In a few years all will be neutrality. The strong, masculine virility of the town; the softer beauty, the richness, the fruitfulness of that mother of men, the countryside, will be debased into one sterile, hermaphrodite beastliness.[28] Thomas Sharp (1932:11)

Sharp's regret, written in 1932 as a polemical defense of the distinction between city and countryside against the advance of suburban sprawl, serves as an unexpected but productive introduction to a discussion of integration. His anxiety was the opposite of ours: while he feared the dissolution of urban and rural peculiarities into a shapeless uniformity, the challenge addressed in this section is almost the opposite, namely the persistence of an artificial separation between urban and rural territories that has structured urban thinking, policy frameworks, and development discourse for much of the twentieth century, to the detriment of both.

In spatial planning, an integrated approach to urban–rural linkages refers to a planning and governance philosophy that treats cities and their surrounding rural areas not as separate, opposing entities but as components of a single, functionally interdependent territorial system—one in which a constant bidirectional flow of people, resources, goods, services, and ecological functions takes place.[29] This approach rejects both the urban-centric model, which treats rural territories as passive hinterlands serving urban demand, and the ruralist romanticization that imagines rural areas as self-sufficient and separable from the urban economies on which they structurally depend. Instead, it calls for planning and governance frameworks capable of managing the complex, asymmetric, and often conflictual relationships between urban and rural territories in ways that produce equitable and sustainable outcomes for both.

It is in precisely this sense that integrated urban–rural development must be understood—not as the homogenization Sharp feared, but as the governance of interdependence between territories that remain distinct in character, function, and development logic. Because the way urban and rural territories are understood, classified, and governed—and the terms on which their interdependencies are managed—carries direct consequences across the economic, social, and environmental dimensions of sustainable development.

[28] Sharp, T. (1932). Town and Countryside: Some Aspects of Urban and Rural Development. London: Oxford University Press.

[29] The foundations of an integrated approach to urban–rural territorial planning were laid long before the concept entered mainstream discourse by Patrick Geddes, in his work *Cities in Evolution* (1915). Geddes was among the first thinkers to argue systematically that the city could not be understood, planned, or governed in isolation from its regional hinterland.

## 4.5 Conclusions

The longstanding divide between urban and rural has deep historical roots, but its modern form was consolidated by industrialization and accelerated urbanization, which—combined with a structural demographic shift toward cities—produced two territories understood as distinct, functionally opposed, and governed accordingly. This institutional separation was not merely descriptive: it generated the planning frameworks, policy silos, and development hierarchies that this chapter has examined.

Reading that separation through the lens of critical scholarship reveals a different picture. The consistent lesson of both theory and official policy frameworks is that every dichotomous category—urban and rural, core and periphery, local and global—becomes an obstacle to sustainable development the moment it is treated as two separate entities rather than two poles of a single relational system. This chapter has not aimed to be comprehensive; its purpose has been purposefully selective: to unsettle the assumption of separation and build the analytical case for an integrated approach grounded in urban–rural linkages.

It is in this light that the urban–rural relationship must be rethought: not as a hierarchy in which rural areas aspire to urban conditions, nor as a romantic opposition in which rural authenticity is preserved against urban contamination, but as a system of interdependence in which both urban and rural territories have legitimate claims, responsibilities, and development trajectories that must be recognized, negotiated, and governed together.

**Mersida Ndrevataj** is an urban scholar with a background in architecture, urban planning, and environmental psychology. She holds a PhD in Urban Planning and Public Policy from the IUAV University of Venice, where her research focused on the implementation of public policies in cultural heritage management. She is currently a lecturer at the IUAV University of Venice and collaborates with several COST Action research networks focused on urban studies. Her work brings together academic research and field-based inquiry by adopting audiovisual and sensory approaches to investigate urban spaces, everyday socio-spatial practices, and lived experiences.

# Part III

# Global Institutions to Foster Sustainable Cooperation and Improve Communication

# 5 Impactful Leadership for the Sustainable Development of a Country: Insights from the Unique Model of Qatar

Rabih El-Haddad and Sabrina Illiano

**Abstract**

This chapter examines the pivotal role of leadership in driving sustainable development, using Qatar as a case study. In moving beyond conventional analyses that equate development with political systems, it argues that certain leadership qualities can decisively shape national trajectories irrespective of governance structures. Through an in-depth exploration of Qatar's rapid economic diversification, strategic investments in human capital and infrastructure, labor and social reforms, and environmental initiatives, this chapter demonstrates how deliberate, long-term planning and adaptive governance can translate national resource wealth into enduring societal and economic outcomes. Qatar's experience highlights the importance of continuity, intergenerational planning, and culturally grounded modernity. It, therefore, offers insights for states and organizations seeking to achieve sustainable development goals. By reframing leadership as a catalyst for measurable, durable progress, this analysis contributes to a broader understanding of how strategic leadership can generate tangible and long-term development impacts in diverse political and cultural contexts.

R. El-Haddad (✉) · S. Illiano
United Nations Institute for Training and Research (UNITAR), Division for Multilateral Diplomacy, Geneva, Switzerland
e-mail: rabih.haddad@unitar.org; sabrina.illiano@unitar.org



P. Wollmann et al. (eds.), *The Sustainable Organization – Volume 2*, Future of Business and Finance, https://doi.org/10.1007/978-3-032-23802-3_5

## 5.1 Introduction

***Plans are not drawn up for the purpose of completing a procedure as a formality only, plans are drawn up to be implemented.—Sheikh Tamim bin Hamad Al Thani.***[1]

How can the impact of leadership be measured beyond politics? When examining pathways toward the achievement of sustainable development goals, the role of leadership can be an elusive factor, difficult to identify yet essential to understand. Often, leadership is conflated with political power, which limits the ability to account for its role in either advancing or constraining development. However, to separate the wheat from the chaff, it is leadership, distinct from the mechanics of political systems, that ultimately provides the vision, direction, and mobilization that is necessary to convert policies and strategies into measurable outcomes. It is through leadership where the power to shape priorities and create the conditions necessary for progress can be most effective. This chapter focuses on leadership as a key driver of economic growth and sustainable development. The meaning of leadership in the context of this discussion extends far beyond the exercise of political authority to encompass the capacity to envision, mobilize, and guide resources toward long-term, lasting goals. This chapter examines leadership not as a function of governance but as a critical element for the creation of a safe environment where sustainable growth, positive peace, and individual happiness prosper. Through this approach, leadership is reframed as a catalyst for sustained economic and social advancement that lasts beyond the immediacy of political cycles.

In the global discourse of sustainable development, there has been relatively little attention paid to the decisive role of leadership style in shaping national trajectories. The case of Qatar, a small monarchical country that has quickly risen from a relatively insular, carbon-dependent state to a global and regional powerhouse in key industries, offers an exemplary opportunity to explore how visionary and strategic leadership can drive rapid development while simultaneously laying the foundations for long-term stability and sustainability. Over the course of just a few decades, Qatar, under the guidance of successive Emirs Sheikh Hamad bin Khalifa Al Thani and Sheikh Tamim bin Hamad Al Thani, has successfully transformed itself into a dynamic, increasingly diversified economy as well as an international hub for diplomacy, education, and technology alongside facilitating consistent progress in sustainable development. This transformation has been neither incidental to, nor has it been purely the result of, resource-driven wealth. Seneca once said, "no man was ever wise by chance." Similarly, no country is prosperous and stable by chance. If wisdom is to be equated with success, then it is our firm belief that the Qatari successful transformation has been intentionally pursued by coherent leadership that is deliberate, future-oriented, and adaptive to changes in the global political landscape.

---

[1] Administrative Control and Transparency Authority: "Words of his highness, the emir of the state of Qatar". https://www.acta.gov.qa/en-us/media/Pages/words-of-his-highness-the-emir-of-the-state-of-qatar.aspx. State of Qatar, © 2014.

Conventional assessments of development often rely on Western political trajectories as the benchmark for progress, with a particular emphasis on democratic governance models as necessary, or, at the very least, desirable preconditions for successful development. Such frameworks tend to assume that the hallmarks of these models, including electoral pluralism and institutional separation of powers, are inherently the most compatible with critical development metrics such as human rights, social well-being, and economic security. Yet, Qatar challenges this assumption with a governance structure rooted in centralized leadership and the enduring legitimacy of the Al Thani ruling family that diverges sharply from the idealized models but that has nevertheless delivered outcomes that align with, and, in many areas, advance, the objectives of the United Nations Sustainable Development Goals (SDGs). To understand Qatar's achievements in this area therefore requires situating the country within its own historical, cultural, and regional context rather than judging it through an external lens that overlooks the particularities of its positioning. In doing so, it is not to suggest that more countries should relinquish their democratic systems in favor of absolute monarchies if they want to advance their development. Rather, it is to call attention to the ways in which Qatar's model of governance, when combined with a leadership style that prioritizes sustainability, forward-thinking, and investment in citizen well-being, has produced quick, effective, and tangible progress toward the SDGs and has enabled Qatar to become a world leader in several areas. Qatar's success reveals that leadership style is a crucial element of sustainable development regardless of political infrastructure, and therefore that it is imperative to pay close attention to how leadership can be leveraged as a tool of progress.

Qatar's leadership model is built on several interrelated pillars that will be explored throughout this chapter. It is visionary, articulating a long-term plan that aligns national development with global sustainability agendas, most evident in the Qatar National Vision 2030. It is transformational in its strategic deployment of its hydrocarbon wealth to build institutions, improve infrastructure, and invest in human capital, which has resulted in significant structural changes to Qatar's economy and society. It is adaptive, demonstrating a willingness to respond effectively to external pressures while maintaining internal resilience, cohesion, and tradition. It is modern in its capacity to provide for political stability and a sense of gradual evolution toward a higher good. Finally, it is a sustainable model that intelligently embeds diversification and long-term success into national planning in line with both domestic goals and international sustainability agendas. Together, these pillars describe the unique way in which Qatar has navigated the challenges of globalization and development while remaining firmly situated within its own cultural traditions and values. Central to the success of this model has been leadership that moves deliberately, one step at a time. Economic development in Qatar has progressed extraordinarily quickly owing to natural resource wealth that has been carefully managed and reinvested through strategic planning. Political reforms, by contrast, have proceeded more gradually. However, this slower advancement should not be interpreted as stubbornness or an unwillingness to conform to international standards but instead should be understood to reflect a conscious decision to prioritize stability and continuity over abrupt transformation, evolution over revolution in an Arab region fraught with political instability. The smooth transition of power from Sheikh Hamad bin Khalifa all Thani to Sheikh Tamim bin Hamad Al Thani in 2013 illustrates this unique principle and highlights how Qatar's leadership is committed to ensuring consistent progress and to building on the legacies of achievements that have come before. In not pushing for political reforms to race ahead of economic foundations or allowing economic progress to become disrupted by political fragmentation, Qatar's approach has struck a careful balance.

As a result, the international community has increasingly recognized Qatar's achievements in this regard. By hosting United Nations conferences such as the 2012 UN Climate Change Conference (COP18) and the 2023 UN conference on Least Developed countries (LDC5), Qatar has signaled its intention and preparedness to contribute not only as a regional leader but as a partner in advancing the global sustainable development agenda. Its consistent improvements across social, economic, and environmental dimensions have thus distinguished it within the Gulf Cooperation Council (GCC) and beyond. This chapter will explore Qatar's leadership model as a driving force of sustainable development, focusing primarily on the economic dimension of Qatar's development while situating these efforts in relation to the broader goals of the SDGs. Structured around the five key qualities of Qatar's leadership that have been identified—visionary, transformational, adaptive, modern, and sustainable—the analysis here will trace how Qatar has forged their own path of development and successfully positioned itself at the forefront of regional and global progress. Furthermore, by considering how far the Qatar of today has come from the Qatar of its early years in such a short period of time, the discussion will highlight this nation's remarkable trajectory to reveal the essential link between leadership and enduring sustainable development. While there are other examples of Qatar's transformative development that have been explored in depth elsewhere, the important takeaway for this chapter is that each point of success on Qatar's trajectory illustrates how impactful leadership that is rooted in long-term planning, strategic investment, and adaptive governance can drive substantial progress toward sustainable development even in different political systems. Though not without its share of controversy or contradictions, Qatar's model therefore offers a compelling case for how leadership, legitimacy, and sustainability can intersect to produce efficient and effective development.

## 5.2 Visionary Leadership: Strategic Vision as a Driver of Development

***There is no passion to be found in playing small and settling for a life that is less than the one you are capable of living.—Nelson Mandela.***[2]

[2] Mckenzie, S.: "Nelson Mandela Inspirational Quotes". https://worldcertification.org/nelson-mandela-inspirationalquotes. World Certification Institute, © 2013.

Visionary leadership is one of the defining features of Qatar's national development narrative, by which it is meant that Qatar's development is facilitated by a leadership style that is committed to long-term, future-oriented strategic frameworks that articulate national development goals. Developing a vision requires leaders with a capacity to dream, to imagine a bold destination for their nation, and to invite their citizens to implement that vision while assuming the risks that be. Most importantly, visionary leadership is an invitation to rise, to challenge oneself and one's own nation to greatness.

The essence of Qatar's visionary leadership lies in thinking beyond immediate resource rents, preventing its natural resource windfall from fostering dependency or rentier stagnation. The Qatar National Vision 2030 (QNV 2030), which was unveiled in 2008, clearly articulates this leadership ethos in the way that it sets out a long-term blueprint to guide progress across economic, human, social, and environmental dimensions. In this way, the visionary approaches of Qatar's leadership have involved anticipating future challenges and opportunities, constructing a trajectory for the country beyond its immediate social and political goals, and mobilizing both resources and institutions to consistently move forward along this path. In just over half a century, having only achieved independence from Britain in 1971, Qatar has undergone a dramatic transformation from a small economy on the margins of the Gulf to one of the most prosperous nations in the world—and, indeed, is on track to soon become one of the most influential actors in the GCC region. While Qatar was certainly aided by its wealth from hydrocarbon revenues, a fact that is undeniable, it would still be reductive and overly simplistic to merely attribute this transformation to its natural resources. After all, there are plenty of resource-rich states that have struggled to do the same. Rather, it is imperative to acknowledge the role of leadership in Qatar's achievements. In doing so, it is possible to garner essential insights into the leadership qualities that are critical to achieving controlled and lasting progress toward sustainable development.

Historically, Qatar's economic base was modest and primarily relied on the pearling and fishing industries. Even with the discovery of oil in the 1940s, Qatar remained relatively underdeveloped with limited growth or progress in infrastructure well into the 1990s. Despite providing a new economic foundation on which to build, the presence of oil alone was not sufficient to usher in a wave of prosperity. It was, in fact, not until 1995, when Sheikh Hamad bin Khalifa Al Thani assumed power, that Qatar truly began to enter a visionary era characterized by modernization, diversification, and global engagement. Alongside aggressively developing Qatar's oil and gas industries, and contrary to the advice of senior international advisors, Sheikh Hamad decided to invest in liquefied natural gas as a new technology that would enable his country to transport and sell gas more easily. He took a huge risk on behalf of his nation that paid off and enabled Qatar to punch above its weight.

He had the foresight to establish the Qatar Investment Authority in 2005 as a sovereign wealth fund dedicated to managing Qatar's oil and natural gas revenues not only for the present, but with an eye toward ensuring future generations would continue to benefit from this wealth. The QIA has since grown to manage an estimated $475 billion in assets with a portfolio spanning multiple continents and sectors including real estate, technology, infrastructure, energy, and finance. Through this strategy, Qatar has leveraged its resource base to simultaneously secure domestic economic stability, protecting it from the volatility of oil and gas markets, and embed itself in key networks of international trade and capital. According to World Bank data, Qatar's GDP annual growth during this period jumped from a mere 3.4% in 1990 to 8% in 2000, ultimately peaking at an astounding 26.2% in 2006.[3] Since then, Qatar has maintained steady growth year after year and is currently projected to be the fastest growing economy of the GCC states in 2026.[4] While the impressive scale of its holdings alone makes the QIA worth significant attention when discussing the conditions that have facilitated this growth, it is also significant in terms of what it represents for Qatar's visionary approach to sustainable development. In this sense, the QIA works in tandem with QNV 2030 by providing the material foundation to support sustained investments in human capital, social development, and infrastructure. In other words, it ensures that the goals and visions for the country that are outlined in QNV 2030 can become a reality.

The remarkably smooth and, notably, voluntary transfer of authority to Sheikh Tamim bin Hamad Al Thani in 2013, who has continued to carry his father's vision forward, reinforces the positive impact of the principles of continuity and stability that are central to Qatar's governance. This transition firmly communicated that the vision for Qatar established by the previous Sheikh was not tied to or dependent on the motivation of a particular individual and was instead embedded into the fabric of Qatar's leadership and governance model. The voluntary abdication of Sheikh Hamad was itself a powerful statement of institutional continuity and assurance that Qatar's trajectory of long-term planning and investment would not be disrupted by political succession. Since assuming leadership, Sheikh Tamim has consolidated and expanded the strategic role of the QIA while delivering on commitments made during the rule of his father. For example, the successful delivery of the 2022 FIFA World Cup, which served not only as a demonstration of the country's organizational capacity and diplomatic strength, but also as an opportunity to test and scale its capacity to achieve sustainable urban development. Qatar won the bid to host this World Cup in 2010, beating out several much larger countries and becoming the first in the Middle East to do so. The event was explicitly framed from its inception by the Government of Qatar as a way of accelerating progress toward the environmental and urban objectives outlined in QNV 2030.[5] By the time the World Cup took place, however, both the domestic and international contexts had changed considerably from when Sheikh Hamim had initially succeeded in the tournament bid. Qatar

---

[3] World Bank: "Qatar GDP and Development Statistics." Washington, DC: World Bank, © 2021–2025.

[4] World Bank: "Qatar GDP and Development Statistics." Washington, DC: World Bank, © 2021–2025.

[5] Scharfenort, N.: "Urban Development and Social Change in Qatar: The Qatar National Vision 2030 and the 2022 FIFA World Cup." *Journal of Arabian Studies* 2 (2), 209–230, © 2012.

had navigated the breakdown of regional diplomatic relations, there was heightened scrutiny of its internal labor practices, and global demands for sustainability criteria in major events had ushered in new expectations of environmentally conscious urban planning. In this context, Sheikh Tamim's leadership was crucial in continuing to realize his father's vision in a way that showcased Qatar's capacity to a global audience and advanced its long-term development goals. With firm adherence to QNV 2030, the event revealed the importance of continuity in vision across generations of leadership: Sheikh Hamad had laid the foundations and set bold ambitions while Sheikh Tamim ensured their successful realization despite a more complex global context.

Much has been written already about the 2022 World Cup as a landmark event for Qatar and, more broadly, the Middle East, and it is not necessary to reiterate the same points here. Its significance for the purposes of this chapter lies instead in what it reveals about Qatar's leadership model and strategic positioning on the global stage from a successful bid in 2010 to an innovative, standard-setting event 12 years later. The 2022 World Cup marked a turning point in seeing GCC states as destinations for mega events, that is, events which are measured not only by their size but by their significant and multidimensional impact on the people, environment, infrastructure, and reputation of the places in which they take place.[6] These require a careful balance of short-term spectacle and long-term legacy building. While they can generate global visibility and national pride in the immediate term, when executed successfully their true developmental value lies in the systems and sociopolitical capital they leave behind. In Qatar's case, this meant integrating sustainability criteria into the planning from the outset of the project, as a core pillar of delivery rather than an inconvenient requirement or an afterthought. The 2022 tournament was promoted as the first "carbon–neutral" World Cup, a claim whose veracity has been debated but that undeniably led to concrete advances such as stadiums built to LEED environmental standards[7] and the expansion of the Doha Metro. These achievements are directly related to policies implemented by Qatar's leadership and reinforced across regimes that laid the foundation for improvements that can be seen in the measurable progress of Qatar's economy—progress that has not only challenged stereotypes of resource-rich economies, but that has highlighted Qatar's resilience and ability to continue to thrive in the face of global challenges. Measured against itself over time, Qatar's trajectory exhibits consistent growth that is intrinsically tied to the steady commitment of its visionary leadership to development and progress.

Taken together, the dimensions discussed here—historical foresight, long-term planning, intergenerational investment, and institutional continuity—illustrate the visionary quality of Qatar's leadership. Qatar has avoided the pitfalls of overreliance on hydrocarbons and sustained incredible economic growth even in

---

[6] Müller, M.: "Mega Events and Their Multidimensional Impact." Cham: Springer Nature, © 2018.

[7] LEED = Leadership in Energy and Environmental Design that represents a globally recognized green building certification system developed by the U.S. Green Building Council (USGBC), see https://www.usgbc.org/leed.

challenging circumstances by committing, both politically and financially, to the goals outlined in QNV 2030. The establishment of the QIA under Sheikh Hamad bin Khalifa Al Thani exemplified in particular the commitment to safeguarding Qatar's prosperity for future generations, while the subsequent voluntary transition of power to Sheikh Tamim bin Hamad Al Thani reaffirmed that this vision was not tied to a single leader but embedded in Qatar's political system and approach to governance. Sheikh Tamim's tenure has built upon the economic growth initiated by his father's policies to demonstrate how visionary goals can continue to be consolidated and achieved despite shifting global contexts, from maintaining growth during regional and global crises to delivering on the ambitious promise of hosting the first FIFA World Cup in the Middle East. The tournament, first envisioned under Sheikh Hamad, became a culmination of this visionary strategy as a platform for advancing sustainable infrastructure, national development goals, and positive global visibility. The key takeaway from this trajectory is that the visionary quality of Qatar's model of leadership sets aside short-term gains or immediate political clout in favor of carefully and intentionally building capacity to support long-term visions. It is this sustained and forward-looking approach that has enabled Qatar to achieve such rapid improvements while also building the resilience needed to meet future challenges.

## 5.3 Transformational Leadership: Reshaping Structures for Economic and Social Success

***The world as we have created it is a process of our thinking. It cannot be changed without changing our thinking.—Albert Einstein.***[8]

Where the visionary dimension of Qatar's leadership model focuses on setting ambitious goals and ensuring long-term stability, the transformational quality lies in the capacity to translate these aspirations into tangible structural changes that inspire Qataris and the cosmopolitan world citizens, especially those who transit in Doha. Transformation starts with an instinctive decision and an act of will in which one decides to leave the convenient present for a brighter unknown future.

A central feature of Qatar's transformation has been the development of its domestic infrastructure as a result of leadership's recognition early on that the prosperity derived from exports of natural gas had to be reinvested into projects that would expand the country's productive capacity and improve the quality of life for its citizens. The construction of Hamad International Airport, for example, in conjunction with the significant expansion of Qatar Airways into one of the world's leading airlines, provided world-class facilities and also turned Doha into a major global transit hub. The Hamad International Airport is now consistently ranked among the best in the world and serves over 40 million passengers annually, further

[8] Sanford, C.: "Heeding Einstein's admonition: A White Paper on regeneration's significance." https://carolsanford.medium.com/heeding-einsteins-admonition-a-white-paper-on-regeneration-s-significance-e4bbbe702670. Medium, © 2023.

boosting Qatar's reputation in the global aviation industry.[9] This project not only connected the citizens of Qatar to worldwide destinations, but also helped to turn Qatar itself into a highly desirable destination itself. The value of such transformative initiatives goes beyond superficial optics, financial gains, or pure utility, as the legitimacy of Qatar's political structure and governance is intrinsically tied to the *perception* of its legitimacy. That is, while the ruling family's legitimacy is not derived from an electoral process, it still relies on being perceived as credible by its citizens; in addition to the tribal values that bind members of the Qatari society together, this perception is firmly rooted in the government's demonstrated capacity to deliver tangible improvements in living standards, economic opportunities, and social services. This arrangement creates an informal social contract wherein political participation is limited, but citizens receive extensive benefits in return such as heavily subsidized (or free) education and healthcare, readily available employment opportunities in profitable public sectors, and generous social support systems. Qatar's model has proven remarkably effective in delivering high levels of quality public goods, social stability, and citizen well-being. Linking economic growth to tangible benefits has enabled the state to maintain a positive image among the citizen body, thereby effectively insulating the leadership from political disruption.

Qatar's leadership has clearly recognized that actual, positive change is a prerequisite for a satisfied population, one whose material needs are met and who thus accepts the political governance structure, and they have consequently built an economy that meets this expectation. The scale of Qatar's economic transformation is striking. Between 2000 and 2022, the GDP per capita rose from $34,000 to more than $84,000, which places Qatar among the wealthiest countries in the world.[10] Though much of this growth was indeed initially fueled by Qatar's exports of liquified natural gas (LNG), the leadership has deliberately reinvested oil and gas revenues back into infrastructure, services, and social development. Far beyond just becoming a high-income economy, Qatar's citizens consequently enjoy a high standard of living wherein welfare services are largely subsidized by the government and employment opportunities are readily accessible. Data on employment in Qatar provides a clear picture of Qatar's success in this area: Qatar's citizen unemployment rate has remained extremely low, coming in at under 1% in recent reports.[11] Qatari nationals have a choice between taking jobs in the public sector, which are prestigious, secure, and well-compensated, or taking advantage of the many available private sector positions in finance, logistics, and services. Furthermore, Qatar's leadership has gone to great lengths to establish initiatives aimed at encouraging entrepreneurship among its younger population as part of the recognition that a dynamic labor market is an essential component of long-term diversification. In this way, Qatar's orientation toward transformation is particularly strategic as it does not

[9] Skytrax: "Hamad International Airport World Rankings." London: Skytrax, © 2022.

[10] World Bank: "Qatar GDP and Development Statistics." Washington, DC: World Bank, © 2021–2025.

[11] Planning and Statistics Authority: "Employment Statistics for Qatari Citizens." Doha: Planning and Statistics Authority, © 2021.

rely on simply propping up domestic industries with its resource revenues but instead actively seeks to equip its population with the capacity to engage with global sectors and build economic resilience. These efforts include programs headed by the Qatar Development Bank and innovation hubs such as the Qatar Science and Technology Park, which provide funding and incubation for small and medium enterprises. Ensuring employment security alongside nurturing opportunities for innovation maintains citizens' confidence in the state's ability to deliver prosperity and creates a citizen base that can contribute significantly and productively to economic growth.

Though the focus in this chapter is on the success of Qatar's leadership model for sustainable economic development, it is important to note this synergy between Qatar's economic goals and its consistent focus on citizen well-being as a crucial measure of progress. This is explicitly laid out in QNV 2030, which identifies human development as one of its four central pillars. The connection between economic growth and well-being can be seen in concrete outcomes, including the rise in life expectancy and the significant decline in maternal and infant mortality since 2000.[12] For the purposes of analyzing the transformative quality of Qatar's leadership, it is necessary to acknowledge that Qatar's social and economic transformation has not required sacrificing traditional values or cultural identity. The measurable evidence of Qatar's progress makes it clear that modern infrastructure, excellent services, and high levels of human well-being can coexist with deeply rooted social and cultural systems. The continuity of leadership through the Al Thani family, the preservation of traditions such as shura-style consultation, and the central role of family and community in social life, alongside investments and global partnerships that further economic growth, illustrate this balance in practice. Economic progress, especially when accomplished within such a short period of time, can easily run the risk of creating expectations that political development must move at the same pace. However, Qatar has followed a different path wherein reform unfolds gradually in step with national capacity and priorities. This approach has helped to prevent economic transformation from generating destabilizing pressures, ensuring that it instead strengthens the legitimacy of leadership. Citizens see tangible benefits, which are consistently being updated and expanded with new technologies, and their trust in the state's ability to navigate global challenges and protect their economic interests is reinforced.

One example of how economic growth has been tied to transformational capacity-building is in the establishment of Education City under the Qatar Foundation.[13] Now home to more than a dozen global universities, research institutes, and innovation centers, Education City embodies Qatar's recognition of the value of creating channels through which its citizens can develop the skills and knowledge to

[12] World Bank: "Qatar GDP and Development Statistics." Washington, DC: World Bank, © 2021–2025.

[13] Al-Musalmani, F.; Maalouf, S.: "Qatar Foundation: A Sustainability Innovator." In: Cochrane, L.; Al-Hababi, R. (Eds.): *Sustainable Qatar: Social, Political and Environmental Perspectives*. Singapore: Springer Nature, © 2022.

participate fully and freely in diverse economic opportunities. Education City hosts branch campuses of leading institutions, including Georgetown University, Carnegie Mellon University, and University College London, allowing Qatar to leverage partnerships with existing universities to create access to world-class education for Qatari citizens without even having to leave the country. Doing so enables students to train in both classic and emerging fields while building the foundation for entrepreneurship and the growth of new industries in Qatar. Through their transformational approach, Qatar's leadership has anticipated the demands of economic diversification for a skilled workforce and culture of innovation and has invested in creating the conditions to meet those demands. Investing in education and research at home lays the groundwork for a knowledge-based economy that can thrive even if hydrocarbons decline and sets the stage for Qatar to become a destination for much more than just oil and gas. This initiative, as part of a systematic reorientation of Qatar's development trajectory toward innovation, human capital, and long-term sustainability, thus reflects transformational leadership in practice. These investments in education will undoubtedly transform Qatar as much as, or even more than, formal policy could by preparing Qatari citizens to sustain and expand economic diversification.

The transformational pillar of Qatar's leadership has produced a remarkable integration of economic and social development to deliver on tangible structural changes that set up citizens for success. Resisting the impulse to consume resource wealth in the short term, Qatar has channeled revenues into creating a resilient economy both enjoyed and supported by a productive, educated population. In this way, the legitimacy that Qatar's political leadership derives not from democratic procedures but from the consistent delivery of growth, security, and well-being offers a critical insight into the value of transformational leadership for sustainable development. The paradox of a country with a unique political governance model that has had such irrefutable success in this area introduces a nuanced understanding of effective leadership beyond the Western liberal-democratic paradigm, especially in the context of the Gulf region's specific sociopolitical fabric and history. Beyond diversifying industries or constructing new cities for the short-term enjoyment of the elite, Qatar's transformational approach to leadership ensures that development means citizens experience the benefits of progress in their everyday lives. It is not enough to merely be visionary; impactful leadership must deliver on the vision through transformation.

## 5.4 Adaptive Leadership: Navigating Change and Building Resilience

***The impediment to action advances action. What stands in the way becomes the way.—Marcus Aurelius.***[14]

[14]Aurelius, M. and Studios, D.: *Meditations—Marcus Aurelius: The New English Translation for Modern Readers*. Julian Everett, © 2024.

The next defining feature of Qatar's leadership model is its capacity for adaptation, remaining responsive to shifting domestic needs as well as to evolving global conditions while still maintaining a consistent trajectory toward long-term goals. Whereas the visionary dimension sets the direction, and the transformational dimension delivers structural change, adaptive leadership reflects the flexibility and resilience that ensure Qatar's goals can remain viable in quick-changing, and even volatile, environments. One of the most obvious tests of Qatar's leadership model has been its ability to adapt to the demographic transformation that has accompanied the rise of its global reputation and influence. In just a few decades, Qatar has grown from a small, sparsely populated state into a hub that attracts workers, students, and professionals from around the world. The population has increased from fewer than 300,000 people in 1970 to nearly three million today, with noncitizens now making up the overwhelming majority.[15] This population growth has been a true test for Qatar's model of leadership as it is one thing to use natural resource wealth to support a small citizen population, but quite another to expand that strategy out to include a large, majority noncitizen base. While there are many societies where such a demographic shift could have easily been destabilizing, Qatar's leadership has instead managed this change in a way that preserves social stability, economic growth, and national identity. In successfully adapting to such a markedly different dynamic, Qatar has demonstrated that it does not owe its success entirely to the good fortune of natural resources but rather deserves credit for its approach to leadership that can accommodate change without losing sight of its priorities.

How leadership responds to changing social realities can be a critical test of its strength, especially when those changes are driven by demographic shifts that require domestic policies and infrastructure to account for a large number of outsiders from a variety of different backgrounds. The recent, and now consistently high, influx of noncitizens into Qatar—whether expatriates, foreign professionals, construction workers, service staff, or students—has had a particularly dramatic effect on such a small state. To put the scale of this shift in perspective, as of 2025 expatriates now make up 88.4% of the population while only the remaining 11.6% are Qatari citizens. The proportion of expatriates in the labor force is similarly high, with estimates that migrants make up an astonishing 95% of the total workforce in Qatar. While Qatari citizens remain the majority in lucrative public sector jobs, and citizen unemployment remains low, it would be naïve to think that such a disparity would be sustainable without a willingness to adapt. Confronting the demographic reality of Qatar is unavoidable, particularly in the context of its growing reputation as a regional and international leader. In the face of this challenge, Qatar's leadership has embraced the new opportunities offered by its status as a sought-after destination by adapting policies, institutions, and services to integrate this diverse population while maintaining social cohesion and avoiding destabilizing threats to its political structure.

[15] World Bank: "Qatar GDP and Development Statistics." Washington, DC: World Bank, © 2021–2025.

One adapted policy area has been the expansion of public healthcare and insurance for noncitizen residents resulting from the recognition that a large expatriate population requires accessible and affordable health services to thrive and productively contribute to Qatar's growth. The National Health Strategy was updated to include mandatory basic health insurance by employers for expatriate workers and their eligible dependents, a critical step toward systematic protection. This move reduces the risk of public health burdens and helps to integrate foreign workers into Qatari systems, which can mitigate potential sources of discontent. The National Health Strategy 2018–2022 framed these changes precisely as a necessary response to accommodate Qatar's new demographic landscape and to build a resilient health system capable of scaling to meet Qatar's ambitious development goals. Qatar is now on the second iteration of the National Health Strategy for 2024–2030, with aims of building on the previous strategy to empower the population through health consciousness. Notably, one of the three principles around which the updated strategy was designed is "supporting Qatar's economic diversity agenda," explicitly linking economic development to health as both an industry and as an essential factor in human social welfare.[16] The new policy ensures that the noncitizen majority is not excluded from receiving the essential care that enables them to continue being productive members of Qatar's society and economy and simultaneously reduces the potential strain on the public healthcare system by shifting some of the onus of responsibility onto private healthcare provider networks.

Closely tied to this decision has been the leadership's willingness to adjust labor policies in response to domestic needs and international expectations. Qatar's development relies heavily on expatriate workers, which historically created possible risks to its standing in terms of worker protections and international criticism. Specifically, the acceleration of Qatar's development and the parallel rise of its reputation in global forums has led to increased scrutiny of the controversial kafala (sponsorship) system.[17] The kafala system, widely used in GCC countries, has long been the subject of criticism. Such critiques stem from the system's structure that ties a worker's legal status in the country to the sponsor, effectively delegating the responsibility for migrant governance to private citizens rather than the state. Consequently, migrants working within the kafala system can be vulnerable to exploitation as they cannot change employment or even decide to leave the country without the permission of their sponsor and this arrangement creates a dependent relationship that can easily be abused. The ethics of the kafala system's purpose has also been called into question with some analyses linking it not only to the desire to exploit migrant workers' labor, but also to ensure that migrant workers are in a continuous state of precariousness, unable to permanently settle and easily expelled when no longer beneficial.[18]

[16] Ministry of Public Health: "National Health Strategy 2024–2030." Doha: Ministry of Public Health, © 2024.

[17] See https://en.wikipedia.org/wiki/Kafala_system.

[18] Damir-Geilsdorf, K.: "The Kafala System and Migrant Labor Governance in the Gulf." Cham: Springer Nature, © 2016.

Rather than dismissing the criticisms or defending the kafala system, Qatar's leadership chose to engage with international institutions, particularly the International Labor Organization (ILO), to initiate a series of reforms. Indeed, this was done under some pressure, but it was also reflective of strategic foresight. Beginning in 2017, Qatar committed to dismantling core features of kafala, which required challenging long-standing practices embedded in the Gulf labor market and in the expectations of employers. Among these were reforms to abolish exit permits and the introduction of a nondiscriminatory minimum wage in 2021—the first of its kind in the Gulf region. Though it is inherently difficult to disrupt systems that have long been deeply embedded in regional practices, Qatar's leadership recognized that achieving their development goals requires demonstrating a capacity to adapt to accommodate norms and standards beyond the domestic or regional context. While enforcement remains a persistent concern that Qatar must continue to monitor, and while human rights organizations have continued to question the pace and consistency of implementation, the institutionalization of new protections marks a distinct departure from the Gulf norm, positioning Qatar as a potential regional leader in labor governance and human rights. This departure is particularly notable given that the reforms were achieved through visionary statecraft that understands the geopolitical value of alignment with global standards. Furthermore, the labor reforms revealed that the legitimacy of leadership in Qatar does in fact extend beyond citizen satisfaction to include international credibility and the effective management of risks to Qatar's reputation. Working with organizations like the ILO and explicitly framing the reforms in the language of human rights and development further reinforced their aspiration to not just align with global decision-making processes but to be seen as a norm-setting actor capable of balancing national sovereignty with shared international standards.

Taken together, Qatar's expansion of healthcare protections for non-nationals and reform of labor practices demonstrate how adaptive leadership is not merely about being reactive to potential criticism but about strategically anticipating and responding to the pace of change—whether in terms of measurable population changes or more subtle cultural shifts. In Qatar, the growing disparity between the proportion of the country who is a Qatari citizen and the proportion of expatriates or foreign workers is a driving force behind the need to adapt, as Qatar's development goals, particularly those related to economic growth, are tied to the central role of noncitizens in its economy and society. Thus, Qatar has adapted and reshaped policies to ensure that such changes become a source of resilience rather than fragility, decisions that were made independent of an electoral cycle or popular protest even if they were partially motivated by international attention. Again, this serves to highlight the qualities of leadership that produce effective, meaningful progress in order to demonstrate that sustainable development can be achieved when such qualities are leveraged in service of development goals.

## 5.5 Modern Leadership: Leveraging Cultural Values for Development Goals

***The future depends on what we do in the present.—Mahatma Gandhi.***[19]

The modern dimension of Qatar's leadership model perhaps requires the most careful clarification, as the idea of what is "modern" can be burdened with connotations and assumptions that inhibit understanding how it relates to Qatar's approach to leadership. The concept of "modernity" is often invoked in ways that assume a linear trajectory of progress that converges toward Western liberal democratic models. The presumed equivalence between modernity and political liberalization, or even the wholesale adoption of Euro-American norms, would seem to preclude a country that is an absolutist monarchy from being described as such. Yet, modernity is a complex and contested term that is inherently shaped by context, history, and culture. Modern leadership in Qatar is about crafting a distinctly Qatari vision of modernity where substantial economic and social development are pursued within frameworks that resonate within local traditions, values, and historical trajectories. Qatar has demonstrated that meaningful, measurable progress can be achieved through economic diversification, institutional reforms, and global engagement without replicating Western governance systems or sacrificing cultural identity. Modern leadership here therefore refers to a model of governance that combines traditional systems with forward-looking economic and institutional reforms.

Qatar's modern leadership is clearly visible in the ways that the state has developed its economy and financial systems to ensure that prosperity is supported by institutions and practices that are consistent with global standards and that prioritize national priorities such as diversification. The selective integration of international norms in its financial and commercial structures is a strategy that strengthens legitimacy and contributes to a more durable economy less susceptible to disruption, producing a world-class financial industry. The creation of the Qatar Financial Centre (QFC) in 2005 was a manifestation of this outlook, establishing modern regulatory and legal frameworks that attracted hundreds of firms in banking, insurance, asset management, and professional services to Qatar. Offering an independent legal environment and tax incentives, this move positioned Qatar as an inviting destination for global finance. In doing so, Qatar has been able to quickly diversify their revenue streams and integrate modern commercial practices that extend beyond the oil and gas industries. By 2022, the QFC hosted over 1400 registered companies managing assets worth billions of dollars, a growth that reflected leadership's success in building a modern financial services sector from the ground up in such a short period of time. The construction of leading trade and investment networks has complimented this transformation as Qatar has climbed in global competitiveness rankings after initiating reforms designed to ensure reliability in contract enforcement, property rights, and ease of doing business; in just the short period from 2022 to 2025, Qatar's global competitiveness index according to the International Institute

[19] Universities Scotland: "The future depends on what we do in the present." https://www.universities-scotland.ac.uk/future-depends-present/. Universities Scotland, © 2016.

for Management Development (IMD) report has risen from ranking of 18th to ranking 9th.[20] This impressive rise to one of the top 10 most competitive global economies, recognized by reputable third-party observers, confirms that Qatar's approach to economic modernization has successfully turned Qatar into an attractive and market where potential investors can confidently seek growth and security. In this sense, it is important to understand modernization as an ongoing process or mindset rather than an end goal, one that is concerned primarily with achieving consistent, demonstrable improvement in the context of shifting global conditions over seeking the superficial performance of democratic governance.

Following from this understanding, Qatar's model of modern leadership emerges as a case study in how economic growth and ambitious financial strategies can be designed in ways that do not erode or threaten traditional cultural identity. Indeed, Qatar's approach to modern leadership demonstrates that it is not only possible, but effective to modernize economically and institutionally in a way that remains culturally grounded. A key example of this is the expansion of Islamic finance, a financial system that aligns deeply with Qatar's identity as an Islamic country by operating in accordance with the principles of Sharia law, including the prohibition of interest-based lending and an emphasis on risk sharing and ethical investment. According to the Qatar Financial Centre's 2025 report, Islamic banking now accounts for 29% of Qatar's total banking assets and has outpaced the growth of conventional banks since 2020.[21] This growth is not incidental to Qatar's strategic focus on the financial sector but has instead been intentionally enabled by the Third Financial Sector Strategic Plan, which was launched by the Qatar Central Bank in 2023 and which integrates Islamic finance throughout its analysis of Qatar's economic goals.[22] The development of the Islamic finance sector in Qatar is much more than a consequence of virtue-signaling the country's Islamic identity and culture. It is a significant, growing, and financially impactful component of the economy that has been cultivated alongside other growth strategies and that demonstrates the compatibility of cultural authenticity, economic modernization, and institutional reform.

Analyzing the success of this combination in Qatar helps to elaborate on what modern leadership means in the Qatari context. It means using cultural norms as assets for development rather than limitations and leveraging them to produce lasting, robust growth and development. Qatar's construction of a globally credible financial sector that incorporates religious and cultural values is the result of utilizing available modern mechanisms strategically to pursue and achieve its visionary goals. Legal and regulatory frameworks for Islamic banking that attract capital, sukuk issuance regimes that are competitive on not only a regional, but an international scale, and innovations in FinTech driven by Islamic banking investment are

---

[20] IMD World Competitiveness Center: "World Competitiveness Ranking 2025." Lausanne: IMD Business School, © 2025.

[21] Qatar Financial Centre: "Qatar Islamic Finance Report 2025 – Expanding Horizons." Doha: Qatar Financial Centre, © 2025.

[22] Qatar Central Bank: "Third Financial Sector Strategic Plan." Doha: Qatar Central Bank, © 2023.

all emblematic of the modern nature of Qatar's economic growth and approach to development. The impact of Qatar's modern leadership on its financial sector, as exemplified by its continued inclusion and support of Islamic finance alongside other global finance models, provides several key insights that are essential for the purposes of this chapter. The first is that the modern quality of Qatar's leadership does not refer to a conformity to Western models or strategies—it is modern because it actively redefines what progress looks like and what it can be driven by within a given cultural, historical, and social context. The second is that it draws on global structures without being bound by them by translating those structures into strategies that are both locally successful and internationally relevant. The third is that modern leadership can and should be understood as a capacity to move a nation's trajectory forward by mobilizing resources and institutions in ways that sustain growth, strength, and adaptability. Taken together, these insights from Qatar's modern leadership offer an inspiring understanding of modernity as something that does not require a fixed template but that can be variously manifested depending on history, values, and economic structure. Rather than pursuing political similarity with Western governance structures in order to be perceived as "modern," true modern leadership should strive to enhance the culturally relevant systems and tools at its disposal in order to achieve development outcomes.

## 5.6 Sustainable Leadership: Integrating Economic and Environmental Security for the Future

***You simply have to put one foot in front of the other and keep going.—George Lucas.***[23]

The final dimension of Qatar's leadership model that will be explored here is its commitment to sustainability as a principle that extends beyond environmental concern to encompass economic resilience, social well-being, and intergenerational prosperity. In the Qatari context, sustainable leadership means using today's wealth and networks to create opportunities for structures, institutions, and resources that will ensure growth and success for future generations. This consideration is particularly critical for a state that is dependent on hydrocarbons where the question of "what comes after oil and gas" is inevitably a central concern of long-term planning. It would be all too simple for Qatar's leadership, with its control over national wealth, to operate with a short-term mindset of consumption now at the expense of long-term gains; in resource-rich states, leadership is often tested by the tension between immediate wealth and long-term stability. The sheer abundance of national revenue derived from such a profitable sector as oil and gas can unfortunately incentivize the postponement of structural reforms to prepare for the potential of a more environmentally conscious future. What distinguishes sustainable leadership is therefore its ability not just to invest in more environmentally sustainable industries,

[23] Gul, Z: "10 powerful quotes by George Lucas to motivate you." https://medium.com/@zaharagul/10-powerful-quotes-by-george-lucas-to-motivate-you-9cc407ebaaa6. Medium, © 2024.

but to resist the pull of resource dependency altogether to instead leverage wealth to build enduring systems of growth.

Qatar's trajectory demonstrates this quality of sustainable leadership and its success beyond the diversification of energy sources shows just how important a sustainable approach is to successful development outcomes. The choice of Qatar's leadership to emphasize investments in institutions, infrastructure, and industries that will provide continued prosperity across generations reflect an understanding that sustainability requires embedding durability and adaptability into the economic and social fabric of the nation. An essential element of Qatar's approach to sustainability is its emphasis on tangible outcomes as a measure of progress. Sustainable development cannot remain at the level of strategies or governance frameworks, though, as discussed, such tools can be useful for guiding national goals. It must instead be evidenced in results that show demonstrable improvement and secure long-term growth. In Qatar, these outcomes are visible in rising life expectancy, declining unemployment rates, expanded healthcare systems, and the growth of non-hydrocarbon sectors that now contribute a steadily increasing share to Qatar's GDP. For example, according to the World Bank, life expectancy at birth in Qatar has risen from about 72 years in 1990 to over 80 years by 2021 while nonresource sectors such as finance, construction, and service industries have become key engines of growth.[24] These are not abstract policy achievements but measurable changes in citizens' well-being and economic diversification that have been driven by visionary and sustained leadership that executes policy, mobilizes resources, and maintains consistency over decades. This dynamic is what makes Qatar's model so distinctive and worthy of deconstructing: it demonstrates that durable, outcome-oriented progress requires leadership capable of aligning vision with results.

Economically, the Qatar Investment Authority discussed earlier is one of the most obvious expressions of sustainable leadership as a mechanism for transforming natural resource revenues into diversified global assets that contribute to Qatar's prosperity in their own right. However, Qatar's leadership has also emphasized that sustainable development cannot be limited to the economic domain alone, particularly for a country whose prosperity is so closely tied to the production and export of oil and gas. Though sustainability in the context of Qatar's leadership has not meant an exclusive focus on environmental concerns, economic sustainability and environmental sustainability inevitably go hand in hand to secure their future. For this reason, Qatar has sought to integrate the two in ways that reinforce one another. In the QNV 2030, commitment to responsible environmental stewardship is identified as one of the four pillars guiding Qatar's development. On the surface, such a commitment might seem strange for, or even incompatible with, a state associated with oil and gas wealth—why would Qatar's leadership be interested in promoting industries or principles that appear to directly challenge the primary source of their prosperity? This apparent contradiction is precisely why sustainability here does not only refer to renewable energy or green policies. Despite it seeming antithetical to

[24] World Bank: "Qatar GDP and Development Statistics." Washington, DC: World Bank, © 2021–2025.

support industries that might one day replace those that have served as the foundation for Qatar's success, Qatari leadership has recognized that, for economic development to be truly sustainable, the reality is that environmental stability must be addressed as well. Where this could be seen primarily as a challenge or an obstacle, Qatar has instead embraced sustainability as an opportunity to further support their goals; this is the foundation of real sustainable leadership.

For Qatar, sustainability of course does not mean reducing dependence on hydrocarbons overnight, and it certainly does not mean abandoning them altogether. Similarly to how Qatar's social and political progress has remained slow and controlled rather than sudden and disrupted, even as economic growth has skyrocketed, sustainability is about ensuring that revenues from the energy sector are used carefully in the service of the diversified economy, whether through funding for green infrastructure or investment in research and development that prepares the country for a post-carbon future. In practice, this has involved pursuing projects that simultaneously generate substantial economic returns and reduce environmental costs, such as the development of carbon capture initiatives in the energy sector and the adoption of sustainable building standards in urban planning. Qatar has even explored solar power as a complementary source of national energy—one of its most significant recent projects in this area is the 2022 commissioning of the Al Kharsaah solar plant, which is projected to support Qatar's growing electricity demand and significantly reduce carbon dioxide emissions.[25] While not yet nearly on a scale that could entirely replace reliance on non-renewable energy sources, the important point to emphasize here is that if Qatar's leadership had only a narrow, short-term vision, there would really be no immediate need to invest in such projects at all; Qatar has more than enough oil and gas for the ruling class to easily maintain their wealth and, while the international community continues to stress the importance of solutions to climate change, oil and gas are not really going anywhere any time soon. But these kinds of projects are part of broader national commitments to not just enjoy prosperity now, but to sustain long-term growth. Investing in renewable energy now allows leadership to reduce future energy costs and insulate the economy from potential climate-related sources of volatility.

In tandem to its exploration of the potential of renewable energies, Qatar has continued to capitalize on and expand its nonrenewable energy industry through projects like the North Field East LNG project, which involves constructing additional facilities to increase Qatar's production of liquid natural gas (LNG).[26] While, for some, this might undermine the value of Qatar's investment in solar energy, it is simply unrealistic to expect that Qatar would not be continuing to utilize these resources, and Qatar has not pursued such projects without consideration of the risks to the environment and its reputation. To complement the expansion, it has initiated additional carbon capture and storage projects to mitigate the environmental impact and has plans to ensure that these projects, as well as existing carbon

---

[25] Qatar Electricity & Water Company: "Al Kharsaah Solar Plant Commissioning Report." Doha: Qatar Electricity & Water Company, © 2022.

[26] Qatar Petroleum: "North Field East LNG Project Report." Doha: Qatar Petroleum, © 2022.

capture infrastructure, scale alongside LNG development. The economic benefit here is thus twofold in that this strategy preserves Qatar's position in global gas markets by making the product cleaner and less carbon-intensive and also anticipates future carbon regulation or carbon pricing that could threaten fossil fuel revenue streams. By aligning economic and environmental strategies in this way, Qatar presents a model for leadership that does not see sustainability as a constraint on national prosperity but that embraces it as an essential condition for its continuation. Sustainability does not have to mean perfection; it means employing a strategy of long-term preparedness and resilience, whether in the face of environmental concerns or other challenges. For a country whose revenues are still, despite impressive diversification, overwhelmingly tied to oil and natural gas, the realistic approach to sustainability is not to abruptly turn away from hydrocarbons but to create a strong economic and energy system that enables growth and improvement. In this sense, like modernity in the previous section, sustainable leadership should not be viewed as a final endpoint. Instead, it should be understood as a process for building security for future generations, a responsibility of leadership to anticipate challenges, use (all) resources wisely, and safeguard the social and environmental foundations on which long-term success depends.

## 5.7 Conclusion: Impactful Leadership

***Lead Me, follow Me, or get Out of My way.—General George S. Patton***[27]

Qatar's developmental trajectory demonstrates that leadership, rather than governance structures in isolation, can be a decisive force in advancing sustainable development. The Qatari case complicates conventional assumptions that democratic institutions are a necessary precondition for progress toward the Sustainable Development Goals (SDGs). Instead, it highlights how the particular qualities of leadership discussed in this chapter—visionary, transformational, adaptive, modern, and sustainable—can be mobilized to generate development outcomes that are measurable and durable. To reiterate, this is not to suggest that governance systems are irrelevant, but rather that the role of leadership as a driver of direction, coherence, and long-term commitment deserves greater analytical attention. The reality of impressive progress and growth in Qatar, particularly its economic success, illustrates that impactful leadership is an essential and effective component of development. The qualities of Qatar's leadership identified here constitute a model that has enabled Qatar to transition from an oil-dependent, peripheral economy into a diversified and globally engaged state capable of promoting economic growth, encouraging social stability, and maintaining cultural continuity. While no single one of these qualities is necessarily unique or distinctive by itself, the manner in which they have been brought together to reinforce each other by Qatar's leadership is what has made them so effective. Visionary planning provides direction, transformational

[27] Killick, C.: "Lead me, follow me, or get out of my way." https://makingteamswork.co/2017/11/28/lead-me-follow-me-or-get-out-of-my-way/. MakingTeamsWork, © 2017.

initiatives reshape (but do not disrupt) economic and social structures, adaptive measures ensure resilience to changing conditions, modern strategies support innovation without erasing tradition, and sustainability anchors growth in long-term security.

Deconstructing Qatar's leadership in this way offers important, and, indeed, essential, insights for global debates on sustainable development. Too often, discussions of the SDGs focus narrowly on governance systems, international financing, or technical capacity at the expense of considering whether any of these things can be truly effective without the right mindset. Qatar's development is a productive context for this analysis because it shows that even where financial resources are abundant, and citizen populations are small, positive outcomes are not automatic. What matters is leadership's ability to channel those resources coherently and deliberately, maintain momentum over decades, and balance international expectations with domestic legitimacy. This perspective reframes sustainable development as not only a question of "what policies are pursued" but of "how leadership mobilizes vision and authority to make them effective over time," crucially shifting the focus from political intent to measurable outputs. For other countries or organizations seeking to advance sustainable development or better align their work with development goals, the lesson is not to replicate Qatar's governance model but to appreciate the deeper qualities of leadership that enable lasting progress. Though it is important to acknowledge that Qatar's model of governance as an absolutist monarchy is not what would typically be considered ideal for achieving sustainable development goals, the central aim of this chapter is not merely to celebrate Qatar's success but to bring attention to qualities that can inspire more efforts to encourage leadership styles that produce tangible outcomes.

It would be understandable to want to, at first sight, dismiss the insights from Qatar's development trajectory as not instructive or valuable for entities with fewer resources. After all, much of what this chapter has focused on is the economic progress that Qatar has achieved through strategic use of its abundant national wealth. But what makes Qatar's case useful to reflect on is not simply the fact that it *is* wealthy but the way that that wealth has been directed. Hydrocarbon revenues could have easily been consumed in ways that entrenched dependency, corruption, and inequality, but rather than going down that path, Qatar's leadership made deliberate choices to invest in economic diversification and social well-being. From establishing the Qatar Investment Authority to constructing Education City, from advancing labor reforms to commissioning one of the region's largest solar power plants, these are choices that reflect a commitment to translating short-term advantages into long-term resilience. In the context of the SDGs themselves, these decisions are consistent with the concrete goals outlined by the SDGs including quality education, economic growth, industry and innovation, sustainability, and climate action. And while Qatar is certainly still faced with challenges, the point to takeaway is that leadership has played a critical role in directing national wealth and institutional capacity toward outcomes that map onto the global development agenda and create domestic prosperity. Such a trajectory is the outcome of leadership decisions that are therefore informative for any organization or state that wants to pursue sustainable development. Leadership for this purpose is about the qualities that guide resource use and institutional design in a positive direction, and it is of the utmost importance for decision-makers to understand how to nurture and leverage those qualities to make deliberate, future-oriented choices.

Qatar's experience therefore contributes to a necessary rethinking of sustainable development discourse, especially as the global community approaches the final stage of the 2030 Agenda and will soon be confronted with the need to consider how to move forward into the coming decades and the new challenges they will bring. As institutions, states, and organizations look toward building a sustainable collective future not only for their own populations but for humanity as a whole, the role of leadership in shaping trajectories, balancing short-term profit with long-term success, and maintaining trust and legitimacy must be taken seriously. In a way, impactful leadership is itself the most critical resource for sustainable development. Recognizing the value of this resource and learning how it can be leveraged for positive growth and improvement will be essential for achieving development outcomes that are both enduring and truly transformative.

At the level of individual Sustainable Development Goals, Qatar's leadership model has facilitated outcomes that demonstrate a remarkable degree of alignment with the global sustainability agenda, which is a direct result of a comprehensive and integrated vision of national progress (Fig. 5.1).

Substantial achievements have been recorded across SDGs 1, 2, 3, 4, 6, 8, 9, 14, 15, and 16 where policies aimed at ensuring a minimum standard of living, providing food and water security, enhancing health and education systems, and

**Fig. 5.1** UN SDG (*source*: https://www.un.org/sustainabledevelopment/), used with friendly permission of UN. The content of this publication has not been approved by the United Nations and does not reflect the views

promoting the development of infrastructure have produced tangible, lasting results. SDG 17, in particular, stands out as a defining feature of Qatar's approach. By hosting major international conferences, building partnerships with UN agencies, and supporting multilateral platforms for knowledge exchange, Qatar has positioned itself as a driver of global cooperation for sustainable development.

Progress is also visible in SDGs 5, 7, 10, 11, 12, and 14 where ongoing reforms are addressing gender equality, the expansion of renewable energy, social inclusion, sustainable urbanization, responsible production and consumption, and climate adaptation. That Qatar continues to work toward these goals is reflective of the understanding that doing so in a gradual manner is more effective and stable than attempting to institute drastic and disruptive changes.

Crucially, as discussed throughout this chapter, Qatar's leadership has adopted a pragmatic and forward-looking approach to development that uses the revenues from its still carbon-based economy to invest strategically in research, education, innovation, and a post-oil transition. This model helps to sustain economic resilience so that today's prosperity can become the foundation of a diversified and sustainable future. In this way, Qatar's leadership demonstrates that a carefully managed resource-based economy can evolve into a platform for broad, inclusive, and globally engaged sustainable development.

## 5.8 At the End: A Brief Reference to the Three-Pillar Model

The Three-Pillar Model (or 3-P-Model)[28] mentioned in the book's subtitle is a preferred thinking and mental style for challenging large transformations, especially in the context of achieving sustainability and coverage of the UN SDGs on a global and local scale, which is described in detail in Chaps. 1 and 2. It can also be explicitly applied to the transformation that was triggered by Qatar's unique leadership model. In brief, the model is based on three key building blocks (pillars) for organizational design and leadership:

(1) The First Pillar: **Sustainable Purpose**

Sustainable Purpose is the raison d'être of an organization—or even of a state—because having a guiding objective means bringing orientation and certainty to the people for their joint endeavors and success. The defined vision for the transformation of Qatar and the resulting successful activities to achieve

[28] Developed in: Wollmann, P. et al. (Eds.): Transforming Public and Private Sector Organizations—Implementing Sustainable Purpose, Travelling Organization and Connectivity for Resilience. Cham: Springer Nature, © 2022; Wollmann, P. et al. (Eds.): Organization and Leadership in Disruptive Times—Design and Implementation of the 3-P-Model. Cham: Springer Nature, © 2021; Wollmann, P. et al.(Eds.): Three Pillars of Organization and Leadership in Disruptive Times—Navigating Your Company Successfully through the twenty-first Century Business World. Cham: Springer Nature, © 2020; Wollmann, P. et al. (Eds.): Reading an Unknown Organization - Leaders in New Roles Thriving in Challenging Environments with the 3-P-Model Mindset. Cham Springer Nature, © 2024.

rapid economic diversification, make strategic investments in human capital and infrastructure, institute labor and social reforms, and start environmental initiatives are all components of a very convincing sustainable purpose for the country. In addition to its national impact, this model is also deeply relevant in an international context, particularly given Qatar's active role in SDG 17.

(2) The Second Pillar: **Traveling Organization**

Traveling Organization means for an organization—or even for a state—accepting to be in a permanent state of flux and transformation, while interacting with the external requirements of its journey with rapid learning and adaptivity. It is obvious that the transformation of Qatar in the last few decades has the character of a prudent change process under impactful leadership that challenges the stakeholders without overwhelming them and takes them along on the journey.

(3) The Third Pillar: **Connectivity**

Connectivity refers to the ability to interconnect on the development journey with all the needed parties and resources both inside and outside the silos of individual areas of development, thereby creating high efficacy and consistency during this process. Remaining connected enables organizations—and states—to build shared learning networks that support growth and improvement. It is obvious that Connectivity is a key success factor for Qatar's proactive transformation as a state and as a significant player on both the regional and global levels.

To summarize: the unique leadership model for Qatar's development and transformation is not only an enlightening use case for sustainable development but also for the application of the 3-P-Model on the level of a whole country.

**Disclaimer** The views, opinions, findings, interpretations, and conclusions expressed in this chapter are those of the author alone and do not reflect the views or official position of the United Nations (UN), the United Nations Institute for Training and Research (UNITAR), or any of their affiliated organizations, staff members, or governing bodies.

## Further Reading

World Bank: "Qatar GDP and Development Statistics". Washington, DC: World Bank, © 2021–2025.

Scharfenort, N.: "Urban Development and Social Change in Qatar: The Qatar National Vision 2030 and the 2022 FIFA World Cup." *Journal of Arabian Studies* 2 (2), 209–230, © 2012.

Müller, M.: "Mega Events and Their Multidimensional Impact". Cham: Springer Nature, © 2018.

Skytrax: "Hamad International Airport World Rankings". London: Skytrax, © 2022.

Planning and Statistics Authority: "Employment Statistics for Qatari Citizens". Doha: Planning and Statistics Authority, © 2021.

Al-Musalmani, F.; Maalouf, S.: "Qatar Foundation: A Sustainability Innovator." In: Cochrane, L.; Al-Hababi, R. (Eds.): *Sustainable Qatar: Social, Political and Environmental Perspectives*. Singapore: Springer Nature, © 2022.

Ministry of Public Health: "National Health Strategy 2024–2030". Doha: Ministry of Public Health, © 2024.

Damir-Geilsdorf, K.: "The Kafala System and Migrant Labor Governance in the Gulf". Cham: Springer Nature, © 2016.

IMD World Competitiveness Center: "World Competitiveness Ranking 2025." Lausanne: IMD Business School, © 2025.
Qatar Financial Centre: "Qatar Islamic Finance Report 2025—Expanding Horizons." Doha: Qatar Financial Centre, © 2025.
Qatar Central Bank: "Third Financial Sector Strategic Plan". Doha: Qatar Central Bank, © 2023.
Qatar Electricity & Water Company: "Al Kharsaah Solar Plant Commissioning Report". Doha: Qatar Electricity & Water Company, © 2022.
Qatar Petroleum: "North Field East LNG Project Report". Doha: Qatar Petroleum, © 2022.

**Mr. Rabih El-Haddad** brings a deep understanding and knowledge of multilateral diplomacy, negotiation, and United Nations history cultivated through his extensive experience supporting governments and senior decision-makers in complex international processes. He currently leads the Division for Multilateral Diplomacy at the United Nations Institute for Training and Research (UNITAR). His work focuses on diplomacy, negotiation, and collaborative leadership as tools for advancing national interests, managing global interdependence, and fostering cooperative solutions across policy areas that include climate change, trade, human rights, sustainable development, and peace and security. His professional practice is grounded in a strong commitment to dialogue, informed decision-making, and the peaceful settlement of disputes as core principles of effective global governance.

**Dr. Sabrina Illiano** is an anthropologist who specializes in cultural diplomacy, international development, and the politics of heritage. Her research explores how widespread discourses about justice and history influence social perceptions of representation, narrative truth, and identity. As an expert on cultural restitution, Sabrina has frequently worked with embassies, museums, libraries, nonprofit organizations, and government agencies on projects related to negotiations surrounding contested collections, ownership, and repatriation. In her current role, she supports the research pillar of the Division for Multilateral Diplomacy at the United Nations Institute for Training and Research (UNITAR). She holds a doctorate in Social and Cultural.

# 6 Rule of Law and Strong Institutions: The Importance of SDG 16 in Post-Conflict Settings

Anna Sabrina Wollmann and Faisal Aminollah

**Abstract**

The rule of law is the cornerstone of political stability, and, as such, it is unequivocally linked to sustainable development, as well as the prevention and resolution of conflicts. It protects individual's rights and fundamental freedoms by creating a system of accountability at the national and international levels. While UN Sustainable Development Goal (SDG) 16 "Rule of Law and Strong Institutions" seeks to establish clear targets for countries to strengthen and achieve the rule of law, recent international developments have diminished the progress made toward the goal. This article examines the rationale between SDG 16, why it is key especially in post-conflict settings, and explores how different initiatives have contributed to the building of strong institutions using real-life examples. It, furthermore, takes stock of the progress made toward the achievement of SDG16 and links the attainment of the rule of law to the Three-Pillar Model (first published in 2020 at https://link.springer.com/book/10.1007/978-3-030-23227-6).

A. S. Wollmann (✉)
United Nations Institute for Training and Research (UNITAR), Division for Multilateral Diplomacy, Geneva, Switzerland
e-mail: anna.wollmann@unitar.org

F. Aminollah
Longpont sur Orge, Essonne, France
e-mail: aminollah.faisal02@gmal.com





P. Wollmann et al. (eds.), *The Sustainable Organization – Volume 2*, Future of Business and Finance, https://doi.org/10.1007/978-3-032-23802-3_6

## 6.1 Introduction

*I believe that every Government that is committed to the Rule of Law at home, must be committed to the Rule of Law abroad. All States have a clear interest as well as a clear responsibility to uphold international law and maintain international order.*

(Kofi Annan, 12 September 2002, Speech to the UN General Assembly[1])

The rule of law is the cornerstone of political stability both within nation states and in the international community. By assuring that natural persons, public and private institutions and entities, as well as the states themselves are accountable to existing laws, the rule of law provides legal certainty. It entails that individuals and their fundamental freedoms are protected by human rights norms, such as the right to a fair trial. Equality before the law and avoidance of arbitrariness is key for private entities, strengthening trade and economic development. Separation of powers, curbing corruption, and restraining the abuse of power create strong and reliable political institutions, which, in turn, are fundamental for international peace and security. Without the rule of law and without strong institutions, the political system within a country can lose its equilibrium. This, in turn, can lead to a political spillover effect, destabilizing entire regions and leading to internal and international conflicts with detrimental effects on civilians and the sustainable development. This article will commence by examining the evolution of the rule of law, peacekeeping, and the emergence of the state-building concept in the twentieth century. Furthermore, important academic critiques and the resulting doctrinal changes will be explored. In the following, the progress made toward the achievement of the UN Sustainable Development Goal (SDG)[2] 16 in general in the past 10 years will be examined, before diving into concrete actions of how the implementation of SDG 16 has been undertaken in post-conflict settings, shedding light on such matters as anticorruption and transitional justice mechanisms. Finally, the linkage between SDG 16 and the Three-Pillar Model[3] will be explored, underlying the paramount importance of the sustainable purpose, traveling organization, and connectivity matrix for successful peacekeeping missions and the achievement of the rule of law.

## 6.2 The Evolution of State-Building Doctrine: Toward a Universal Agenda (SDG 16)

### 6.2.1 The Emergence of the State-Building Concept

#### 6.2.1.1 From Crisis Management to Structural Reconstruction (1948–1992)

In the wake of the United Nations' creation in 1945, the practice of peacekeeping was shaped by the direct influence of the ideological and strategic confrontation between the Soviet and Western blocs. During these early decades, international

[1] See https://www.un.org/webcast/ga/57/statements/sgE.htm.

[2] See https://sdgs.un.org/goals and especially https://www.un.org/sustainabledevelopment/peace-justice/.

[3] See https://link.springer.com/book/10.1007/978-3-030-23227-6).

interventions were designed according to a so-called "first-generation peacekeeping"[4] model, which was essentially reactive and military. The first operations, deployed from 1948 onward, did not aim to transform the internal structures of states but acted as ceasefire monitoring and interposition mechanisms between warring parties. This classic practice was built on an unwavering trinity that guaranteed the Organization's neutrality: the consent of the parties, strict impartiality, and the non-use of force except in self-defense.[5] During this period, state sovereignty remained a barrier, limiting UN action to the technical management of military crises. The collapse of the Soviet bloc and the simultaneous lifting of ideological blockades within the Security Council in the late 1980s and early 1990s radically shifted the focus of international intervention. Conflict was no longer viewed simply as a military crisis to be contained, but through its root causes: structural deficits in governance, justice, and legitimacy.

In 1992, at the Security Council's request, Secretary-General Boutros Boutros-Ghali published the seminal report *An Agenda for Peace*, which theorized this paradigm shift. This foundational document introduced four key concepts for strengthening peace: preventive diplomacy, peacemaking, peacekeeping, and, for the first time in a formalized way, post-conflict peacebuilding.[6] The latter was defined as an "action to identify and support structures which will tend to strengthen and solidify peace in order to avoid a relapse into conflict."[7] This more ambitious approach marked a historic turning point. It implied that international and regional actors would now engage in projects aimed at strengthening the capacities of state institutions. The goal was twofold: to restore state authority while guaranteeing citizens' fundamental rights. This new *state-building* doctrine was tested in the major crisis zones of the 1990–2001 decade, notably in Bosnia and Herzegovina, Kosovo, Rwanda, and Sierra Leone.[8] For the United Nations, the task was to implement an approach focused on restoring the formal structure of the state, where rebuilding state capacities became a central, indispensable operational practice for preventing a relapse into violence. From the mid-1990s onward, this shift reflected a renewed understanding of armed conflict, now seen as a symptom of structural state dysfunction rather than merely a security breakdown.[9]

Concretely, this materialized through the direct intervention of United Nations peacekeeping operations in the restoration and extension of state authority, which could include operational assistance to state institutions and, where relevant, capacity-building support or assistance for constitutional or institutional

[4] Oliver P Richmond, *Peace in International Relations* (second edn, Routledge, London 2020).

[5] United Nations Department of Peacekeeping Operations and Department of Field Support, *United Nations Peacekeeping Operations: Principles and Guidelines* (UN Capstone Doctrine, United Nations, New York, 2008) 31–35.

[6] Boutros Boutros-Ghali, *An Agenda for Peace: Preventive Diplomacy, Peacemaking and Peacekeeping*, UN Doc A/47/277–S/24111 (17 June 1992).

[7] Boutros Boutros-Ghali, *An Agenda for Peace*, UN Doc. A/47/277–S/24111 (1992), para. 21.

[8] Roland Paris, *At War's End: Building Peace after Civil Conflict* (Cambridge University Press 2004).

[9] Roland Paris, *At War's End: Building Peace after Civil Conflict* (Cambridge University Press 2004).

processes.[10] The UN notably focused on supporting the reactivation of national legal systems, assisting in the development of transitional legal frameworks, and helping to reconstitute administrations capable of exercising minimal authority over the territory. The goal was not yet the deep normative transformation of societies, but rather the restoration of institutional mechanisms functional enough to manage social and political conflicts through nonviolent means. However, this phase remained largely characterized by an exogenous, technocratic-institutional, and sequential approach.[11] In this model, the formal reconstruction of the state often preceded, and sometimes overshadowed, local dynamics of political legitimization, as illustrated in Kosovo following the 1999 conflict, through the United Nations Interim Administration Mission in Kosovo (UNMIK).[12] While this method achieved rapid gains in stabilization and administrative capacity, it revealed structural limitations: fragile national ownership, prolonged dependence on international assistance, and a disconnect between the reconstructed institutions and social realities. The observed gap between this formal institutional architecture and local social realities spurred the development of major theoretical critiques, centered on questioning the very content of these reforms.

#### 6.2.1.2 Academic Critique: A Conceptual Reconfiguration of Peacebuilding

**_Liberal Peacebuilding_ (Roland Paris): The Challenge of Sequencing and _Institutionalization Before Liberalization_[13]**

The first major school of critique of this disconnect between formal institutional architecture and local social realities, *liberal peacebuilding*, argues that peacebuilding has long been equated with the hasty export of Western liberal institutions.[14] Driven by the predominant influence of NATO countries after the Cold War, this model aimed for comprehensive societal transformation through the simultaneous promotion of representative democracy, market economies, and the *rule of law*. This strategy was based on the conviction that grafting these universal principles would be enough to durably stabilize societies emerging from civil conflict.

Operationally, this approach translated into a race to organize national elections, draft liberal constitutions, and implement drastic economic liberalization pushed by the IMF and the World Bank. The United Nations invested heavily in deep security sector reform, while codifying individual freedoms and promoting transitional

[10] United Nations, *United Nations Peacekeeping Operations: Principles and Guidelines* (UN Department of Peacekeeping Operations 2008).

[11] Roland Paris, 'International Peacebuilding and the "Mission Civilisatrice"' (2002) 28 *Review of International Studies*.

[12] United Nations Security Council, UN Doc S/RES/1244 (10 June 1999).

[13] Roland Paris, *At War's End: Building Peace after Civil Conflict* (Cambridge University Press 2004).

[14] Roland Paris, *At War's End: Building Peace after Civil Conflict* (Cambridge University Press 2004).

justice to break cycles of impunity. The goal was to create, from the top down, a normative framework capable of regulating social tensions through law and the market.

However, Roland Paris's analysis highlights that this premature liberalization can exacerbate tensions rather than ease them. In contexts of extreme fragility, economic openness often benefits predatory elites, while newly proclaimed laws remain unenforceable due to a lack of administrative capacity. To remedy these failures, Paris advocates for the "*Institutionalization Before Liberalization*" approach, a call for pragmatic sequencing where strengthening public institutions must come *before* democratic and economic opening.[15] This critique of the liberal nature of reforms inevitably leads to questions about a more fundamental dimension: the intrinsic strength of the state apparatus itself.

#### ***State Institutionalism* (Francis Fukuyama): State Strength Versus the Scope of Its Functions**

The paradigm of *state institutionalism*, developed by Francis Fukuyama, proposes viewing the state not as a mere vessel for liberal norms, but as a functional organism whose basic capacities need strengthening. Fukuyama argues that the primary cause of reconstruction policy failure lies in the intrinsic weakness of institutions. Therefore, international reforms should focus not on expanding state functions, but on prioritizing the strengthening of administrative, judicial, and coercive capacities.[16]

Central to this thesis is the distinction between scope, the range of functions a state undertakes, and state strength, which denotes its actual capacity to plan, execute policies, and enforce laws transparently.[17] A state can claim ambitious objectives while sorely lacking the material ability to implement them a hallmark of clientelist states. The challenge of *state-building*, therefore, lies in improving bureaucratic quality and ensuring the impersonality of administrative authority.

To make this reconstruction work, Fukuyama identifies several essential pillars: effective law enforcement, predictable public action, and credible judicial institutions. Once consolidated, these dimensions allow the state to be more than a facade; they enable it to act as a neutral and effective regulator of society. The goal is to build a political order capable of resisting corruption and exercising genuine democratic accountability.[18] Yet, a technically efficient bureaucracy is not enough if it lacks a deep bond of trust with the citizens it is meant to serve.

[15] Roland Paris, *At War's End: Building Peace after Civil Conflict* (Cambridge University Press 2004).

[16] Francis Fukuyama, *State-Building: Governance and World Order in the twenty-first Century* (Cornell University Press 2004).

[17] Francis Fukuyama, *State-Building: Governance and World Order in the twenty-first Century* (Cornell University Press 2004).

[18] Francis Fukuyama, *State-Building: Governance and World Order in the twenty-first Century* (Cornell University Press 2004).

**The *Local Turn* (Mac Ginty, Richmond, Lederach): Peace as a Hybrid, "Bottom-Up" Process**

This need for social legitimacy fueled the rise of the "local turn," which conceptualizes peace not as a top-down blueprint, but as a relational and dynamic process. Championed by Mac Ginty and Richmond, this perspective highlights the hybrid nature of post-conflict political orders.[19] Stability does not come merely from applying an international institutional template, but from the continuous interaction between formal state structures and local systems of governance and legitimacy.[20]

The *local turn* argues for a re-evaluation of local actors, traditional authorities, and community leaders as genuine co-producers of peace. This vision aligns with the work of John Paul Lederach, who advocates for a peacebuilding approach "from the bottom up." In his "peace pyramid," Lederach shows that sustainable societal transformation cannot be limited to pacts negotiated by political elites at the top; it must be rooted in the lived realities of people at the grassroots level.[21]

Therefore, this perspective urges international institutions to adopt a "people-centered" approach, where judicial and administrative reforms are calibrated according to the actual needs and perceptions of citizens. It's about empowering local actors to drive deep social and relational change, ensuring the state is seen not as a foreign entity, but as an infrastructure serving the community.[22]

Ultimately, it is the synthesis of these three schools of thought, integrating the sequencing imperatives of liberal peacebuilding, the organic capacity focus of *state institutionalism*, and the social legitimacy imperative of the local turn, that enabled the United Nations to internalize these critiques and durably reshape its intervention doctrine.

### 6.2.2 The Institutional Internalization of Critiques: Toward a New Paradigm for Sustainable Peace

The above-described doctrinal debates of the 2000s, shaped by Roland Paris's analysis of *liberal peacebuilding*, Francis Fukuyama's framework of *state institutionalism*, and the *local turn* advocated by Roger Mac Ginty and Oliver Richmond, gradually informed the internal reflections of the United Nations. This intellectual maturation gained formal institutional recognition in October 2014 with the establishment of the High-Level Independent Panel on Peace Operations (HIPPO),

[19] Roger Mac Ginty and Oliver P Richmond, 'The Local Turn in Peace Building: A Critical Agenda for Peace' (2013).

[20] Oliver P Richmond, *A Post-Liberal Peace* (Routledge 2011).

[21] John Paul Lederach, *Building Peace: Sustainable Reconciliation in Divided Societies* (United States Institute of Peace Press 1997).

[22] Oliver P Richmond, *Peace Formation and Political Order in Conflict-Affected Societies* (Oxford University Press 2016).

mandated by Secretary-General Ban Ki-moon at the request of Member States.[23] This independent panel was tasked with reviewing all peace operations, including peacekeeping missions under the Department of Peacekeeping Operations (DPKO) and special political missions under the Department of Political Affairs (DPA), to assess gaps between mandates and outcomes on the ground.

Published in 2015, the HIPPO report[24] stands as the first high-level Secretariat document to explicitly integrate critiques from the academic literature on *state-building*. The panel acknowledged that the rapid transplantation of liberal institutional models had often produced formal institutions aligned with international standards, yet that were weakly rooted socially and politically. This finding validates Roland Paris's analysis in his 2004 book *At War's End*, which argued that premature liberalization without consolidated institutional capacities heightens the risk of conflict relapse. Furthermore, the report operationalizes Francis Fukuyama's[25] distinction between the *scope* of state functions and the effective *strength* to implement them, lamenting that mandate expansion had not been matched by sufficient investment in administrative and judicial capacities.

By incorporating insights from Mac Ginty, Richmond, and Lederach, the HIPPO report finally recognizes that legitimacy cannot be imposed through formal arrangements but depends on inclusive political processes and local ownership of reforms. The panel stresses the need to prioritize strengthening governance and the rule of law before extending state functions, aligning with the sequential approach championed by Fukuyama in his 2004 and 2014 works. This new understanding of the interdependence among institutions, local legitimacy, and stability laid the groundwork for integrating peace more comprehensively into global development objectives. The integration of peace and the rule of law into the international development agenda stems from the structural limitations of the Millennium Development Goals (MDGs), adopted in 2000. Designed with a technocratic focus, the MDGs sidelined governance and violence issues, assuming that economic growth and improved social indicators would naturally stabilize societies.[26] However, experience gathered by the United Nations throughout the 2000s demonstrated that developmental progress remained fragile and reversible in the absence of institutions capable of ensuring legal security and trust between the state and society.[27]

This institutional learning spurred critical reflection on the artificial separation between development and peace, leading to the adoption of the 2030 Agenda in

---

[23] United Nations, *Secretary-General's statement on the appointment of the High-Level Independent Panel on Peace Operations* (31 Oct 2014) SG/SM/16301-PKO/451.

[24] Report of the High-Level Independent Panel on Peace Operations, *Uniting Our Strengths for Peace: Politics, Partnership and People*, UN Doc A/70/95–S/2015/446 (2015).

[25] Francis Fukuyama, *State-Building: Governance and World Order in the twenty-first Century* (Cornell University Press 2004).

[26] United Nations, *United Nations Millennium Declaration*, UN Doc A/RES/55/2 (8 September 2000).

[27] United Nations Development Programme, *Governance for Peace: Securing the Social Contract* (UNDP 2012).

September 2015. Through resolution A/RES/70/1, the UN General Assembly placed the rule of law at the heart of a universal framework via Sustainable Development Goal 16 (SDG 16). This goal specifically aims to promote peaceful and inclusive societies, provide access to justice for all, and build effective, accountable, and inclusive institutions at all levels.[28] SDG 16 thus affirms that the absence of legitimate institutions is both a root cause of conflict and a cross-cutting obstacle to development.[29]

Far from representing a break, the inclusion of SDG 16 marks the normative culmination of an intellectual trajectory seeking to blur the boundaries between security, governance, and development. By integrating principles debated in *peacebuilding* and *state-building* circles, this goal embodies a vision where peace and the rule of law are no longer external prerequisites, but constitutive elements of long-term development pathways. To translate this normative ambition into practice, the United Nations developed an operational framework capable of turning these principles into concrete actions across the entire conflict cycle.

## 6.3 Toward Strong Institutions: Stocktaking of SDG 16 in 2025

Given that it has now been 10 years since the adoption of the Sustainable Development Goals and Agenda 2030 in September 2015, and 5 years remain until the objectives should be reached, it is important to take stock of the process to see how the above-described principles were translated into concrete actions. According to the Sustainable Development Goals Report 2025: "35 per cent of SDG targets are on track or making moderate progress. Nearly half are moving too slowly and, alarmingly, 18 per cent are in reverse."[30]

This validates Roland Paris's thesis on the risk of rule of law relapse if the underlying institutions are not sufficiently consolidated,[31] as both the Covid-19 pandemic in the early 2020s and the exacerbation of several geopolitical conflicts in recent years placed an additional strain on the achievement of the SDGs. In our context, especially the Sustainable Development Goal 16 with its objective to "Promote peaceful and inclusive societies for sustainable development, provide access to justice for all and build effective, accountable and inclusive institutions at all levels" is of utmost relevance. It covers a plethora of targets, namely (Table 6.1):

In terms of achievement of the above targets, the current statistics are not reassuring and underscore both the accuracy of the High-Level Independent Panel on

---

[28] United Nations General Assembly, *Transforming our world: the 2030 Agenda for Sustainable Development*, UN Doc A/RES/70/1 (25 September 2015).

[29] United Nations and World Bank, *Pathways for Peace: Inclusive Approaches to Preventing Violent Conflict* (World Bank 2018).

[30] United Nations (2025) The Sustainable Development Goals Report 2025, p. 2.

[31] Roland Paris, *At War's End: Building Peace after Civil Conflict* (Cambridge University Press 2004).

**Table 6.1** UN SDG 16 in detail. Overview drafted by the authors. 

| | |
|---|---|
| Target 16.1 | Significantly reduce all forms of violence and related death rates everywhere |
| Target 16.2 | End abuse, exploitation, trafficking, and all forms of violence against and torture of children |
| Target 16.3 | Promote the rule of law at the national and international levels and ensure equal access to justice for all |
| Target 16.4 | By 2030, significantly reduce illicit financial and arms flows, strengthen the recovery and return of stolen assets, and combat all forms of organized crime |
| Target 16.5 | Substantially reduce corruption and bribery in all their forms |
| Target 16.6 | Develop effective, accountable, and transparent institutions at all levels |
| Target 16.7 | Ensure responsive, inclusive, participatory, and representative decision-making at all levels |
| Target 16.8 | Broaden and strengthen the participation of developing countries in the institutions of global governance |
| Target 16.9 | By 2030, provide legal identity for all, including birth registration |
| Target 16.10 | Ensure public access to information and protect fundamental freedoms, in accordance with national legislation and international agreements |
| Target 16. A | Strengthen relevant national institutions, including through international cooperation, for building capacity at all levels, in particular in developing countries, to prevent violence and combat terrorism and crime |
| Target 16.B | Promote and enforce non-discriminatory laws and policies for sustainable development |

Peace Operations (HIPPO) reports findings, and the continued need for evidence-based and targeted actions.

- Target 16.1: Reduce violence everywhere, which seeks to significantly reduce all forms of violence and related death rates globally. However, 2024 saw a surge of 40% increase in loss of lives amid armed conflicts from 2023, marking the third consecutive annual rise.[32] A total of 36.1 thousands of Conflict-Related civilian deaths were recorded globally with the majority of victims being women and children[33] in the world (2024).
- Target 16.2: Protect children from abuse, exploitation, trafficking and violence with the objective to end abuse, exploitation, trafficking, and all forms of violence against and torture of children. Yet, recent trends saw a stark 31% increase in child trafficking since 2019 and an overall increase of human trafficking victims of 25% in 2022.[34]
- Target 16.3: Promote the rule of law and ensure equal access to justice that especially targets the promotion of the rule of law at the national and international levels and aims to ensure equal access to justice for all. The Sustainable Development Goals Report 2025, however, finds that 31 per cent of the global

[32] United Nations (2025) The Sustainable Development Goals Report 2025, p. 40.

[33] Global SDG Database (2025) and last visited on 23 November 2025.

[34] United Nations (2025) The Sustainable Development Goals Report 2025, p. 41.

prison population were held in pretrial detention, with the numbers varying significantly per continent.[35]

- Target 16.5: Substantially reduce corruption and bribery, which seeks to reduce corruption and bribery in all their forms substantially by 2030. Globally 13.2% of firms experienced at least one bribe payment request in 2025.[36]
- Target 16.6: Develop effective, accountable, and transparent institutions at all levels, which is measured by the reliability of budgets and fiscal discipline. According to the Sustainable Development Goals Report 2025: "global budget deviations improved from 12.6 per cent in 2020 to 9.6 per cent in 2023."[37]
- Target 16.7: Ensure responsive, inclusive, and representative decision-making that aims to guarantee responsive, inclusive, participatory, and representative decision-making at all levels. However, the Sustainable Development Goals Report 2025 found that in 2024 only 44% of the people surveyed in 83 countries believe that they have a stake in the political decision-making.[38]
- Target 16.8: Strengthen the participation in global governance, which seeks to broaden and strengthen the participation of developing countries in the institutions of global governance. Yet, the percentage of developing countries in the voting rights at international organizations has remained stagnant in recent years.[39]
- Target 16.9: Provide universal legal identity, which aims to ensure birth registration and related legal identity for all by 2030. While this has been reached in the Global North, the proportion of children under 5 years of age whose births have been registered is merely 51.1% in Sub-Saharan Africa and 25.6 in Oceania (excluding Australia and New Zealand).[40]
- Target 16.a: Ensure public access to information and protect fundamental freedoms that seek to ensure public access to information and protect fundamental freedoms, in accordance with national legislation and international agreements. Significant progress was made toward this target as "a 2024 survey found that 95 per cent of 125 countries have legal guarantees, but enforcement remains uneven."[41]
- Target 16.c: Promote and enforce non-discriminatory laws and policies, which aims to further strengthen non-discriminatory laws and policies for sustainable development. However, recent studies have found that "one in five people report having been discriminated against on at least one ground prohibited by interna-

[35] United Nations (2025) The Sustainable Development Goals Report 2025, p. 41.

[36] Global SDG Database (2025) and last visited on 23 November 2025, United Nations (2025) The Sustainable Development Goals Report 2025—Statistical Annex, p. 213.

[37] The Sustainable Development Goals Report 2025, p. 41.

[38] United Nations (2025) The Sustainable Development Goals Report 2025, p. 41.

[39] United Nations (2025) The Sustainable Development Goals Report 2025—Statistical Annex, p. 217.

[40] United Nations (2025) The Sustainable Development Goals Report 2025—Statistical Annex, p. 218.

[41] United Nations (2025) The Sustainable Development Goals Report 2025, p. 41.

tional law in the past 12 months, with LDCs reporting the highest rates averaging 24.3 per cent."[42]

While this list seems to paint a very bleak picture of the progress made toward the achievement of SDG 16, it is important to take the new and evolving circumstances nobody could have predicted in 2015 into consideration. As mentioned above, the Covid-19 pandemic had a negative impact on the global development, frequently even reversing previously achieved objectives, such as a surge in poverty and a resulting increase in child labor.[43] Moreover the geopolitical landscape has undergone a tremendous transformation with several conflicts (re-) erupting and multilateralism and multilateral institutions being challenged from Brexit to the change in US Foreign Policy Doctrines through the two Trump Administrations,[44] sometimes labeled as Donroe Doctrine.[45] Furthermore, new technological developments, such as the advent of artificial intelligence and the emerging issue of cybercrime, have given rise to new threats to the rule of law, as will be examined below.

## 6.4 Challenges to the Rule of Law in the Twenty-First Century

The Rule of Law originated as a counter point to arbitrary justice systems in which those accused of crimes were at the mercy of the personal whims of the ruling class, and those with titles and funds are above the law.[46] While in the twenty-first century reference to the Rule of Law can be found in many constitutions, it is frequently imperfectly applied.[47] The application of the Rule of Law falls short in cases where political pressure is exerted over judges and investigators, rights are suspended for a prolonged time, and laws are passed by emergency powers outside of the normal legislative process. Such practices hollow out the trust in government institutions, creating a vicious spiral in which people seek to exploit the imperfect system for their own ends, thus further reducing its perceived legitimacy, as pointed out especially for fragile contexts by Roland Paris.

Consequently, as "the breakdown of the rule of law may be both the cause and the consequence of conflict,"[48] this can lead to serious tensions and even internal conflicts. With the ongoing geopolitical conflicts, it is thus evident that the rule of law remains under constant threat of relapse.

---

[42] Global SDG Database (2025) and United Nations (2025) The Sustainable Development Goals Report 2025—Statistical Annex, p. 40.

[43] United Nations (2020) The Sustainable Development Goals Report 2020, p. 3.

[44] National Security Strategy of the United States of America November 2025.

[45] https://edition.cnn.com/2026/01/05/media/trump-donroe-doctrine-ny-post-fox-news.

[46] Brahimi (2002) The Rule of Law at Home and Abroad, p. 9.

[47] Brahimi (2002) The Rule of Law at Home and Abroad, p. 11.

[48] Brahimi (2002) The Rule of Law at Home and Abroad, p. 9.

Moreover, recent developments such as the advent of artificial intelligence (AI) in legal contexts can further hollow out the reliability of institutions and the trust of the public in them. In some countries, AI has now been introduced in the courtroom, which raises the question of how the rule of law and strong institutions are equipped for the transition into the digital sphere. The use of AI in the field has several potential benefits for judicial systems, such as improved inclusion by using Large Language Models (LLMs) to translate decisions into accessible language.[49] However, there is also a plethora of pitfalls and dangers. Given the inability of Large Language Models (LLMs) to not answer a question, and the resulting "hallucinated" answers, one cannot be certain that all responses are factually correct. As found by UNESCO experts: "[...] reports have surfaced from Australia, Brazil, Canada, South Africa, Spain, and the United States regarding judges and lawyers who have issued judicial decisions or submitted legal documents that contained references to non-existent rulings due to the inappropriate use of AI chatbots."[50] This trend could potentially further undermine the rule of law and the entire justice system, as legal reasoning and the administration of justice is outsourced to opaque algorithms. While UNESCO has published the Guidelines for the Use of AI Systems in Courts and Tribunals,[51] it is their application in the national jurisdictions that will determine how the rule of law will be safeguarded against this influence. Akin to the challenges in fragile contexts, it is the premature liberalization of AI while the underlying regulatory institutions remain weak that tests the rule of law.

## 6.5 The Sustaining Peace Framework: An Operational Lens

In the context of conflicts and building on the 2030 Agenda and the normative elevation of SDG 16, the above-described challenges are sought to be addressed by the *Sustaining Peace* framework. This framework now provides the operational lens for integrating justice and the rule of law into differentiated political trajectories of conflict prevention, management, and resolution. Formally adopted by General Assembly resolution A/RES/70/262 and Security Council resolution S/RES/2282,[52] this framework redefines peace not as a static intervention, but as a comprehensive and continuous undertaking. In this regard, the United Nations explicitly defines sustaining peace as "… a goal and a process to build a common vision of a society, ensuring that the needs of all segments of the population are taken into account […]."[53] Through this approach, the United Nations explicitly affirms that strength-

[49] UNESCO (2025) Guidelines for the Use of AI Systems in Courts and Tribunals, p. 7f.

[50] UNESCO (2025) Guidelines for the Use of AI Systems in Courts and Tribunals, p. 8.

[51] UNESCO (2025) Guidelines for the Use of AI Systems in Courts and Tribunals.

[52] United Nations General Assembly, UN Doc A/RES/70/262 (27 April 2016); United Nations Security Council, UN Doc S/RES/2282 (27 April 2016).

[53] United Nations General Assembly and Security Council, *Report of the Advisory Group of Experts on the Review of the Peacebuilding Architecture*, UN Doc A/70/95–S/2015/446 (17 June 2015) para 2; endorsed by UNGA Res 70/262 (27 April 2016).

ening institutions is not about applying a one-size-fits-all model or a one-off intervention, but about a sequenced and flexible process whose priorities must be adapted to existing capacities and each country's specific political context, as reflected in the Review of the United Nations peacebuilding architecture.[54]

This sequential approach rests on a core idea: judicial and institutional functions produce different effects depending on the phase of the conflict cycle. Upstream of violence (pre-conflict), weaknesses in the rule of law, impunity, and legal exclusion are identified as major structural vulnerabilities that can fuel social grievances and violent mobilization. During active conflict, the role of institutions shifts to prioritize limiting arbitrariness, protecting civilians, and preserving a minimum level of legality essential for the social contract's survival. Finally, in post-conflict phases, restoring the credibility of judicial institutions becomes the central pivot for rebuilding trust between state and society, a sine qua non for preventing the resurgence of violence.[55]

Thus, *Sustaining Peace* frames justice not as an end in itself, but as an evolving function, whose strategic priorities must be adjusted over time to ensure lasting stability.[56] Within this framework, emphasis is placed on gradual, targeted, and politically informed interventions that aim to preserve or restore essential institutional functions, rather than immediately rebuilding comprehensive systems. This reflects an explicit recognition of the constraints inherent in contexts of conflict and prolonged fragility: limited resources, fragmented legitimacy, normative pluralism, and contested state authority.[57]

This normative journey from *state-building* to the adoption of SDG 16 and *Sustaining Peace*—thus moves beyond abstract theoretical models toward a pragmatic approach to international institutional intervention. This new consensus now demands the translation of these normative principles into concrete levers for action, structured around five operational pillars tailored to the specificities of post-conflict contexts.[58]

---

[54] United Nations General Assembly and Security Council, *Review of the United Nations peacebuilding architecture*, UN Doc A/RES/72/276–S/RES/2413 (26 April 2018).

[55] United Nations and World Bank, *Pathways for Peace: Inclusive Approaches to Preventing Violent Conflict* (World Bank 2018).

[56] United Nations General Assembly, UN Doc A/RES/70/262 (27 April 2016); United Nations Secretary-General, *Peacebuilding and Sustaining Peace*, UN Doc A/72/707–S/2018/43 (18 January 2018).

[57] Organisation for Economic Co-operation and Development (OECD), *States of Fragility 2018* (OECD Publishing, 2018).

[58] United Nations General Assembly, UN Doc A/RES/70/262 (27 April 2016); United Nations Secretary-General, *Peacebuilding and Sustaining Peace*, UN Doc A/72/707–S/2018/43 (18 January 2018).

## 6.6 Operational Guidance for Building Institutional Capacity in Post-Conflict Contexts: A Practice-Based Approach

Following the normative shift toward the *Sustaining Peace* framework and SDG 16, the central challenge lies in translating normative ambitions into concrete operational strategies adapted to the specific constraints of post-conflict environments.[59] This narrative advocates for a flexible, sequenced approach that moves beyond rigid models to address the root causes of fragility through four interconnected levers: redefining anticorruption as a strategic tool to dismantle war economies[60]; shifting access to justice from physical infrastructure to the functional empowerment of populations[61]; pragmatically integrating informal legal systems to leverage local legitimacy[62]; and positioning transitional justice as a catalyst for structural reform.[63] Through this systemic integration, the framework shifts the focus from isolated projects to building the foundations of legitimacy in spaces where formal authority is absent or contested.

### 6.6.1 Anticorruption in Conflict Situations: War Economy and Legitimacy Reconstruction

While the Sustaining Peace framework laid the groundwork for a political and sequential approach, it is in the field of anticorruption, anchored in the 2003 United Nations Convention against Corruption (UNCAC),[64] that this evolution found one of its operational applications. Since 2016, the UN system has undertaken a fundamental reassessment of the relationship between corruption and armed conflict.[65] Previously seen as a mere administrative pathology to be treated after the fact,

---

[59] United Nations General Assembly, *Transforming our world: the 2030 Agenda for Sustainable Development*, UN Doc A/RES/70/1 (25 September 2015); United Nations General Assembly, UN Doc A/RES/70/262 (27 April 2016).

[60] United Nations Development Programme, *Fighting Corruption in Post-Conflict and Recovery Situations: Learning from the Past* (UNDP, 2015).

[61] United Nations Development Programme, *Global Programme Annual Report 2013:. Strengthening the Rule of Law in Crisis-Affected and Fragile Situations* (UNDP, 2014).

[62] United Nations Development Programme, *Informal Justice Systems: Charting a Course for Human Rights-Based Engagement* (UNDP, 2013).

[63] United Nations Secretary-General, *The Rule of Law and Transitional Justice in Conflict and Post-Conflict Societies*, UN Doc S/2011/634 (12 October 2011); Office of the United Nations High Commissioner for Human Rights, *Rule-of-Law Tools for Post-Conflict States: Transitional Justice* (OHCHR, 2006).

[64] United Nations General Assembly, *United Nations Convention against Corruption*, UN Doc A/RES/58/4 (31 October 2003).

[65] United Nations and World Bank, *Pathways for Peace: Inclusive Approaches to Preventing Violent Conflict* (World Bank 2018).

corruption is now identified as a constitutive mechanism of conflicts, a power infrastructure that shapes the dynamics of violence.[66]

This epistemological shift is grounded in the analytical framework of the United Nations Office on Drugs and Crime (UNODC), which posits that corruption does not merely accompany conflicts but is a condition for their perpetuation. UNODC formulates this causal link explicitly: "Corruption fuels and thrives on conflict. Corruption leads to crises of state legitimacy, failures of governance and economic dislocation. […] Corrupt practices may also cause warring factions to prolong conflict, using it as a cover to loot natural resources, misappropriating or diverting humanitarian aid or deriving other benefits from the war economy."[67]

The publication of the joint UN–World Bank report *Pathways for Peace* in 2018[68] operationalized these theoretical insights by identifying systemic corruption as a structural factor in the recurrence of violence. This finding invalidated the effectiveness of classic technical frameworks designed for stable contexts, forcing the UN to view corrupt practices not as marginal deviations but as central elements of the war's political economy.[69] This theoretical redefinition now makes it possible to highlight the concrete mechanisms through which predation takes root and fuels war.

On the ground, corruption is not a theoretical abstraction but manifests in tangible financial and logistical circuits that allow war to persist. The first function identified by UN research is the direct financing of armed groups. This occurs through the diversion of natural resource revenues, as well as predation on populations via illicit taxation and the monetization of passage at checkpoints.[70]

The second and third functions concern the consolidation of belligerents' power. On one hand, managing loyalties is achieved through diverted funds, which are used to buy the allegiance of local leaders and distribute selective immunities, thereby integrating armed actors into disintegrating state structures.[71] On the other hand, the diversion of development aid provides a framework for laundering war profits,

[66] Sarah Chayes, *Thieves of State: Why Corruption Threatens Global Security* (W.W. Norton & Company, 2015).

[67] United Nations Office on Drugs and Crime, *Corruption: Threats and Trends in the Twenty-First Century*, Eleventh United Nations Congress on Crime Prevention and Criminal Justice (UNODC 2005) para 26.

[68] United Nations and World Bank, *Pathways for Peace: Inclusive Approaches to Preventing Violent Conflict* (World Bank 2018).

[69] United Nations Office on Drugs and Crime, *Corruption: Threats and Trends in the Twenty-First Century*, Eleventh United Nations Congress on Crime Prevention and Criminal Justice (UNODC 2005).

[70] United Nations Security Council, *Final Report of the Panel of Experts on the Illegal Exploitation of Natural Resources and Other Forms of Wealth of the Democratic Republic of the Congo*, UN Doc S/2002/1146 (16 October 2002).

[71] United Nations Security Council, *Final report of the Panel of Experts on the Democratic Republic of the Congo*, UN Doc S/2022/479 (14 June 2022).

notably through fictitious public contracts and the capture of humanitarian aid, depriving populations of the dividends of peace.[72]

Finally, the fourth function is the instrumentalization of the judiciary, where the system becomes a strategic target to entrench impunity. As documented in South Sudan (2018–2022), this instrumentalization is brutal: it involves the monetization of arrest warrants, the use of pretrial detention as an extortion tool, and the initiation of selective prosecutions to eliminate political rivals.[73] These practices consolidate a post-conflict predatory governance where military networks reconstitute themselves as institutional elites.[74] Faced with this predatory governance that instrumentalizes the state, anticorruption efforts must move beyond mere diagnosis to become a proactive tool for securing stability.

From the UN perspective, the fight against corruption has become a direct lever for stabilization and development, aiming to address the root causes of renewed violence.[75] The experience of United Nations engagement in areas liberated from ISIS in Iraq illustrates this broader logic. In post-liberation contexts marked by intense competition over land, resources, and reconstruction rents, United Nations support to institutional recovery and governance has functioned as a stabilizing factor by mitigating disputes linked to property claims and reconstruction processes. By strengthening safeguards around aid delivery and public spending, such interventions contribute to the restoration of basic services and help rebuild trust between state institutions and local communities, while gradually undermining the social legitimacy of actors embedded in war economies.

To operationalize this imperative, the United Nations has increasingly relied on the UNCAC as a normative and technical reference framework, guiding efforts to align national systems with international anticorruption standards. In practice, this has involved the mobilization of provisions relating to judicial integrity under Article 11 and to technical assistance under Articles 60 to 62.[76] This strategy first relies on jurisdictional specialization, illustrated by UNDP support to the Central Anti-Corruption Criminal Court in Iraq. This specialized structure has led to a measurable increase in prosecutions and the adoption of rigorous procedural standards, insulating judicial processes from direct political interference.

This engineering is complemented by two other levers: strengthening investigative capacities and monitoring.[77] UNDP programs train magistrates in financial

---

[72] United Nations Development Programme, *Governance for Peace: Securing the Social Contract* (UNDP 2012).

[73] United Nations Mission in South Sudan and Office of the United Nations High Commissioner for Human Rights, *Deprivation of Liberty in South Sudan, January 2023—May 2024* (Joint UNMISS–OHCHR Report, December 2024).

[74] William Reno, *Warlord Politics and African States* (Lynne Rienner Publishers 1998).

[75] United Nations and World Bank, *Pathways for Peace: Inclusive Approaches to Preventing Violent Conflict* (World Bank 2018).

[76] United Nations General Assembly, *United Nations Convention against Corruption*, UN Doc A/RES/58/4 (31 October 2003), arts 11, 60–62.

[77] United Nations Development Programme, *Iraq Judicial System and Grand Corruption Cases: The Central Anti-Corruption Criminal Court—Trial Monitoring Programme, First Report (August 2022–July 2023)* (UNDP, 2024).

analysis to trace illicit flows. Simultaneously, the establishment of independent judicial monitoring, as implemented by UNDP in Iraq, serves as a real-time diagnostic to identify systemic flaws (delays, access to defense) and transform procedural transparency into a lever for performance. This technical approach, linking integrity and security, was recently the subject of major doctrinal formalization at the international level.[78]

This operational trajectory was formally consolidated in December 2025 in Doha, during the 11th Conference of the States Parties to the United Nations Convention against Corruption (CoSP-UNCAC).[79] On this occasion, UNODC and the Department of Peace Operations (DPO) jointly presented two key documents. The first, titled *Achieving Peace through Integrity: Addressing Corruption in Conflict-Affected Settings*,[80] normatively establishes judicial incapacity to address economic crimes as an autonomous factor for relapse into violence, positioning the credibility of courts as a key indicator of state legitimacy.

The second document, *Practical Guidance for Field Engagement by the United Nations*,[81] provides field missions with the necessary methodological toolkit to sequence these reforms. It offers frameworks for analyzing specific vulnerabilities of courts, such as the capture of appointments or political interference, to tailor capacity-building to the unique security constraints of each context.

The major contribution of this 2025 normative and operational corpus lies in the explicit articulation between anticorruption and judicial capacity-building. Strengthening technical skills and anchoring efforts in the UNCAC[82] are no longer considered parallel objectives but as interdependent components of a political strategy aimed at shielding the judicial sphere from the reconstitution of war economies. By re-establishing the integrity of the judicial apparatus as a condition for its legitimacy, this comprehensive strategy forms the indispensable foundation upon which improving population access to justice through judicial institution capacity-building in conflict contexts can subsequently be built.

---

[78] United Nations Office on Drugs and Crime (UNODC), *Global Programme against Corruption (GPAC)*.

[79] Conference of the States Parties to the United Nations Convention against Corruption (CoSP11), Doha, Qatar, 15–19 December 2025.

[80] United Nations Office on Drugs and Crime (UNODC), United Nations Peacekeeping and United Nations Department of Political and Peacebuilding Affairs (DPPA), *Achieving Peace Through Integrity: Addressing Corruption in Conflict-Affected Settings* (practical guidance presented at the 11th Conference of the States Parties to the UNCAC, Doha, Qatar, 15–19 December 2025).

[81] United Nations Office on Drugs and Crime (UNODC), United Nations Peacekeeping and United Nations Department of Political and Peacebuilding Affairs (DPPA), *Practical Guidance for Field Engagement by the United Nations* (part of *Achieving Peace Through Integrity: Addressing Corruption in Conflict-Affected Settings*, UNODC, 2025).

[82] United Nations General Assembly, *United Nations Convention against Corruption*, UN Doc A/RES/58/4 (31 October 2003).

### 6.6.2 Improving Population Access to Justice Through Judicial Institution Capacity-Building in Conflict Contexts

Access to justice is not merely a technical question of procedure or geographic availability; it represents a cardinal political function in the transition to peace. This understanding transcends an instrumental vision to anchor itself in a normative obligation: the United Nations General Assembly has reaffirmed that "equal access to justice for all"[83] is a fundamental right without which no other civil or political right can be exercised. Since 2016, this approach has aligned with the Sustaining Peace paradigm, enshrined in resolutions A/RES/70/262 and S/RES/2282.[84] The UN has integrated access to justice as a strategic factor in preventing recurrent violence, considering that an accessible judicial system functions as an indispensable social regulation mechanism to channel collective grievances into institutional pathways.

However, implementing these imperative clashes with the brutal reality of contemporary conflicts, which cause multidimensional systemic collapse. On a material level, this translates into the targeted destruction of infrastructure and the elimination of the physical bases of administration, on a human level, through the flight or assassination of qualified personnel. Functionally, widespread insecurity makes holding hearings impossible, while political fragmentation leads to the extreme politicization of remaining institutions. As the Secretary-General established as early as 2004,[85] this institutional vacuum deprives populations of protection and directly fuels cycles of impunity and revenge that perpetuate violence.

This structural failure creates a multitiered justice system where access barriers—prohibitive costs, illiteracy, dangerous travel—de facto exclude vulnerable populations. UNDP notes that it is precisely the groups most exposed to violations, such as women, internally displaced persons, and minorities, who face the most insurmountable barriers.[86] This exclusion exacerbates social fractures and erodes the very possibility of trust in a state under reconstruction. Faced with these disparities and state incapacity, it has become necessary to move beyond the classic institutional approach to adopt a definition of justice centered on the real needs of justice-seekers.

To address these challenges, UNDP developed a functional reading where access to justice is defined as the real and effective capacity of individuals to mobilize the system for specific purposes. This approach identifies two fundamental primary capacities: the capacity to obtain effective redress for serious violations, a sine qua non condition for reconciliation and acknowledgment of wrongs, and the capacity

83 United Nations General Assembly, *The rule of law at the national and international levels*, UN Doc A/RES/67/1 (24 September 2012), para 14.

84 United Nations General Assembly, *Review of the United Nations peacebuilding architecture*, UN Doc A/RES/70/262 (27 April 2016); United Nations Security Council, *Review of the United Nations peacebuilding architecture*, UN Doc S/RES/2282 (27 April 2016).

85 United Nations Secretary-General, *The rule of law and transitional justice in conflict and post-conflict societies*, UN Doc S/2004/616 (23 August 2004), para 5.

86 United Nations Development Programme, *Access to Justice Practice Note* (UNDP, 2004).

to resolve everyday disputes peacefully, such as land or family disputes, thus offering a vital alternative to the informal or violent mechanisms that proliferate in the absence of a state.[87]

This analytical framework includes two other critical dimensions: the capacity to access basic legal services (information on rights, fair trial) and, crucially, the capacity to obtain enforceable decisions. A judgment that cannot be implemented due to corruption or insecurity is deemed an "empty shell" that undermines the credibility of the entire system. This vision is enshrined in the UNDP People-Centered Approach to Justice and Security,[88] which emphasizes the concrete social and political effects of justice rather than traditional quantitative indicators.

It justifies targeted operational intervention to strengthen national institutions, as they fulfill an irreplaceable stabilizing function. To translate this vision into action, the UN system has structured its interventions around an integrated approach combining institutional capacity-building, people-centered justice delivery, and the expansion of inclusive and accessible justice services, as reflected in UNDP's Global Programme for Strengthening the Rule of Law, Human Rights, Justice and Security for Sustainable Peace and Development (Phase IV, 2022–2025). This approach is operationalized primarily through technical support to justice institutions, aimed at restoring the functionality, territorial coverage, and effectiveness of courts and justice services.[89]

The first pillar of UN action aims to reconstitute the professional core of the judicial system. The work of UNDP and peacekeeping missions includes creating integrated justice access centers, designed as one-stop shops for guidance and legal aid, and the technical training of magistrates on critical issues such as sexual violence or land restitution.[90] Simultaneously, United Nations programs have increasingly emphasized the modernization and digitalization of judicial institutions as tools to enhance procedural transparency and institutional efficiency in fragile and conflict-affected settings.[91]

However, when insecurity or distance isolates populations, justice must project itself physically. This logic underpins the deployment of mobile courts, supported by UN missions to hold hearings in remote areas. These itinerant jurisdictions

[87] UN General Assembly, *Universal Declaration of Human Rights* Art 8 (adopted 10 December 1948) (right to an effective remedy for human rights violations); Praia City Group and United Nations, *SDG 16.3.3 Policy Brief* (22 January 2025) (defining access to justice as the ability to defend rights and obtain just resolution of disputes through formal or informal mechanisms).

[88] United Nations Development Programme, *The UNDP People-Centred Approach to Justice and Security: A Policy Framework for Justice and Security Programming* (UNDP, 2025).

[89] United Nations Development Programme, *Global Programme for Strengthening the Rule of Law, Human Rights, Justice and Security for Sustainable Peace and Development, Phase IV (2022–2025)* (UNDP 2022); see also United Nations Development Programme, *UNDP Strategic Plan 2022–2025* (UNDP 2021).

[90] United Nations Development Programme, *The UNDP People-Centred Approach to Justice and Security: A Policy Framework for Justice and Security Programming* (UNDP, 2025).

[91] United Nations Development Programme, *The UNDP People-Centred Approach to Justice and Security: A Policy Framework for Justice and Security Programming* (UNDP, 2025).

handle a massive volume of minor civil and criminal cases, relieving pressure on fixed courts while restoring a symbolic and concrete presence of the rule of law in areas otherwise left to arbitrariness.[92]

Beyond their immediate dispute-resolution function, mobile justice mechanisms have been recognized by the United Nations as contributing to the extension of state authority and institutional presence in post-conflict settings, particularly in relation to land and property disputes, which are recurrent drivers of local instability.[93] These initiatives prove that access to justice is inseparable from the restoration of territorial sovereignty. While reconstructing institutional provision is crucial, it must be accompanied by strengthening the demand for justice through citizen empowerment and their inclusion in governance.[94]

Alongside institutional provision, the UN works to bridge the representation gap through expanded legal assistance. In line with the "Leave No One Behind"[95] principle, this approach includes training and deploying paralegals, supporting legal clinics, and conducting mass awareness campaigns to inform populations of their rights. The objective is to empower marginalized groups in the face of complex systems by providing them with tools to navigate the judicial apparatus.[96]

This approach particularly targets gender inequalities, as illustrated by the UN in Colombia or Sierra Leone. By combining specialized training for judges and support for networks of women lawyers, these programs have tangibly improved women's capacity to obtain redress for gender-based violence.[97]

Finally, to ensure the sustainability of reforms, the UN framework promotes participatory legal governance. The Global Programme 2022–2025[98] emphasizes the necessity for continuous institutional dialogue and systematic community consultation mechanisms to identify access barriers and co-design solutions. This represents a fundamental shift, moving from a vertical service delivery model to a logic of accountability and co-creation, the only guarantee of a justice system perceived as legitimate by the populations it serves. However, to support the goal of effective and equitable access to justice for populations, UN frameworks have gradually

[92] UN General Assembly, *Strengthening and coordinating United Nations rule of law activities*, Report of the Secretary-General, UN Doc A/69/181 (24 July 2014) para 60.

[93] United Nations Secretary-General, *The rule of law and transitional justice in conflict and post-conflict societies*, UN Doc S/2004/616 (23 August 2004) paras 4–6.

[94] United Nations Development Programme, *The UNDP People-Centred Approach to Justice and Security: A Policy Framework for Justice and Security Programming* (UNDP, 2025).

[95] United Nations General Assembly, *Transforming our World: the 2030 Agenda for Sustainable Development*, UN Doc A/RES/70/1 (25 September 2015), paras 4.

[96] United Nations Development Programme, *Access to Justice Practice Note* (UNDP, 30 October 2015).

[97] United Nations, *A Practitioner's Toolkit on Women's Access to Justice Programming* (United Nations, 2018).

[98] United Nations Development Programme, *Global Programme for Strengthening the Rule of Law, Human Rights, Justice and Security for Sustainable Peace and Development, Phase IV (2022–2025)* (UNDP 2022).

recognized the role of informal and local justice mechanisms in extending the territorial reach of judicial systems in post-conflict contexts.

### 6.6.3 The Integration of Informal and Local Justice Mechanisms: Between Operational Pragmatism and the Gradual Reconstruction of Judicial Legitimacy

In legal landscapes devastated by armed conflict, the state's claim to a monopoly on normative production reveals itself to be a theoretical fiction in the face of an observable reality: that of a deeply entrenched legal pluralism. The United Nations system now acknowledges that state justice, often marginalized or discredited, coexists with a dense constellation of local, customary, religious, and community-based mechanisms that continue to regulate social life.[99] This coexistence does not constitute an archaic residue destined to disappear with modernization, but rather represents a structural characteristic of fragile societies, exacerbated by conflict that undermines central authority while reinforcing local solidarities.

This recognition marks a major doctrinal shift, initiated as early as 2004 by the United Nations Secretary-General, who noted with realism that rebuilding the rule of law could not be limited to the physical restoration of infrastructure, as "Additionally, while focusing on the building of a formal justice system that functions effectively and in accordance with international standards, it is also crucial to assess means for ensuring the functioning of complementary and less formal mechanisms, particularly in the immediate term."[100] This observation necessitates a fundamental shift in perspective: informal justice ceases to be perceived as an obstacle to building the rule of law and is instead understood as a resilient social infrastructure for regulation.

Indeed, these mechanisms often prove more accessible, swift, and culturally comprehensible than distant and costly state courts. To ignore this reality would condemn international interventions to building formal institutions devoid of meaning for the majority of justice-seekers, disconnected from local dynamics of legitimization. The resilience of informal systems stems not solely from cultural preference but responds to precise functional imperatives dictated by the crisis environment.

Empirical studies conducted within the United Nations system[101] demonstrate that the widespread recourse to informal mechanisms stems primarily from a practical rationality of institutional and social survival in the face of the collapse of state provision. This choice is compelled by absolute material inaccessibility, due to the

---

[99] United Nations Development Programme, United Nations Children's Fund and UN Women, *Informal Justice Systems: Charting a Course for Human Rights-Based Engagement* (2012).

[100] United Nations Secretary-General, *The Rule of Law and Transitional Justice in Conflict and Post-Conflict Societies*, UN Doc S/2004/616 (23 August 2004) §36.

[101] United Nations Secretary-General, *The Rule of Law and Transitional Justice in Conflict and Post-Conflict Societies*, UN Doc S/2004/616 (23 August 2004).

destruction of infrastructure and insecurity, as well as by paralyzing procedural delays that transform any official judicial process into a risky gamble. This is compounded by a radical distrust of formal institutions perceived as intrinsically corrupt or politicized, contrasting with the social, linguistic, and normative proximity of customary mechanisms whose actors are identifiable within the community.

Beyond these default motives, these systems ensure central and positive operational functions indispensable for maintaining a minimal social order. They often constitute the only mechanism capable of regulating highly flammable daily disputes, land conflicts, inheritance, and debts preventing their degeneration into fratricidal violence through immediate local mediation that prevents escalation. UNDP thus estimates that in some post-conflict contexts, up to 80 percent[102] of civil and family disputes are resolved outside the formal justice system, through traditional chiefs' councils, religious authorities, or community committees.

This regulatory capacity proves particularly vital for managing explosive tensions related to mass forced displacement and population returns, areas where the absence of accessible legal solutions constitutes a major factor in the recurrence of violence. By preserving a minimal normative legitimacy where the state has disappeared, these mechanisms fill a critical security vacuum. Faced with this ambivalence between pragmatic necessity and the imperative of transformation, the UN had to formalize a new strategic approach.

Over the past two decades, the UN has forged a nuanced and explicit operational doctrine that rejects both the substitution's approach and the romantic idealization of traditional justice. The chosen path is that of a conditional, critical, and regulated integration of informal mechanisms into a comprehensive rule of law reconstruction strategy. This position, theorized within the *People-Centered Justice* framework,[103] postulates that any relevant reform must start from the real-life trajectories of justice-seekers to channel and transform the normative energy of the informal sector rather than confronting it head-on.

The ultimate objective is not to institutionalize permanent legal pluralism, but to use this pluralism as a transitional lever for legitimization and capacity-building. The aim is to orchestrate a progressive and normative convergence toward a reformed, accessible, and rights-respecting formal judicial system. This strategy aims to create a hybrid and evolving judicial ecosystem, where different mechanisms mutually reinforce rather than compete with each other, thereby guaranteeing broader jurisdictional coverage without sacrificing international standards.

This approach allows for moving beyond the sterile dichotomy between formal and informal justice to focus on the quality of service delivered to the justice-seeker. By conditioning its support on the evolution of customary practices, the UN engages

[102] United Nations Development Programme, United Nations Children's Fund and UN Women, *Informal Justice Systems: Charting a Course for Human Rights-Based Engagement* (UN Women 2012) 7.

[103] United Nations Development Programme, *The UNDP People-Centred Approach to Justice and Security: A Policy Framework for Justice and Security Programming* (UNDP, 2025).

in a process of profound social transformation that requires specific technical modalities of intervention on the ground.

The operational translation of this approach relies firstly on the creation of institutional bridges and formalized referral mechanisms between customary and judicial authorities. In contexts such as Somalia, Afghanistan (pre-2021), or Sierra Leone,[104] bidirectional protocols compel traditional authorities to refer serious crimes (genocide, sexual violence) to formal courts, while the latter delegate minor disputes to community mediation. This articulation aims to decongest courts while combating impunity, allowing for the mutual legitimization of both systems through a recognized division of labor.

Secondly, the UN deploys targeted training and mentoring programs to progressively align local practices with human rights. Initiatives in South Sudan and Mali[105] train customary chiefs and mediators in principles of gender equality and nonviolent mediation, engaging a process of gradual normalization through learning and dialogue without brutal institutional absorption. The Afghan experience of UNAMA demonstrated the effectiveness of this support to *jirgas* and *shuras* in processing a massive volume of disputes, although the superior legitimacy of informal mechanisms there sometimes highlighted the persistence of distrust toward the central state.[106]

Finally, this integration constitutes a major security issue, notably through its strategic articulation with stabilization agendas. In Mali, support for local mechanisms explicitly aims to prevent their capture by extremist armed groups that instrumentalize justice to establish territorial control. MINUSMA assessments confirm that these mechanisms, when articulated with state jurisdictions, help defuse intercommunal tensions and contain the normative influence of violent groups, thereby validating pragmatic integration as a lever for preventing violent extremism. If the integration of local mechanisms stabilizes the social body by resolving daily disputes, the durable reconstruction of the rule of law requires addressing the legacy of mass violence through transitional justice, the only means capable of transforming this security stabilization into institutional political legitimacy.[107]

---

[104] United Nations Development Programme, United Nations Children's Fund and UN Women, *Informal Justice Systems: Charting a Course for Human Rights-Based Engagement* (UN Women 2012).

[105] United Nations Development Programme, United Nations Children's Fund and UN Women, *Informal Justice Systems: Charting a Course for Human Rights-Based Engagement* (UN Women 2012).

[106] United Nations Assistance Mission in Afghanistan, *Fighting Corruption in Afghanistan: Stepping Up Transparency, Integrity and Accountability* (UNAMA, August 2021).

[107] United Nations Multidimensional Integrated Stabilization Mission in Mali (MINUSMA), *Report of the Secretary-General on the Situation in Mali*, UN Doc S/2018/866 (28 September 2018).

### 6.6.4 Transitional Justice: A Lever for Reconstruction and Legitimation

In societies emerging from mass violence, transitional justice can no longer be considered a mere symbolic ornament or an exceptional mechanism. Since 2016, the United Nations system has theorized it as a structuring political function, indispensable for state reconstruction and the restoration of legitimacy eroded by arbitrariness. This paradigmatic evolution was institutionalized by the adoption of the twin Sustaining Peace resolutions (A/RES/70/262 and S/RES/2282)[108] in 2016, which now explicitly integrate transitional justice into the peacebuilding architecture, no longer as an ex-post reaction, but as a tool for the structural prevention of conflict.

This expanded conception is codified within the UN, which defines this field as encompassing "the full range of processes and mechanisms associated with a society's attempts to come to terms with a legacy of large-scale past abuses."[109] The UN now establishes a direct causal link between the absence of judicial treatment of the past and violent recurrence. As demonstrated in the joint UN-World Bank,[110] persistent impunity is not merely a moral failure; it fuels grievances, legitimizes narratives of revenge, and radically delegitimizes the state, creating fertile ground for radicalization.

Conversely, well-integrated transitional justice breaks this dynamic by offering institutional alternatives to vengeance. By reaffirming that no one is above the law, it transforms justice perceived during conflict as an instrument of domination, into a vector for political stabilization and social cohesion. It thus acts as a strategic investment in future peace by tackling one of its most corrosive enemies: the sense of institutionalized injustice. This political ambition, however, cannot be realized without deep institutional anchoring, requiring a new articulation with the existing judicial apparatus.

One of the critical lessons post-2016 is that transitional mechanisms are doomed to failure if they are not organically articulated with a substantive strengthening of national judicial capacities. In its strategic document, UNDP[111] clearly identified the limits of parachuted mechanisms that, operating in an institutional vacuum, have no lasting impact on the overall judicial apparatus. The UN strategy now favors systemic integration: the aim is to use the political momentum and resources of the transition to catalyze deep reforms and professionalize magistrates, rather than creating disconnected parallel entities.

[108] United Nations General Assembly, UN Doc A/RES/70/262 (27 April 2016); United Nations Security Council, UN Doc S/RES/2282 (27 April 2016).

[109] United Nations Secretary-General, *The rule of law and transitional justice in conflict and post-conflict societies*, UN Doc S/2004/616 (23 August 2004), para 8.

[110] United Nations and World Bank, *Pathways for Peace: Inclusive Approaches to Preventing Violent Conflict* (World Bank 2018).

[111] United Nations Development Programme, *Global Programme for Strengthening the Rule of Law, Human Rights, Justice and Security for Sustainable Peace and Development, Phase IV (2022–2025)* (UNDP 2022).

The operational consequence of this realization is major: the UN approach has evolved toward a systemic integration of transitional justice into the broader rule of law strengthening agenda. It is no longer a matter of creating parallel entities, but of using the political momentum and resources of the transitional period to catalyze deep judicial reforms. The focus is now on the targeted professionalization of magistrates (notably in international criminal and humanitarian law), the creation of specialized jurisdictions integrated into the national system, and the reform of procedural frameworks so they can absorb the complexity of mass crime cases while guaranteeing trial fairness.

This approach aims to fulfill four central judicial functions identified in the Secretary-General's Report on strengthening the rule of law,[112] legally breaking with structural impunity to restore legality; restoring the intrinsic credibility of the system by proving its capacity to judge powerful actors; reconfiguring norms through acculturation to international standards; and finally, producing official recognition of harm. Transitional justice is no longer a justice for the transition, but a justice of the transition, actively participating in the qualitative transformation of the state.

To concretize this integration, the UN favors the model of specialized hybrid chambers anchored within national jurisdictions. The example of the Special Criminal Court (SCC) in the Central African Republic, supported by MINUSCA and UNDP,[113] perfectly illustrates this logic: composed of national and international magistrates, it judges serious crimes while serving as a capacity-building lever. The SCC has helped raise the professional standards of the entire Central African judicial system by demonstrating the feasibility of fair trials for mass crimes.

Concurrently, targeted technical assistance programs, such as those of UNITAD in Iraq, strengthen investigation and prosecution skills. Faced with the complexity of digital evidence or mass graves left by Daesh, UNITAD does not merely collect evidence; it trains Iraqi judges and prosecutors in forensic analysis and the management of massive evidence. According to UNITAD Reports to the Security Council,[114] the aim is to build, in real time, the national system's capacity to absorb and process the judicial legacy of the conflict through advanced training.

These technical mechanisms are complemented by institutional reform processes, notably the "vetting" of judicial personnel. Far from being a blind purge, this process aims at a functional and credible reconfiguration of the judicial apparatus

[112] United Nations Secretary-General, *The rule of law and transitional justice in conflict and post-conflict societies*, UN Doc S/2004/616 (23 August 2004).

[113] See Human Rights Watch, *Central African Republic: First Trial at the Special Criminal Court* (12 April 2022) noting that the Special Criminal Court (SCC) in the Central African Republic is a hybrid national court staffed by both international and national personnel and is operated *"in partnership with the United Nations,"* with support from the United Nations peacekeeping mission MINUSCA and the United Nations Development Programme (UNDP).

[114] United Nations Security Council, *Letter dated 1 May 2021 from the Special Adviser and Head of the United Nations Investigative Team to Promote Accountability for Crimes Committed by Da'esh/Islamic State in Iraq and the Levant addressed to the President of the Security Council*, UN Doc S/2021/419 (1 May 2021).

by removing individuals implicated in past violations. The Secretary-General's Thematic Reports on Transitional Justice and the Rule of Law[115] emphasize that this is an indispensable condition for the depoliticization of justice and the restoration of public trust, allowing for the retention of competence while guaranteeing the integrity of the institution. These technical mechanisms are not an end in themselves, but the necessary foundation for making justice audible to those who need it most: the victims.

A decisive contribution of contemporary transitional justice lies in the central place accorded to victims, no longer as passive objects but as active subjects of the judicial process.[116] The UN insists that the effective access of victims to mechanisms, via the status of civil party or protected testimony, is a key factor of legitimation. When justice recognizes the status of victim and grants reparations, it ceases to be a distant abstraction to become a concrete service rendered to the wronged citizen, thereby rebuilding the social contract from the ground up.[117]

The architecture established in Colombia after 2016, with the Special Jurisdiction for Peace (JEP), embodies this ambition. This sophisticated system, supported by the UN, places victims at the heart of the procedure and provides for collective reparations, while maintaining constant dialogue with the ordinary justice system. The Reports to the Security Council[118] demonstrate how this model can reconfigure citizen expectations and anchor the legitimacy of institutions in their capacity to respond to the recognition needs of affected populations.

In conclusion, transitional justice asserts itself as the final pillar of judicial strengthening. It uses the moral imperative of judging past crimes as a catalytic energy to rebuild state capacity. By compelling the judicial system to reform in order to handle its heaviest caseloads, it completes the cycle of reconstruction: there can be no lasting peace without justice, nor legitimate justice without capable institutions, these two imperatives mutually constructing each other in the fragility of the postwar period. At the conclusion of this analysis of operational levers, from the fight against corruption to transitional justice, it is now necessary to move beyond a sectoral approach and mobilize a transversal analytical framework capable of evaluating the coherence and sustainability of these interventions: the Three-Pillar[119] (3-P) analytical model.

[115] United Nations Secretary-General, *The Rule of Law and Transitional Justice in Conflict and Post-Conflict Societies*, UN Doc S/2004/616 (23 August 2004).

[116] United Nations Secretary-General, *Transitional Justice: A Strategic Tool for People, Prevention and Peace—Guidance Note of the Secretary-General* (United Nations, 2023).

[117] United Nations General Assembly, *Basic Principles and Guidelines on the Right to a Remedy and Reparation for Victims of Gross Violations of International Human Rights Law and Serious Violations of International Humanitarian Law*, UN Doc A/RES/60/147 (16 December 2005).

[118] United Nations Secretary-General, *Report of the Secretary-General on the United Nations Mission in Colombia*, UN Doc S/2017/801 (26 September 2017).

[119] See: https://link.springer.com/book/10.1007/978-3-030-23227-6).

## 6.7 The 3 Pillars as Conceptual Foundations for the Organizational Sustainability of Peace

Analysis of international engagement reveals a dynamic of continuous progress in conceptualizing peace. Since the end of the Cold War, UN thinking has undergone remarkable maturation, evolving from post-conflict peacebuilding formulated in *An Agenda for Peace* (1992)[120] to the contemporary and holistic framework of *Sustaining Peace*.[121] This positive evolution has allowed for the progressive integration of empirical lessons and major doctrinal contributions, such as the critique of liberal peacebuilding, institutionalist approaches, and the local turn.[122] This normative sophistication was crowned by the report of the High-Level Independent Panel on Peace Operations (HIPPO, 2015) and the adoption of the 2030 Agenda with Sustainable Development Goal 16, providing a robust architecture of principles to promote peaceful and inclusive societies.[123]

However, to fully operationalize these principles in complex environments, it is pertinent to mobilize a distinct analytical grid, capable of articulating the political purpose, the adaptive dynamic, and the relational quality of collective action. This is where the independent contribution of the 3P model (Purpose, Traveling, Connectivity), developed by Peter Wollmann, Frank Kühn, Michael Kempf, and Reto Püringer, is situated.[124] This transversal conceptual framework offers an integrative perspective to bring coherence to the operational practices analyzed in this work. It enables the identification of positive organizational conditions for sustainability, providing a method for designing interventions that are simultaneously

---

[120] Boutros Boutros-Ghali, *An Agenda for Peace: Preventive Diplomacy, Peacemaking and Peacekeeping*, UN Doc A/47/277–S/24111 (17 June 1992).

[121] United Nations General Assembly, UN Doc A/RES/70/262 (27 April 2016); United Nations Security Council, UN Doc S/RES/2282 (27 April 2016).

[122] Roland Paris, *At War's End: Building Peace after Civil Conflict* (Cambridge University Press 2004). Francis Fukuyama, *State-Building: Governance and World Order in the twenty-first Century* (Cornell University Press 2004). Roger Mac Ginty, *International Peacebuilding and Local Resistance* (Palgrave Macmillan 2011); Oliver P Richmond, *A Post-Liberal Peace* (Routledge 2011).

[123] United Nations, *Report of the High-Level Independent Panel on Peace Operations on uniting our strengths for peace*, UN Doc A/70/95–S/2015/446 (17 June 2015). United Nations General Assembly, *Transforming our world: the 2030 Agenda for Sustainable Development*, UN Doc A/RES/70/1 (25 September 2015), esp. Sustainable Development Goal 16.

[124] Peter Wollmann, Frank Kühn and Michael Kempf (eds), *Three Pillars of Organization and Leadership in Disruptive Times: Navigating Your Company Successfully Through the twenty-first Century Business World* (Cham: Springer Nature, Cham 2020). Wollmann, P. et al. (Eds.) (2022) Transforming Public and Private Sector Organizations—Implementing Sustainable Purpose, Travelling Organization and Connectivity for Resilience. Cham: Springer Nature; Wollmann, P. et al. (Eds.) (2021) Organization and Leadership in Disruptive Times—Design and Implementation of the 3-P-Model. Cham: Springer Nature; Wollmann, P. et al. (Eds.) (2020) Three Pillars of Organization and Leadership in Disruptive Times—Navigating Your Company Successfully through the twenty-first Century Business World. Cham: Springer Nature; Wollmann, P. et al. (Eds.): (2024) Reading an Unknown Organization - Leaders in New Roles Thriving in Challenging Environments with the 3-P-Model Mindset. Cham Springer Nature.

politically anchored (Purpose), agile in their implementation (Traveling), and relationally connected (Connectivity).

### 6.7.1 Purpose: Social Legitimacy as the Engine of Resilience

The *Purpose* pillar places the social relevance and ultimate aim of action at the heart of institutional success. It formalizes a fundamental lesson: to be sustainable, an institution must be perceived as intrinsically legitimate and useful by the populations it serves. This concept positively synthesizes academic contributions by recalling that state authority does not rest solely on its administrative capacity, but on a shared narrative of justice and utility. It invites structuring interventions around a foundational question: for whom and for what purpose does this institution act? This approach allows for transforming the perception of the rule of law from an external constraint into an accepted social resource. Thus, by underscoring the impact and social purpose of an institution of an action, institutions can strengthen their public backing and legitimacy.

This requirement for meaning permeates all the operational levers studied. Anticorruption thus becomes a political struggle for public trust; access to justice asserts itself as a vector for reparation and recognition; the integration of local mechanisms anchors the legal order in social reality; and transitional justice acts as a catalyst for re-founding the social contract. *Purpose* therefore acts as the transversal condition for resilience, ensuring that technical reforms, even the most sophisticated ones, produce living and credible institutions, socially useful, rather than formal structures disconnected from citizen expectations.[125]

### 6.7.2 Traveling: Institutional Action as a Dynamic of Adaptation

The *Traveling* pillar highlights the capacity of interventions to evolve positively within fluid environments. Far from a rigid linear trajectory, it conceptualizes peace consolidation as a living, contingent, and adaptive process. The observed experiences demonstrate that the success of judicial reforms relies on their ability to navigate changing local political dynamics. This concept introduces a logic of organizational reflexivity, where continuous strategic adjustment, learning from experience, and flexible sequencing are major assets for transforming uncertainty into an opportunity for consolidation.

This dynamic approach allows interventions to maintain their internal coherence while adjusting to the realities on the ground. Within this framework, the dilemmas

[125] United Nations, *Report of the High-Level Independent Panel on Peace Operations*, UN Doc **A/70/95–S/2015/446** (17 June 2015); United Nations General Assembly, *Transforming our world: the 2030 Agenda for Sustainable Development*, UN Doc A/RES/70/1 (25 September 2015), SDG 16; United Nations Secretary-General, *Peacebuilding and Sustaining Peace*, UN Doc A/72/707–S/2018/43 (18 January 2018).

encountered, for example, between the demand for justice and stability in anticorruption efforts, are no longer obstacles, but parameters for navigation. *Traveling* thus underscores the crucial importance of context monitoring mechanisms and feedback loops. Sustainable strategies are those that manage to progressively integrate social and institutional constraints, making permanent evolution a condition for the performance and anchoring of reforms.

### 6.7.3 Connectivity: Peace as an Integrated Relational Ecosystem

The *Connectivity* pillar highlights the relational dimension as a key factor of stability. It is based on the observation that post-conflict societies are plural ecosystems where governance relies on the quality of links between a multitude of actors: the state, customary authorities, civil society, and international partners. The sustainability of peace therefore depends on the density, fluidity, and trust characterizing these interactions. This model explains the success of initiatives like mobile courts, which re-establish a geographic and human link, or formalized bridges between state justice and customary mechanisms, which create mutual recognition.

By fostering this relational coherence, *Connectivity* allows for overcoming institutional fragmentation. It values hybrid architectures, where collaboration between national and international magistrates strengthens the overall credibility of the system. In summary, the joint application of the 3Ps, Purpose, Traveling, Connectivity, offers a strategic compass for realizing SDG 16.[126] It demonstrates that reconstructing the state through justice is not a juxtaposition of technical silos, but a continuous organic process aimed at weaving a resilient, legitimate, and connected governance network.

## 6.8 Conclusion

This article commenced by tracing not only the development of peacekeeping and the rule of law since the mid-twentieth century, but also by shining light on its three major critiques as voiced by Paris, Fukayama, as well as Mac Ginty, Richmond, Lederach respectively. As examined, the shift from the Agenda for Peace (1992) to the contemporary and holistic framework of Sustaining Peace, as well as the adoption of the 2030 Agenda with Sustainable Development and especially Goal 16, showcases an institutionalization of the framework. This framework with its many targets creates a strong supporting architecture for the achievement of the rule of law and the strengthening of institutions. In the post-conflict context, this means:

1. Strengthening institutions and the public trust therein by implementing anticorruption measures.

[126] United Nations General Assembly, UN Doc A/RES/70/262 (27 April 2016); United Nations Security Council, UN Doc S/RES/2282 (27 April 2016).

2. Ensuring accountability by providing access to justice for all, including through the employment of informal and local justice mechanisms.
3. Enhancing capacity of the institutions through targeted capacity-building programs.
4. Securing legitimacy through participatory governance and including marginalized groups by providing them with tools to navigate the judicial apparatus.

As outlined above, the work of the United Nations through UNDP and the various peacekeeping missions uses a plethora of programs, such as the creation of creating integrated justice access centers, and the technical training of magistrates on critical issues, to achieve these aims.

It is important to bear in mind that the rule of law and legal certainty is not only undermined in conflict and post-conflict situations, but that the strengthening of public and private institutions, which seek to achieve SDG 16 globally, is essential to maintain the rule of law at home and abroad—or rather domestically and internationally.

Taking the lessons learnt from conflict and post-conflict contexts and applying the Three-Pillar Matrix, those wishing to strengthen the rule of law and prevent both internal and external conflicts need to commence:

1. Re-establishing legitimacy of institutions (or the *sustainable purpose*) by building trust in the institutions, fighting corruption, and ensuring accountability.
2. Adapting institutions to the new context (or a *traveling organization*), by remaining agile and flexible to address the issues at hand, and by constantly monitoring the system for trends, changes, and challengers.
3. Re-connecting with all stakeholders (or connectivity) and involving especially marginalized groups through participatory processes.

**Disclaimer** *The views and opinions expressed in this publication are those of the authors and do not necessarily reflect the views, policies, or positions of the United Nations Institute for Training and Research (UNITAR) or the United Nations. The author contributed to this publication in a personal capacity and not in an official role as a representative of UNITAR. Responsibility for the content of this chapter rests solely with the author.*

## Further Reading

National Security Strategy of the United States of America November 2025 available via https://www.whitehouse.gov/wp-content/uploads/2025/12/2025-National-Security-Strategy.pdf and last visited on 31 January 2026

United Nations (2025) The Sustainable Development Goals Report 2025 available via https://unstats.un.org/sdgs/report/2025/ and last visited on 23 November 2025

United Nations (2020) The Sustainable Development Goals Report 2025 available via https://unstats.un.org/sdgs/report/2020/ and last visited on 23 November 2025
United Nations (2023) New Vision of the Secretary-General for the Rule of Law available via https://www.un.org/ruleoflaw/wp-content/uploads/2023/10/Rule-of-Law-New-Vision-English.pdf and last visited on 23 November 2025
Global SDG Database (2025) available via https://unstats.un.org/UNSDWebsite/undatacommons/sdgs and last visited on 23 November 2025
United Nations (2025) The Sustainable Development Goals Report 2025—Statistical Annex available via https://unstats.un.org/sdgs/files/report/2025/E_2025_62_Statistical_Annex_I_and_II.pdf and last visited on 23 November 2025
UNESCO(2025) Guidelines for the Use of AI Systems in Courts and Tribunals.
Annan, Kofi (2002) Speech to the United Nations General Assembly on 12 September 2002, retrieved from https://press.un.org/en/2002/ga10044.doc.htm and last visited on 25 January 2026
Brahimi, Lakhdar (2002) The Rule of Law at Home and Abroad. Dag Hammarsköld Lecture 2002.
Wollmann, P. et al.(Eds.) (2020) Three Pillars of Organization and Leadership in Disruptive Times—Navigating Your Company Successfully through the twenty-first Century Business World. Cham: Springer Nature.
Wollmann, P. et al. (Eds.) (2021) Organization and Leadership in Disruptive Times—Design and Implementation of the 3-P-Model. Cham: Springer Nature.
Wollmann, P. et al. (Eds.) (2022) Transforming Public and Private Sector Organizations—Implementing Sustainable Purpose, Travelling Organization and Connectivity for Resilience. Cham: Springer Nature.
Wollmann, P. et al. (Eds.): (2024) Reading an Unknown Organization - Leaders in New Roles Thriving in Challenging Environments with the 3-P-Model Mindset. Cham Springer Nature,
CNN. Donroe Doctrine, see https://edition.cnn.com/2026/01/05/media/trump-donroe-doctrine-ny-post-fox-news

**Anna Sabrina Wollmann** holds a PhD in International Law from Maastricht University and has over a decade of experience in leading international and multistakeholder projects in the field of international law, capacity-building, and legal education. She has been working at the United Nations Institute for Training and Research (UNITAR) since 2017, leading international law projects for the Division for Multilateral Diplomacy since 2018. Prior to joining UNITAR, Sabrina worked as a Researcher for Maastricht University and the European Commission DG: Internal Market.

**Faisal Aminollah** is an expert in international law and governance, specializing in institutional strengthening and development strategies in fragile and post-conflict environments. Currently serving as a University Lecturer in Law based in Qatar, he holds an LL.M. in Business Law from the Sorbonne Law School in Paris. He has also completed coursework in International Public Law at Hamad Bin Khalifa University (HBKU) in Qatar. With extensive experience within United Nations institutions, notably with UNITAR, he has coordinated and advised on capacity-building programs focused on peace, judicial systems, and governance. In this capacity, he has led strategic initiatives related to the 2030 Agenda, contributing concretely to the implementation of SDG 16 (peace, justice, and effective institutions) in Africa, Latin America, the Middle East, and Asia. His work centers on institutional legitimacy, transparency, and the protection of fundamental rights in transitional contexts. Author of the book *Les mutations du système éducatif dans les pays du Conseil de coopération du Golfe* (2020), he regularly contributes to academic and diplomatic discourse on multilateralism and public policy.

# Toward a Multilateral State-to-State Dispute Resolution Mechanism Within the WIPO Framework to Protect Fundamental Right and Enhance Access to Justice in Line with SDG 16

7

Ahmed Al-Sulaiti

**Abstract**

The governance of international intellectual property (IP) is at a crossroads. With the paralysis of the WTO Appellate Body and the proliferation of fragmented forums such as free trade agreements and investor–state arbitration, the resolution of state-to-state IP disputes suffers from incoherence, unpredictability, and inequity. These deficiencies undermine legal certainty, discourage innovation, and disproportionately disadvantage developing economies seeking knowledge-based growth. This chapter argues that the absence of a centralized, specialized mechanism calls for the establishment of a permanent state-to-state dispute resolution body within the World Intellectual Property Organization (WIPO). By leveraging WIPO's technical expertise and treaty-based legitimacy, such a mechanism could restore predictability, ensure inclusivity, and embed public-interest

**Remark from the Editors**
This is a special chapter of the book; on the one hand, as it touches a crucial precondition for all research and development efforts to achieve sustainability and UN SDG coverage in general—legal certainty and transparent procedures for research and development efforts protection and enforcement. We are convinced that innovation based on research and development is a key factor for sustainability and SDG coverage—and many local, regional, national, and international ecosystems were brought in place for this in the last couple of years. On the other hand, the chapter represents a really fundamental concept for a framework to solve open questions and issues in this field and to create the ecosystem needed also in this field, written by somebody from the inner circles of the respective UN agencies. It is an honor for us that these considerations are first published in our book.

A. Al-Sulaiti (✉)
Doha, Qatar
e-mail: ahalsulaiti@hbku.edu.qa



P. Wollmann et al. (eds.), *The Sustainable Organization – Volume 2*, Future of Business and Finance, https://doi.org/10.1007/978-3-032-23802-3_7

considerations often sidelined in privatized adjudication. The chapter explores institutional design options, emphasizing diversity in judicial appointments, transparent procedures, and capacity-building through a proposed WIPO Judicial Academy. Ultimately, it situates the proposal within the broader framework of the UN Sustainable Development Goals, particularly SDG 16 and SDG 17, as a pathway to strengthen multilateralism and promote equitable participation in the global knowledge economy.

## 7.1 Introduction

The current era is marked by the emergence of a global economy increasingly driven by knowledge and innovation. Intellectual property (IP) has evolved from a peripheral legal concern into a central determinant of economic growth, competitiveness, and national strategic advantage.[1] In knowledge-based economies, intangible assets such as patents, trademarks, copyrights, and trade secrets often constitute the core of value creation, necessitating a comprehensive system for their protection and enforcement.[2] Effective IP regimes should incentivize research and development (R&D) by ensuring adequate returns for innovators, attract foreign direct investment (FDI), facilitate technology transfer, and underpin participation in global value chains. Yet, the cross-border nature of modern trade and the integration of digital technologies have heightened the exposure of states to IP-related disputes that transcend their territorial boundaries, creating frictions in international economic relations and revealing the limitations of existing dispute settlement mechanisms.[3]

The rise in inter-state IP disputes reflects the structural realities of a globalized knowledge economy. States are now more frequently confronted with conflicts over the scope and enforcement of IP rights, particularly in sectors where innovation cycles are short and commercial stakes are high, such as pharmaceuticals, biotechnology, and information technology. The financial burden of these disputes can be significant, both in direct litigation costs and in the economic uncertainty generated by protracted or inconsistent adjudication.[4] For developing and mid-sized economies seeking to transition toward knowledge-based development models, these inefficiencies risk undermining the incentives for innovation and investment that IP

[1] Peter K Yu, 'Three Megatrends in the International Intellectual Property Regime' (2023) 41 Cardozo Arts & Entertainment Law Journal 457.

[2] Graham Room and others, *The European Challenge: Innovation, Policy Learning and Social Cohesion in the New Knowledge Economy* (first edn, Bristol University Press 2005) <https://www.cambridge.org/core/product/identifier/9781847421357/type/book> accessed 31 July 2025.

[3] ibid.

[4] Gürkan Çapar, '(Il)Legitimacy of International Intellectual Property Regime?' (2023) 36 Leiden Journal of International Law 721.

systems are meant to foster.[5] Within this context, the coherence and predictability of international IP dispute resolution have become essential components of both economic strategy and legal governance.

Despite the centrality of IP to global economic activity, the institutional architecture for the resolution of inter-state IP disputes is fragmented and underdeveloped. The World Trade Organization (WTO) has, since the entry into force of the TRIPS[6] Agreement, served as the primary multilateral forum for state-to-state IP disputes through its Dispute Settlement Understanding (DSU).[7] For much of its existence, the WTO's dispute settlement mechanism (DSM) was regarded as a model of legalized, rules-based adjudication. Its quasi-automatic adoption of panel and Appellate Body reports provided states with confidence that trade and IP conflicts could be resolved with a high degree of predictability. Yet, this system has entered a profound institutional crisis.[8] Since 2019, the Appellate Body has been rendered nonfunctional due to the United States' persistent blocking of judicial appointments, leaving the multilateral trading system without a functioning appellate mechanism. This paralysis has produced a cascading effect on legal certainty: losing parties can now "appeal into the void," effectively stalling the enforcement of panel rulings, and undermining the credibility of the entire dispute settlement framework.[9]

For IP disputes, the consequences of this paralysis are acute. These cases often involve technically complex questions and high-stakes industries, where delayed or inconsistent rulings carry significant economic and policy costs. Moreover, smaller states are disproportionately disadvantaged in the absence of a functional multilateral adjudicatory forum, as their ability to compel compliance from larger economies is largely dependent on the legitimacy and enforceability of multilateral rulings.[10] The temporary Multi-Party Interim Appeal Arbitration Arrangement (MPIA), created under Article 25 of the DSU, has offered a partial stopgap for willing participants.[11] Yet its voluntary nature, limited membership, and lack of universal coverage render it inadequate as a long-term solution for the coherent and authoritative resolution of inter-state IP conflicts.

The vacuum left by the WTO has been filled, in part, by fragmented alternative forums, particularly free trade agreements (FTAs) and investment arbitration. FTAs

---

[5] Peter K Yu, 'The Second Transformation of the International Intellectual Property Regime' in Jonathan Griffiths and Tuomas Mylly (eds), *Global Intellectual Property Protection and New Constitutionalism* (first edn, Oxford University Press, Oxford 2021) <https://academic.oup.com/book/41133/chapter/350649782> accessed 31 July 2025.

[6] **T**rade-**R**elated Aspects of **I**ntellectual **P**roperty Rights.

[7] Peter K Yu, 'State-to-State and Investor-State Copyright Dispute Settlement' [2019] Texas A&M University School of Law Legal Studies Research Paper Series 421.

[8] Siyu Feng and Sik Cheng Peng, 'From Unilateralism to Multilateralism: WTO Mediation of Legal and Financial Challenges in International Intellectual Property Disputes' (2024) 10 China and WTO Review 101.

[9] Phumelele Dhlamini, 'The World Trade Organization (WTO) Appellate Body Crises: A Critical Analysis' (University of Cape Town 2020).

[10] Yu, 'The Second Transformation of the International Intellectual Property Regime' (n 5).

[11] Dhlamini (n 8).

have proliferated with increasingly robust IP chapters, often containing TRIPS-plus obligations that advance the interests of rights holders in developed states.[12] While these agreements create alternative avenues for enforcement, they also exacerbate forum shopping and fragmentation, as the same dispute may be addressed under different treaty regimes with divergent substantive and procedural rules.[13] This multiplicity of forums undermines the coherence of international IP law, threatening core principles of the rule of law such as predictability, stability, and equality of treatment among states.

Investor–state dispute settlement (ISDS) has further complicated the landscape. In many bilateral investment treaties, IP rights are classified as protected investments, allowing multinational corporations to bring claims directly against host states.[14] ISDS offers certain procedural advantages such as confidentiality, relative expedition, and the possibility of direct monetary compensation, but its application to IP disputes raises systemic concerns. Arbitral awards lack precedential value, producing inconsistent and unpredictable outcomes; proceedings are often prohibitively expensive, with average costs exceeding eight million USD and some cases reaching 30 million USD; and the shift of IP disputes into the investment law domain risks displacing the public-interest considerations embedded in IP regimes.[15] For developing countries, the inequities are particularly stark, as the asymmetry of resources and expertise often leaves them disadvantaged in defending against complex claims brought by multinational rights holders. The proliferation of ISDS in the IP context has, therefore, contributed to incoherence, inconsistency, and inequity in the global governance of IP, fragmenting a system historically grounded in multilateralism.[16]

Within this fractured landscape, the World Intellectual Property Organization (WIPO) occupies a paradoxical position. As the custodian of foundational IP treaties, WIPO has a longstanding mandate to promote international cooperation, harmonization, and capacity-building in the IP field.[17] Its Arbitration and Mediation Center has seen a dramatic increase in private ADR cases, with a record 548 cases filed in 2022, reflecting growing demand for specialized dispute resolution.[18] Yet, WIPO lacks a binding inter-state dispute settlement mechanism. Moreover, while the Paris and Berne Conventions theoretically allow recourse to the International Court of Justice, this avenue has never been invoked.[19] This institutional limitation has forced states to rely on a patchwork of WTO, FTA, and ISDS mechanisms,

---

[12] Çapar (n 4).

[13] ibid.

[14] Yu, 'The Second Transformation of the International Intellectual Property Regime' (n 5).

[15] Yu, 'State-to-State and Investor-State Copyright Dispute Settlement' (n 6).

[16] Yu, 'The Second Transformation of the International Intellectual Property Regime' (n 5).

[17] Kat Gritsenko, 'Arbitration as a Dispute Resolution Mechanism for Cross-Border Intellectual Propery Disputes' (2024) 15 Cybaris: An Intellectual Property Law Review 371.

[18] ibid.

[19] Yu, 'Three Megatrends in the International Intellectual Property Regime' (n 1).

perpetuating fragmentation and weakening the normative coherence of the multilateral IP regime.[20]

The absence of a centralized, specialized forum for inter-state IP disputes has profound implications for both the rule of law and economic governance. Fragmentation encourages forum shopping, incentivizes strategic behavior by powerful states, and erodes the predictability that is essential for innovation-driven investment.[21] It also diminishes the ability of international law to reconcile the dual nature of IP: as a private right incentivizing creativity and as a public instrument serving broader socioeconomic and cultural objectives.[22] Without a coherent adjudicatory framework, there is a growing risk that IP enforcement will be driven by ad hoc investment or trade considerations rather than a balanced and principled multilateral system.

Given these circumstances, the case for developing a dedicated, WIPO-based mechanism for inter-state IP dispute resolution is compelling. Such a mechanism could provide a coherent and authoritative forum capable of reconciling the technical complexity of IP law with the imperatives of international economic governance. By leveraging WIPO's specialized expertise, a new adjudicative structure could ensure faster, higher-quality decisions, reduce forum fragmentation, and reinforce multilateralism at a moment when the WTO's credibility is under strain.[23] Crucially, a state-centric mechanism housed within WIPO would also be better positioned to integrate public-interest considerations into its rulings, avoiding the excessive privatization of IP norms that has characterized ISDS practice.

The objective of this chapter is to critically examine the institutional and normative deficiencies of the current system and to articulate a policy-driven proposal for the creation of a WIPO-based adjudicatory mechanism for inter-state IP disputes. In doing so, it situates the discussion within the broader context of the global shift toward knowledge economies, the paralysis of existing dispute settlement structures, and the urgent need to restore coherence, predictability, and legitimacy to the governance of international intellectual property law.

## 7.2 Inclusivity, Representation, and Judicial Capacity-Building

The legitimacy of any international adjudicatory body rests upon the quality of its jurisprudence and the representativeness and inclusivity of its composition. As proposals for a dedicated WIPO-based mechanism for state-to-state intellectual property disputes gain traction, the question of who sits on the bench becomes as critical as the legal mandate itself. Representation in judicial institutions has long been a point of contestation, reflecting broader debates about sovereignty, democracy, and

[20] Çapar (n 4).

[21] ibid.

[22] Yu, 'The Second Transformation of the International Intellectual Property Regime' (n 5).

[23] Çapar (n 4).

global governance. Critics frequently argue that international courts impose Western-centric values under the guise of universal norms, undermining national sovereignty and local legal traditions. Populist movements across various regions frame international judicial bodies as encroachments on national identity, portraying them as elitist institutions far removed from domestic publics.[24] Within this context, inclusivity in the composition and functioning of such a mechanism is a substantive requirement to ensure legitimacy, effectiveness, and global ownership of the adjudicatory process.

Historically, the composition of international courts has prioritized geographic representation, ensuring that the major world regions are visibly present. This model, however, is increasingly inadequate. While it may address balance among states, it fails to capture structural inequalities that persist within states and across regions.[25]Women, minority ethnic groups, and underrepresented legal traditions continue to be marginalized, despite the growing emphasis on pluralism in global governance. For example, despite women comprising over 60% of domestic judges in France, the country has historically refrained from nominating women to international judicial posts. Such disparities reinforce systemic exclusion and highlight that geographic balance alone cannot secure genuine inclusivity.[26]

Judicial diversity must therefore be understood expansively. It encompasses the nationality of judges as well as their gender, racial, cultural, and professional backgrounds. The "functional argument" for diversity emphasizes that judges' backgrounds shape their approaches to decision-making. A bench comprised of individuals from varied social, cultural, and legal traditions is likely to deliberate more rigorously, identify blind spots, and reach decisions that are better attuned to global realities.[27]At the same time, the "legitimacy argument" argues that diversity enhances both the normative legitimacy (the right of an institution to rule) and the sociological legitimacy (public acceptance and trust in its rulings).[28] A homogeneous bench risks reinforcing perceptions of bias, exclusion, and neo-colonialism, especially in a field as sensitive as IP, where disputes often implicate asymmetries between developed and developing states.

The dangers of insufficient diversity are not abstract. Empirical research suggests that judges may demonstrate unconscious bias, sometimes favoring their nominating states or approaching disputes with perspectives shaped by their socio-legal

[24]Almqvist J, 'The Impact of Cultural Diversity on International Criminal Proceedings' (2006) 4 Journal of International Criminal Justice 745.

[25]De Carvalho JS and Uriburu J, 'Problematising Diversity: The Change That International Lawyers (Do Not) Want for International Courts' (2023) 10 London Review of International Law 391.

[26]Føllesdal A, 'How Many Women Judges Are Enough on International Courts?' (2021) 52 Journal of Social Philosophy 436.

[27]Maučec G and Dothan S, 'The Effects of International Judges' Personal Characteristics on Their Judging' (2022) 35 Leiden Journal of International Law 887.

[28]Greenwood L, 'Tipping the Balance – Diversity and Inclusion in International Arbitration' (2017) 33 Arbitration International 99; Torbisco-Casals N, 'The Legitimacy of International Courts: The Challenge of Diversity' (2021) 52 Journal of Social Philosophy 491.

backgrounds. In the context of IP disputes, which often intersect with broader socio-economic and cultural rights, such biases can distort outcomes. Consider disputes over access to medicines or traditional knowledge: a bench dominated by judges trained in Western intellectual property paradigms may undervalue claims rooted in indigenous traditions or public health concerns.[29] Similarly, underrepresentation of women and minorities risks entrenching blind spots in legal reasoning, particularly in cases with gendered implications for innovation and knowledge production.

Moreover, opacity in judicial selection processes aggravates these issues. In many international courts, political loyalty or diplomatic bargaining dominates over merit and impartiality. The absence of transparent selection procedures fuels perceptions that appointments serve the interests of powerful states rather than the integrity of the judicial process. This undermines trust in international adjudication and weakens compliance with its rulings. If a WIPO-based dispute resolution mechanism is to command authority across diverse member states, its design must directly address these concerns by embedding inclusivity into its institutional structure.[30]

Existing international courts and tribunals offer valuable lessons on mechanisms to enhance inclusivity. The International Tribunal for the Law of the Sea (ITLOS) explicitly mandates equitable geographic distribution and requires that its composition reflect the principal legal systems of the world. The International Court of Justice (ICJ), while less prescriptive, has developed a tradition of ensuring representation from all five UN regional groups. These practices, though imperfect, highlight the role of formal quotas and informal traditions in embedding diversity.[31]

The International Criminal Court (ICC) provides perhaps the most instructive example. Since its establishment in 2003, the ICC has maintained close to 50% female representation, a notable achievement compared to other international courts.[32] This outcome stems from its unique election procedure, which incorporates minimum gender quotas and a multiround voting system designed to ensure balance across gender, geography, and legal traditions. States are required to nominate at least six male and six female candidates, and judges are elected in rounds until diversity thresholds are satisfied. This procedural innovation demonstrates that deliberate institutional design can overcome entrenched biases and deliver a more representative bench without compromising merit.[33]

---

[29] Greenwood L, 'Specialised IP Judiciary: What Are the Key Elements to Consider When Establishing or Reforming an Effective IP Court?' (2022) 71 GRUR International 206.

[30] Gurgula O, Padamczyk M and Shemtov N, 'Specialised IP Judiciary: What Are the Key Elements to Consider When Establishing or Reforming an Effective IP Court?' (2022) 71 GRUR International 206.

[31] Torbisco-Casals N, 'The Legitimacy of International Courts: The Challenge of Diversity' (2021) 52 Journal of Social Philosophy 491.

[32] International Criminal Court marks International Women's Day" (ICC, 8 March 2022) https://www.icc-cpi.int/news/international-criminal-court-marks-international-womens-day-0.

[33] Rosemary Grey, Kcasey McLoughlin & Louise Chappell, 'Gender and Judging at the International Criminal Court: Lessons from "Feminist Judgment Projects"' (2021) 34(1) *Leiden Journal of International Law* 247–264.

These precedents suggest that a WIPO-based mechanism should go beyond rhetorical commitments to inclusivity. Instead, it should adopt binding rules on gender, regional, and legal system diversity, reinforced by transparent selection procedures. In this regard, the ICC model of minimum nomination criteria and multiround elections offers a promising blueprint.

A WIPO-based dispute resolution mechanism must strike a delicate balance between inclusivity and expertise. Intellectual property disputes are technically complex, often requiring specialized legal and scientific knowledge. Therefore, inclusivity cannot be pursued in isolation from competence. Rather, the challenge lies in crafting selection procedures that simultaneously guarantee diversity and ensure high levels of professional qualification.

Several complementary mechanisms can achieve this balance. First, mandatory quotas should be embedded in the founding statute, requiring states to nominate candidates across gender lines and from all UN-defined regional groups. This would prevent systemic exclusion and ensure equitable geographic representation. Second, legal system diversity should be explicitly mandated. Given that IP law draws from both civil law and common law, the bench must include judges with familiarity across these traditions. This requirement would enhance the mechanism's capacity to adjudicate disputes with sensitivity to diverse legal cultures.

Third, screening committees composed of independent experts should vet nominees before they are considered for election. The Court of Justice of the European Union (CJEU) has pioneered such an approach through its Article 255 Panel, which assesses candidates' qualifications and impartiality. This practice minimizes the influence of political bargaining and ensures that only qualified candidates reach the ballot.[34] For WIPO, the screening committee could include representatives from civil society, academia, and professional associations, thereby broadening participation beyond states and strengthening the perceived impartiality of the process.

Fourth, the voting system should incorporate multiple rounds and diversity thresholds, modeled after the ICC. A single-round voting procedure risks producing a homogeneous bench dominated by candidates from powerful states. Multiround voting ensures that underrepresented candidates have a fair opportunity to be elected, and diversity thresholds guarantee that the final bench reflects a balance of perspectives.

Finally, provisions should explicitly prohibit the simultaneous appointment of multiple judges from the same nationality. This prevents overrepresentation by powerful states and reinforces the principle of equality among member states.

Beyond inclusivity in appointments, the legitimacy and effectiveness of a WIPO-based mechanism will hinge on its ability to build global judicial capacity in IP law. At present, many national judiciaries lack the expertise required to adjudicate complex IP disputes, particularly in developing countries where judicial training and resources remain limited. This creates asymmetries in global adjudication: while

[34] Mowbray A, 'The Consideration of Gender in the Process of Appointing Judges to the European Court of Human Rights' (2008) 8 Human Rights Law Review 549.

judges from developed states arrive with deep expertise and institutional support, their counterparts from less developed regions often lack comparable exposure.

Capacity-building initiatives can address this imbalance. A dedicated training program under WIPO's auspices could provide specialized instruction in both substantive IP law and the technical fields, such as biotechnology, pharmaceuticals, and digital technologies, that frequently underpin disputes. Such programs would enhance the expertise of prospective judges and create a global pool of trained professionals capable of contributing to national and international adjudication alike. For instance, Ukraine has recommended that judges in specialized IP courts possess both legal and technical expertise, ensuring that rulings in patent disputes are accurate and consistent. A similar approach could be adopted at the international level, requiring candidates for a WIPO bench to demonstrate either pre-existing technical qualifications or successful completion of WIPO-administered training. This would mitigate the risk of judicial inconsistency and enhance confidence in the quality of rulings.

Capacity-building also has broader systemic benefits. By investing in the training of judges from underrepresented regions, WIPO can foster greater participation of developing states in global IP governance. This would counter the perception that international adjudication privileges the interests of developed economies, thereby strengthening the sociological legitimacy of the system. Moreover, enhanced judicial expertise across regions would reduce the asymmetry that currently incentivizes forum shopping and reliance on bilateral or regional dispute settlement forums.

The pursuit of inclusivity in judicial appointments inevitably confronts the question of merit. Critics argue that quotas and representation rules compromise the appointment of the "best" candidates by prioritizing diversity over excellence. However, such critiques often overlook that merit itself is a socially constructed standard, shaped by the dominance of powerful states and legal traditions. A narrow definition of merit risks entrenching existing inequalities rather than rectifying them.

Empirical evidence demonstrates that diverse judges can meet merit standards while also bringing unique perspectives to the bench. The ICC's record of female judges, many of whom have distinguished themselves in both national and international law, challenges the assumption that diversity undermines competence. On the contrary, diversity enhances deliberative quality and reduces the risk of bias, thereby strengthening the substantive quality of judgments.

For a WIPO-based mechanism, inclusivity and merit should not be framed as competing imperatives. Instead, the design of selection procedures should operationalize both: requiring demonstrated expertise in IP law and related fields while also mandating diverse representation. In this way, diversity becomes a dimension of merit, reflecting the recognition that legitimacy and quality are interdependent.

Designing a WIPO-based adjudicatory mechanism with inclusivity provisions is a matter of institutional design and political will. States must be willing to cede some control over judicial appointments, subjecting nominees to independent screening and binding diversity rules. This may provoke resistance, particularly from powerful states accustomed to dominating appointments. Yet, the alternative is a mechanism perceived as biased, exclusionary, and illegitimate that results in an outcome that would undermine its effectiveness and durability.

Moreover, inclusivity should be framed as a global public good. A representative bench enhances the legitimacy of international adjudication, promotes compliance with rulings, and fosters trust among stakeholders. It also contributes to the diffusion of knowledge and expertise, strengthening national judiciaries and reducing asymmetries in global governance. For developing states, participation in an inclusive adjudicatory system offers a pathway to greater voice and influence in global IP governance, countering the marginalization that often accompanies bilateral and regional arrangements.

Finally, judicial capacity-building must be institutionalized as a continuous process. Rather than ad hoc training, WIPO should establish a permanent judicial academy for IP law, providing ongoing education, networking, and resources for judges worldwide. This would not only ensure the quality of adjudication within the WIPO mechanism but also contribute to the broader strengthening of IP governance globally.

## 7.3 Legal and Institutional Framework: A Way Forward?

Before outlining a forward-looking design for a WIPO-based state-to-state dispute resolution mechanism, it is useful to take stock of the diverse, and often underutilized, dispute settlement provisions already embedded across international IP treaties. While many of the foundational conventions—such as the Berne and Paris Conventions[35]—contain clauses permitting referral of disputes to the International Court of Justice (ICJ), these have never been invoked, largely due to widespread reservations and the political reluctance of states to escalate IP conflicts to the ICJ. More recent treaties, such as the Convention on Biological Diversity or the International Treaty on Plant Genetic Resources, rely on a mix of negotiation, mediation, arbitration, and conciliation, but remain practically dormant. By contrast, the TRIPS Agreement stands out as the most operationalized system, having linked IP disputes to the WTO's Dispute Settlement Understanding (DSU), though this mechanism itself is now weakened by the paralysis of its Appellate Body.

Regional frameworks also illustrate this fragmentation (Table 7.1). The EU's Unified Patent Court represents a bold move toward centralized adjudication, while the Bangui Agreement has institutionalized arbitration and mediation within OAPI.[36] By contrast, the Gulf Cooperation Council (GCC) framework highlights the limitations of regional approaches: although the GCC Economic Agreement (2001) contains general provisions for dispute settlement and the GCC adopted common patent and trademark laws, its mechanisms are weak and indirect. IP disputes fall only within the general economic dispute provisions, which require

[35] See Table 7.1.

[36] Agreement on a Unified Patent Court (entered into force 1 June 2023) OJ C 175, 20 June 2013.

mutual consent for arbitration and have never been tested for IP-specific cases.[37] The GCC Patent Office itself had stopped accepting new applications in 2021 (for the interim), underscoring the fragility of regional IP institutions.[38] The comparative table (Table 7.1) below illustrates these divergences and highlights the systemic fragmentation that any new WIPO-based mechanism must overcome.

Against this backdrop, the proposed WIPO-based state-to-state dispute resolution mechanism must be grounded in a treaty-based legal foundation that both accommodates the consensual nature of international adjudication and responds to the pressing governance deficits exposed by the paralysis of the WTO's Appellate Body. The design challenge is twofold: first, to create a legal instrument with the requisite normative authority and procedural specificity to withstand political contestation; and second, to embed institutional guarantees that will ensure the mechanism's longevity and credibility within the broader corpus of international intellectual property law.

From a doctrinal perspective, there are two primary vehicles through which such a mechanism could be established. The first is the negotiation of a stand-alone multilateral treaty, open to all WIPO members, that expressly creates a permanent adjudicative body with jurisdiction over disputes arising under WIPO-administered agreements. This approach, comparable to the creation of the WTO Agreement itself, would allow for a self-contained set of jurisdictional and procedural rules insulated from the interpretive constraints of pre-existing instruments. However, it faces the practical hurdle of achieving consensus among 193 WIPO members, many of whom have divergent views on the appropriate balance between IP protection and public-interest considerations.

The second option is the adoption of a protocol or amendment to the WIPO Convention of 1967 under Article 17 of that instrument, which permits the General Assembly to adopt measures necessary to attain WIPO's objectives. A protocol would have the advantage of nesting the mechanism within WIPO's existing institutional structure, thereby benefiting from its Secretariat, administrative expertise, and network of national IP offices. Yet, this route would require careful drafting to ensure that the protocol's provisions override or harmonize with the dispute settlement clauses in specific WIPO-administered treaties, some of which, like the Paris and Berne Conventions, currently provide only for recourse to the International Court of Justice—a path that has never been taken in practice.

In either formulation, the principle of *pacta sunt servanda* under Article 26 of the Vienna Convention on the Law of Treaties would ensure that states accepting the new instrument would be legally bound to comply with its jurisdiction and

[37] Economic Agreement Between the Gulf Cooperation Council States (signed 31 December 2001, entered into force 1 January 2003) https://www.gcc-sg.org/en us/CognitiveSources/DigitalLibrary/Lists/DigitalLibrary/EconomicCooperation.aspx accessed 1 October 2025.

[38] *Operational Status of the GCC Patent Office (GCCPO)—2025 Summary*, Kadasa (16 September 2024) https://kadasa.com.sa/the-gcc-patent-office-stops-receiving-new-patent-applications/ accessed 1 October 2025. As of 2025, the new GCCPO examination system is applied only in Bahrain, Kuwait, and Qatar, while Saudi Arabia, the UAE, and Oman continue to process patent applications exclusively through their respective national offices.

**Table 7.1** Comparative table: international and regional IP treaties and dispute settlement clauses; created by the author, used with permission, all rights reserved

| Treaty/instrument | Year | Dispute settlement clause | Practical usage |
|---|---|---|---|
| Berne convention for the protection of literary and artistic works | 1979 (latest act) | Art. 33: Disputes not settled by negotiation may be brought before the ICJ, unless states opt out by reservation | Never used; many states filed reservations excluding ICJ jurisdiction |
| Paris convention for the protection of industrial property | 1979 (latest act) | Art. 28: Same ICJ route as Berne convention; opt-out possible | Never invoked; ICJ mechanism remains dormant |
| Universal copyright convention | 1952 | Art. XV: Disputes not settled by negotiation shall go to the ICJ unless states agree otherwise | Not used; ICJ mechanism remains theoretical |
| Convention on biological diversity (CBD) | 1992 | Art. 27: Negotiation, mediation, or third-party good offices; failing that, arbitration or ICJ (if accepted); otherwise conciliation | Flexible system, but inter-state disputes have not reached arbitration/ ICJ |
| International treaty on plant genetic resources for food and agriculture (ITPGRFA) | 2001 | Art. 22: Negotiation, mediation; optional arbitration or ICJ (if declared); default to conciliation | No known cases; provisions remain untested |
| TRIPS agreement (WTO) | 1994 | Art. 64: Incorporates WTO DSU (binding, enforceable panel and appellate body rulings) | Most active mechanism; several IP disputes litigated under DSU. Currently weakened by appellate body crisis |
| Washington treaty on intellectual property in respect of integrated circuits | 1989 | Art. 14: Consultations → good offices/mediation/arbitration → panel procedure under WIPO assembly | Not yet widely operationalized |
| Bangui agreement (OAPI) | 2015 | Art. 4, 34: Jurisdiction of national courts; arbitration and mediation via OAPI arbitration Center | Used regionally; promotes ADR but limited to member states |
| EU agreement on a unified patent court | 2013 | Establishes a permanent unified patent court for disputes over European patents | Operationalized (since 2023), first cases already being heard |
| GCC economic agreement & GCC IP framework | 2001 (Economic Agreement); GCC Patent/ Trademark Laws (2014–2021) | General dispute clause: Amicable settlement → GCC arbitration Center (by consent) → judicial commission (not IP-specific). No state-to-state IP clause in GCC IP laws | Weak/indirect: No known IP disputes resolved under GCC framework; GCC patent office stopped new filings in 2021, highlighting institutional fragility |

rulings. However, the drafters must decide whether jurisdiction should be compulsory upon ratification or optional, as in the case of the International Court of Justice's optional clause under Article 36(2) of its Statute. Given the failures of purely voluntary submission in the ICJ context, and the need for legal certainty in the IP domain, the stronger case lies with compulsory jurisdiction—albeit with narrowly tailored opt-outs for certain sensitive subject matters, such as national security-related IP disputes.

The institutional design must avoid reproducing the pathologies that have afflicted other international adjudicatory bodies. A key lesson from the WTO is that appellate review, while valuable for legal consistency, can also become a focal point for political obstruction, as demonstrated by the US blockade of Appellate Body appointments since 2017. One solution is to adopt a single-tier adjudicatory structure, as in the International Tribunal for the Law of the Sea (ITLOS), but with an internal mechanism for en banc review of panel decisions to ensure jurisprudential coherence. This model would preserve consistency without creating a second layer vulnerable to appointment politics.

Selection of adjudicators must operationalize the inclusivity principles discussed in Sect. 7.2, but here those principles intersect with the functional demands of IP adjudication. WIPO-administered disputes frequently turn on highly technical issues, such as the patentability of biotechnological inventions under the Patent Cooperation Treaty or the scope of performers' rights under the Beijing Treaty on Audiovisual Performances. Adjudicators must therefore combine demonstrable expertise in IP law with familiarity in relevant scientific or technological fields. This dual-competence requirement mirrors the approach of specialized national IP courts, such as the German Federal Patent Court, which pairs legally trained judges with technically qualified members. At the international level, the requirement could be implemented through a two-stage selection process: an independent vetting committee—modeled on the Article 255 panel used by the Court of Justice of the European Union—to assess qualifications, followed by election by WIPO's General Assembly under strict diversity quotas for geography, gender, and legal tradition.

Procedurally, the mechanism must reconcile efficiency with the due process guarantees that underpin the legitimacy of adjudication. Strict timelines for each stage of proceedings—such as the nine-month period for panel reports in WTO disputes—should be coupled with safeguards for adequate party participation, including opportunities for written submissions, oral hearings, and rebuttals. Given the prevalence of scientific evidence in IP disputes, panels should be empowered to appoint neutral technical experts under procedures analogous to Rule 50 of the International Tribunal for the Law of the Sea (ITLOS) Rules. To enhance transparency, hearings should be public by default, with confidentiality reserved for trade secrets or security-sensitive information, a balance increasingly favored in investor–state arbitration reforms under the United Nations Commission on International Trade Law (UNCITRAL) Working Group III.

One of the most legally sensitive issues is the mechanism's relationship with existing forums. Under the current regime, many disputes that could implicate WIPO treaties are litigated under the WTO's TRIPS Agreement, often because the DSU provides a binding, enforceable process absent in WIPO. To avoid duplicative or conflicting rulings, the new legal instrument should contain a *lis pendens* clause preventing parallel proceedings, combined with a "choice of forum" provision requiring parties to elect a single forum at the outset of a dispute. More radically, states could agree to amend the TRIPS Agreement under Article X of the WTO Agreement to refer certain IP disputes—particularly those with limited trade impact—to WIPO's jurisdiction, thereby formalizing the division of labor between the two institutions.

Enforcement mechanisms must also move beyond the declaratory nature that undermines compliance in some international courts. Borrowing from the WTO's model, the WIPO tribunal could authorize countermeasures, such as suspension of IP-related treaty benefits, against noncompliant states. While the legality of cross-retaliation across different treaty regimes raises complex questions under general international law, the practice has been accepted in the WTO context, as in the EC–Bananas III (Ecuador) arbitration. A compliance review procedure akin to DSU Article 21.5 would allow panels to determine whether remedial actions taken by the losing party satisfy the ruling, preventing indefinite defiance.

Crucially, these legal and institutional design choices are not ends in themselves; they are instrumental to correcting the systemic weaknesses identified in the current architecture. By embedding the mechanism in a treaty framework with compulsory jurisdiction, equipping it with procedures tailored to the technical and public-interest dimensions of IP, and aligning its enforcement tools with established international practice, WIPO would not only fill an institutional void but also reassert the multilateral governance of IP as a domain of law rather than of political bargaining. This is particularly vital in light of the Sustainable Development Goals, which frame effective institutions not as abstract ideals but as preconditions for equitable economic participation in a globalized knowledge economy.

## 7.4 Alignment with Sustainable Development Goals 16 and 17

The proposed WIPO-based state-to-state dispute resolution mechanism is not merely an institutional innovation within the field of intellectual property; it is an operationalization of specific obligations and policy commitments embedded in the United Nations 2030 Agenda for Sustainable Development. In particular, its institutional design and intended functions speak directly to the normative and practical content of SDG 16—Peace, Justice and Strong Institutions and SDG 17—Partnerships for the Goals. These goals are interlinked: SDG 16 articulates the substantive requirements of effective governance, while SDG 17 prescribes the cooperative and resource-mobilizing conditions necessary to realize those requirements.

### 7.4.1 SDG 16—Peace, Justice, and Strong Institutions

From a legal standpoint, SDG 16's relevant targets are not merely programmatic; they are expressions of an emerging consensus on the elements that constitute effective and legitimate governance at both the national and international levels. Target 16.3 calls upon states to "promote the rule of law at the national and international levels and ensure equal access to justice for all," while Target 16.6 requires the development of "effective, accountable and transparent institutions."[39] The proposed WIPO mechanism would instantiate these obligations in the IP domain through its binding jurisdiction, transparent procedures, and inclusivity guarantees.

The link between binding jurisdiction and the rule of law is particularly salient in the wake of the WTO Appellate Body's paralysis. Without a functioning multilateral forum, states have increasingly relied on ad hoc, bilateral solutions to resolve IP-related disputes arrangements that often reflect disparities in bargaining power rather than principled legal reasoning. In doctrinal terms, this shift undermines the *erga omnes partes* character of multilateral IP treaties, which are intended to create uniform rights and obligations among all parties. By reintroducing a compulsory, impartial adjudicatory forum, the WIPO mechanism would restore the expectation that disputes will be resolved according to law rather than power, thereby advancing Target 16.3 in a concrete, treaty-based form.[40]

Target 16.7's emphasis on "responsive, inclusive, participatory and representative decision-making" intersects directly with the design choices outlined in Sect. 7.2. Representation across gender, geography, and legal systems is not merely a question of optics; it is a functional necessity in a body that will interpret and apply treaties binding on a culturally and economically diverse membership. Comparative experience from bodies such as the International Criminal Court—which embedded gender and geographic quotas in its election rules—demonstrates that without binding diversity obligations, institutional inertia will tend toward overrepresentation of a narrow set of states and legal traditions. Embedding such requirements in the constitutive treaty of the WIPO mechanism would transform inclusivity from a political aspiration into a legal guarantee, thereby operationalizing Target 16.7 within the specific context of IP governance.

The mechanism also has a direct bearing on Target 16.4, which aims to "significantly reduce illicit financial and arms flows, strengthen the recovery and return of stolen assets and combat all forms of organized crime." While this target is often discussed in relation to transnational criminal law, the proliferation of counterfeit

---

[39] Winfried Huck and Jennifer Maaß, 'Gaining a Foot in the Door: Giving Access to Justice with SDG 16.3?' (2021) Cambridge Centre for Environment, Energy and Natural Resource Governance Working Paper 2021–5, 1–34; Catherine Dixon, 'From the Director General: How can ADR promote access to justice?' (2022) 2022 The Resolver: The Quarterly Magazine of The Chartered Institute of Arbitrators 1, 6.

[40] Winfried Huck and Jennifer Maaß, 'Gaining a Foot in the Door: Giving Access to Justice with SDG 16.3?' (2021) Cambridge Centre for Environment, Energy and Natural Resource Governance Working Paper 2021–5, 8–10; United Nations General Assembly, Transforming our world: the 2030 Agenda for Sustainable Development, preamble and para 35, UN Doc A/RES70/1 (2015).

goods, large-scale trademark infringement, and state-tolerated piracy operations are forms of transnational organized crime that undermine both economic stability and public welfare. Existing multilateral tools for addressing such conduct are fragmented, with enforcement often dependent on bilateral pressure or private litigation. By granting states a binding forum to challenge systemic failures in IP enforcement—particularly where such failures facilitate organized criminal networks—the WIPO mechanism would bring a presently under-regulated area of transnational crime within the purview of a multilateral judicial process.

### 7.4.2 SDG 17—Partnerships for the Goals

Where SDG 16 defines the institutional qualities to be achieved, SDG 17 outlines the cooperative modalities required to achieve them. Its targets include the mobilization of resources (Target 17.3), enhancement of policy coherence (Target 17.14), and strengthening of multistakeholder partnerships (Target 17.16).[41] The proposed WIPO mechanism can be understood as both a product of and a catalyst for such partnerships.

First, the creation of the mechanism would itself be an exercise in global partnership,[42] requiring negotiation among all WIPO members and the alignment of their respective interests in IP protection, enforcement, and access. The negotiation process would necessarily involve other international organizations—notably the WTO, WHO, and UNESCO—whose mandates intersect with IP in areas such as trade, health, and cultural heritage. Such cross-institutional collaboration is a direct instantiation of Target 17.16, which calls for multistakeholder partnerships "mobilizing and sharing knowledge, expertise, technology, and financial resources."

Second, the mechanism's envisaged WIPO Judicial Academy for IP Law—as discussed in Sect. 7.2—would be an instrument of capacity-building in line with Target 17.9. This target recognizes that effective participation in global governance requires not only formal membership but also the technical and institutional capacity to engage meaningfully in decision-making. By providing sustained training to judges, legal advisors, and technical experts from underrepresented regions, the Judicial Academy would enhance the substantive equality of states in the dispute settlement process, thereby reinforcing the legitimacy of the mechanism's outputs.

Third, the mechanism's role in promoting policy coherence aligns with Target 17.14, which urges the enhancement of policy coherence for sustainable development. Intellectual property sits at the intersection of multiple policy domains—trade, investment, public health, and environmental protection—each governed by

[41] United Nations, *'The 17 Goals: Goal 17—Strengthen the means of implementation and revitalise the Global Partnership for Sustainable Development'* (Sustainable Development Goals, UN, undated) https://sdgs.un.org/goals.

[42] Komal Kalha, 'How IP Enables Accelerated Progress on the Sustainable Development Goals' (IFPMA, 25 April 2024) https://www.ifpma.org/insights/how-ip-enables-accelerated-progress-on-the-sustainable-development-goals/.

its own treaty regime. Fragmented dispute settlement processes risk producing contradictory interpretations of similar or identical treaty language, undermining legal certainty and creating compliance dilemmas for states. By providing a single, specialized forum for the adjudication of disputes under WIPO-administered treaties, the mechanism would contribute to the harmonization of IP norms across different regimes, thereby advancing both legal certainty and sustainable development objectives.[43]

### 7.4.3 Interdependence of SDG 16 and SDG 17 in the IP Context

The interaction between SDG 16 and SDG 17 is particularly evident in the IP field.[44]Effective, transparent, and inclusive adjudication (SDG 16) cannot be achieved in the absence of the partnerships, resource mobilization, and policy coherence mechanisms envisaged by SDG 17. Conversely, partnerships under SDG 17 gain normative credibility when they are embedded in the rule-of-law structures promoted by SDG 16.

Consider, for example, a hypothetical dispute in which a developing country alleges that a developed state has failed to prevent the export of counterfeit pharmaceuticals into its market, in violation of the Anti-Counterfeiting Trade Agreement (ACTA) and the TRIPS Agreement. Under the current system, such a dispute might be addressed—if at all—through informal bilateral consultations or WTO litigation. However, if the developed state is outside the WTO's compulsory jurisdiction for such matters (or if WTO enforcement is politically blocked), the dispute may remain unresolved. Under the proposed WIPO mechanism, the case could be brought before a specialized panel with the technical expertise to assess both the legal and factual dimensions of the claim. The adjudication process would embody SDG 16's rule-of-law and inclusivity principles, while the implementation of the ruling could be facilitated by SDG 17-style partnerships involving WIPO's enforcement programs, WHO's pharmaceutical regulation initiatives, and targeted development assistance from donor states.

### 7.4.4 Avoiding "SDG-Washing"

For the mechanism to make a genuine contribution to the 2030 Agenda, it must guard against the risk of "SDG-washing"—the tendency to invoke SDG language as a legitimizing veneer without measurable institutional change. This risk is acute in the context of international adjudication, where high-minded principles often

[43] Ahmed Abdel-Latif and Pedro Roffe, 'The Interface Between Intellectual Property and Sustainable Development' in Irene Calboli and Maria Lillà Montagnani (eds), *Handbook of Intellectual Property Research: Lenses, Methods, and Perspectives* (OUP 2021).

[44] Brazil, 'Compilation of Member States Inputs on SDGs to WIPO's Work' (CDIP/18/4, WIPO, 8 August 2016) https://www.wipo.int/meetings/en/details.jsp?meeting_id=40504.

coexist with opaque appointment processes, limited access, and inconsistent compliance.[45]

To mitigate this risk, the constitutive treaty should incorporate explicit performance indicators linked to relevant SDG targets. For example, annual reporting could include the proportion of adjudicators meeting diversity criteria (Target 16.7), average time to resolution of disputes (Target 16.6), number of states receiving judicial capacity-building assistance (Target 17.9), and instances of cross-institutional collaboration in dispute resolution (Target 17.16). These indicators would not only provide a basis for internal assessment but would also allow external stakeholders—including civil society and academia—to hold the mechanism accountable for its SDG-related commitments (Table 7.2).

By embedding the qualities of SDG 16 into the institutional DNA of the mechanism, and by mobilizing the cooperative and resource-oriented strategies of SDG 17, the proposed WIPO dispute resolution system has the potential to be more than a procedural fix to a fragmented governance landscape. It could become a model for how specialized international organizations operationalize the SDGs within their treaty regimes, moving beyond rhetorical endorsement to measurable, enforceable practice. In doing so, it would reaffirm the centrality of multilateralism in addressing transnational challenges that no single state—however powerful—can resolve alone.

## 7.5 Conclusion and Policy Recommendations as Takeaways

The analysis presented throughout this chapter has demonstrated that the absence of a centralized, specialized mechanism for the resolution of state-to-state intellectual property disputes poses a significant challenge to the stability, coherence, and legitimacy of the global IP system. As states increasingly rely on knowledge-based economic growth, and as intellectual property rights play a pivotal role in shaping patterns of trade, investment, and technology, the need for predictable, authoritative, and fair dispute settlement becomes crucial.

The paralysis of the WTO Appellate Body since 2019 has fundamentally altered the state-to-state dispute settlement landscape, specifically in terms of IP. Once considered a model of rules-based adjudication, the WTO dispute settlement system now suffers from a loss of credibility and enforceability, with the "appeal into the void" phenomenon undermining legal certainty and creating opportunities for strategic noncompliance. In its wake, alternative venues such as IP chapters in FTAs have proliferated. While these may offer ad hoc solutions, their fragmented nature, overlapping jurisdictions, and varying obligations have exacerbated forum shopping and procedural inconsistencies. The resulting institutional vacuum erodes the normative coherence of the multilateral IP regime and threatens core rule-of-law principles such as predictability, stability, and equality of treatment.

[45] V Ferrón Vílchez, 'SDGwashing: A Critical View of the Pursuit of SDGs and Its Relationship with Environmental Performance' (2022) 34(4) *Development Policy Review* 539–556,

**Table 7.2** Alignment of the proposed WIPO dispute resolution mechanism with SDG 16 and SDG 17; created by the author, used with permission, all rights reserved

| SDG and target | Relevant UN indicator | Operational link to the proposed mechanism |
|---|---|---|
| SDG 16.3—Promote the rule of law at the national and international levels and ensure equal access to justice for all | 16.3.1—Proportion of victims of violence in last 12 months who reported to competent authorities; 16.3.2—Unsentenced detainees as proportion of prison population *(proxy for access to justice)* | By creating a compulsory, impartial, and technically competent forum for IP disputes, the mechanism restores multilateral rule-of-law pathways in the IP field, ensuring all member states—Regardless of economic power—Have equal access to adjudication |
| SDG 16.4—Significantly reduce illicit financial and arms flows, strengthen the recovery and return of stolen assets, and combat all forms of organized crime | 16.4.1—Total value of inward and outward illicit financial flows | Extends multilateral enforcement against systemic counterfeiting, piracy, and state-tolerated IP crime, which are recognized forms of transnational organized crime |
| SDG 16.6—Develop effective, accountable, and transparent institutions at all levels | 16.6.2—Primary government expenditures as a proportion of original approved budget *(proxy for institutional efficiency)* | Embeds procedural safeguards (public hearings, reasoned rulings, compliance review) that create a transparent, rules-based IP governance structure under WIPO auspices |
| SDG 16.7—Ensure responsive, inclusive, participatory, and representative decision-making at all levels | 16.7.1—Proportions of positions in public institutions compared to national distributions | Implements binding diversity quotas for adjudicator selection (geography, gender, legal tradition), institutionalizing inclusivity in international IP adjudication |
| SDG 17.3—Mobilize additional financial resources for developing countries from multiple sources | 17.3.1—Additional financial resources mobilized for development | Uses mixed funding model (state contributions, case fees, voluntary donations) to sustain operations and support developing state participation |
| SDG 17.9—Enhance international support for implementing effective and targeted capacity-building | 17.9.1—Dollar value of financial and technical assistance committed to developing countries | Establishes a permanent WIPO judicial academy for IP law, delivering continuous capacity-building for judges and legal experts from underrepresented regions |
| SDG 17.14—Enhance policy coherence for sustainable development | No global indicator specified *(measured qualitatively)* | Provides a single, specialized forum for interpreting WIPO treaties, reducing contradictory rulings across trade, investment, and cultural regimes |
| SDG 17.16—Enhance the global partnership for sustainable development | 17.16.1—Number of countries reporting progress in multistakeholder development effectiveness monitoring frameworks | Facilitates cooperation among states, WIPO, WTO, WHO, UNESCO, civil society, and industry through amicus participation, technical expertise, and coordinated enforcement follow-up |

In this context, the creation of a permanent, WIPO-led dispute settlement body emerges as a structural necessity for the future of international IP governance. WIPO, as the sole specialized UN agency devoted exclusively to IP, possesses the institutional expertise, technical capacity, and convening power to fill this gap. A WIPO-based dispute resolution mechanism would serve as a coherent, authoritative forum capable of reconciling the technical complexity of IP with the imperatives of international economic governance. Crucially, it would contribute to the realization of Sustainable Development Goal 16, which calls for effective, accountable, and inclusive institutions at all levels. Access to justice in the IP context contributes to enabling equitable participation in the global knowledge economy and the protection of the rule of law in cross-border innovation systems.

However, the legitimacy and effectiveness of such a body will depend heavily on its institutional design. As discussed, the selection of adjudicators must embody both inclusivity and competence. Geographic, gender, and legal system diversity are essential for ensuring representativeness and legitimacy, but these must be balanced with the need for deep technical expertise in IP law and related fields. Binding rules on diversity, coupled with transparent, merit-based selection procedures, should form the backbone of the appointment process. Lessons from the International Criminal Court's gender quotas and multiround elections can provide a valuable template, ensuring that diversity is operationalized rather than left to rhetorical commitment.

Furthermore, judicial capacity-building should be institutionalized through a permanent WIPO Judicial Academy for IP Law. This body would provide continuous training, peer exchange, and resources for judges, fostering high-quality adjudication and contributing to sustainable judicial capacity-building.

The proposed mechanism should be grounded in a clear and binding legal instrument, preferably in the form of a new treaty or protocol under WIPO's auspices. Core design principles such as neutrality, independence, and transparency must be embedded into the instrument to prevent politicization and ensure fair treatment of all parties.

The jurisdictional scope should be carefully defined to include disputes concerning the interpretation and application of multilateral IP treaties administered by WIPO, as well as relevant provisions in other international agreements where states consent to WIPO jurisdiction.

### 7.5.1 Normative and Practical Rationale

From a normative standpoint, a WIPO-led dispute settlement body would reaffirm multilateralism at a time when the fragmentation of global governance institutions is widespread. By institutionalizing rules-based dispute settlement of state-to-state IP disputes, such a mechanism would promote legal certainty, deter opportunistic behavior, and strengthen compliance with international obligations. It would also contribute to the depoliticization of IP disputes by providing states with a neutral, specialized forum, thus reducing the reliance on politically charged, bilateral negotiation or power-asymmetric settlement approaches.

From a practical perspective, WIPO's technical expertise in IP law, its administrative infrastructure, and its capacity to mobilize a global network of stakeholders uniquely position it to operate an adjudicative mechanism that is both procedurally efficient and substantively robust. A specialized mechanism could deliver faster, higher-quality decisions compared to general trade bodies, thereby reducing transaction costs for states and improving the overall functionality of the IP system.

### 7.5.2 Roadmap for Implementation

Given the political and technical complexity of establishing such a body, a phased approach is advisable:

- WIPO should commission comprehensive studies examining institutional models, procedural frameworks, and resource implications. These studies should draw on comparative experiences from other international adjudicatory bodies.
- A multidisciplinary panel of legal scholars, practitioners, diplomats, and technical experts should be convened to draft a preliminary statute and procedural rules, incorporating feedback from member states and stakeholders.
- Formal consultations should be held between governments, intergovernmental organizations, and the private sector to build consensus on the scope, design, and mandate of the mechanism.
- Targeted diplomatic initiatives should be undertaken to secure commitments from key member states, particularly those with significant stakes in IP governance, to lead the initiative.
- Include mandatory diversity requirements in the judge selection process and establish a permanent WIPO Judicial Academy for IP Law to ensure sustainable capacity-building.

In conclusion, the establishment of a WIPO-led permanent state-to-state dispute settlement mechanism is both a response to an urgent institutional deficit and a forward-looking investment in the stability of the global knowledge economy. In filling the void left by the WTO's dispute settlement paralysis, WIPO has the opportunity to enhance legal certainty, foster inclusive and equitable participation in global innovation markets, and contribute meaningfully to the achievement of SDG 16. The choice before the international community is clear: either allow fragmentation and unpredictability to continue undermining the integrity of the international IP regime, or take decisive, coordinated action to institutionalize a comprehensive, efficient, and inclusive mechanism for resolving disputes.

The policy path outlined in this study offers a pragmatic yet ambitious framework for realizing the latter. Moreover, advancing this initiative will require further studies to refine the design and assess the feasibility of such a mechanism.

## 7.6 A Brief Reference to the Three-Pillar Model

The relevance of the Three-Pillar Model[46] as a preferred thinking and mental style for challenging transformations, especially also in the context of achieving sustainability and coverage of the UN SDGs, is already described in detail in Chap. 1. In brief, the model is based on three key building blocks (pillars) for organization design and leadership for the VUCA[47] world:

(1) The First Pillar: **Sustainable Purpose**

Sustainable Purpose is the raison d'être of an organization, having a guiding objective means bringing orientation and certainty to the people for their joint endeavor and success. Contributing to sustainability is for sure a convincing sustainable purpose of an organization. It is obvious that striving for legal certainty and transparent procedures for research and development efforts protection and enforcement is a very convincing sustainable purpose, a key building block to reach the crucial SDGs 16 and 17.

(2) The Second Pillar: **Traveling Organization**

Traveling Organization means for an organization accepting to be in a permanent state of flux and transformation, while interacting with the external requirements of its journey with rapid learning and adaptivity. It is obvious that the realization of the establishment of a WIPO-led permanent state-to-state dispute settlement mechanism will be a fundamental transformation that requires the special mindset of a Traveling Organization on the part of all stakeholders.

(3) The Third Pillar: **Connectivity**

Connectivity is all about the ability to interconnect on the journey with all needed parties and resources inside and outside the silos, thereby creating high efficacy and consistency and during this process building also shared learning network. It is obvious that Connectivity is a key success factor for the required transformation to the establishment of a WIPO-led permanent state-to-state dispute settlement mechanism.

## 7.7 UN SDG Overview

The mainly addressed SDGs 16 and 17 are an important part of the overall set (Fig. 7.1).

---

[46] Wollmann, P. et al. (Eds.): Transforming Public and Private Sector Organizations—Implementing Sustainable Purpose, Travelling Organization and Connectivity for Resilience. Cham: Springer Nature, © 2022; Wollmann, P. et al. (Eds.): Organization and Leadership in Disruptive Times—Design and Implementation of the 3-P-Model. Cham: Springer Nature, © 2021; Wollmann, P. et al. (Eds.): Three Pillars of Organization and Leadership in Disruptive Times—Navigating Your Company Successfully through the twenty-first Century Business World. Cham: Springer Nature, © 2020; Wollmann, P. et al. (Eds.): Reading an Unknown Organization - Leaders in New Roles Thriving in Challenging Environments with the 3-P-Model Mindset. Cham Springer Nature, © 2024.

[47] VUCA = Volatile, Uncertain, Complex, Ambiguous.

**Fig. 7.1** UN SDG (source: https://www.un.org/sustainabledevelopment/), used with friendly permission of UN. The content of this publication has not been approved by the United Nations and does not reflect the views

More details on the UN SDGs and their historical development are covered in detail in Chaps. 1 and 3 and also in Chap. 9 of Volume 1 of *The Sustainable Organization—How Organizations Address the 17 UN SDGs Using the 3-P-Model.*[48]

**Acknowledgments** The author wishes to express his sincere appreciation to Saniya Khanna and Tasneem Muftah for their dedicated research assistance and their invaluable contribution to the organization of the conference entitled "The Role of Alternative Dispute Resolution (ADR), Intellectual Property (IP) Protection, and Innovation in Advancing Foreign Direct Investment (FDI) for Sustainable Development," held in Doha on September 1–2, 2025.

## Further Reading

Çapar G, '(Il)Legitimacy of International Intellectual Property Regime?' (2023) 36 Leiden Journal of International Law 721

Dhlamini P, 'The World Trade Organization (WTO) Appellate Body Crises: A Critical Analysis' (University of Cape Town 2020)

Feng S and Peng SC, 'From Unilateralism to Multilateralism: WTO Mediation of Legal and Financial Challenges in International Intellectual Property Disputes' (2024) 10 China and WTO Review 101

Gritsenko K, 'Arbitration as a Dispute Resolution Mechanism for Cross-Border Intellectual Propery Disputes' (2024) 15 Cybaris: An Intellectual Property Law Review 371

[48] See: https://link.springer.com/book/10.1007/978-3-031-89549-4.

Room G and others, *The European Challenge: Innovation, Policy Learning and Social Cohesion in the New Knowledge Economy* (first edn, Bristol University Press 2005) <https://www.cambridge.org/core/product/identifier/9781847421357/type/book> accessed 31 July 2025

Yu PK, 'State-to-State and Investor-State Copyright Dispute Settlement' [2019] Texas A&M University School of Law Legal Studies Research Paper Series 421

Yu PK, 'The Second Transformation of the International Intellectual Property Regime' in Jonathan Griffiths and Tuomas Mylly (eds), *Global Intellectual Property Protection and New Constitutionalism* (first edn, Oxford University Press, Oxford 2021) <https://academic.oup.com/book/41133/chapter/350649782> accessed 31 July 2025

'Three Megatrends in the International Intellectual Property Regime' (2023) 41 Cardozo Arts & Entertainment Law Journal 457

Almqvist J, 'The Impact of Cultural Diversity on International Criminal Proceedings' (2006) 4 Journal of International Criminal Justice 745

De Carvalho JS and Uriburu J, 'Problematising Diversity: The Change That International Lawyers (Do Not) Want for International Courts' (2023) 10 London Review of International Law 391

Føllesdal A, 'How Many Women Judges Are Enough on International Courts?' (2021) 52 Journal of Social Philosophy 436

Greenwood L, 'Tipping the Balance—Diversity and Inclusion in International Arbitration' (2017) 33 Arbitration International 99

Gurgula O, Padamczyk M and Shemtov N, 'Specialised IP Judiciary: What Are the Key Elements to Consider When Establishing or Reforming an Effective IP Court?' (2022) 71 GRUR International 206

'Specialised IP Judiciary: What Are the Key Elements to Consider When Establishing or Reforming an Effective IP Court?' (2022) 71 GRUR International 206

Holst C and Langvatn SA, 'Descriptive Representation of Women in International Courts' (2021) 52 Journal of Social Philosophy 473

Larsson O and others, 'Selection and Appointment in International Adjudication: Insights from Political Science' (2023) 14 Journal of International Dispute Settlement 134

Maučec G and Dothan S, 'The Effects of International Judges' Personal Characteristics on Their Judging' (2022) 35 Leiden Journal of International Law 887

Mowbray A, 'The Consideration of Gender in the Process of Appointing Judges to the European Court of Human Rights' (2008) 8 Human Rights Law Review 549

Schmitt B, 'Legal Diversity at the International Criminal Court' (2021) 19 Journal of International Criminal Justice 485

Torbisco-Casals N, 'The Legitimacy of International Courts: The Challenge of Diversity' (2021) 52 Journal of Social Philosophy 491

Widłak T, 'Rule of Law and the Criteria for Appointment of Judges: A Case for Judicial Virtues' (2025) 17 Hague Journal on the Rule of Law 199

Zhang Y, 'Equitable Representation on International Benches and the Appointment of Tribunal Members in Investor–State Dispute Settlement: A Historical Perspective' (2023) 14 Journal of International Dispute Settlement 428

*Vienna Convention on the Law of Treaties* (adopted 23 May 1969, entered into force 27 January 1980) 1155 UNTS 331.

*Statute of the International Court of Justice* (adopted 26 June 1945, entered into force 24 October 1945).

WTO, *Understanding on Rules and Procedures Governing the Settlement of Disputes* (15 April 1994) Marrakesh Agreement Establishing the World Trade Organization, Annex 2.

International Tribunal for the Law of the Sea, *Rules of the Tribunal* (28 October 1997).

Consolidated Version of the *Treaty on the Functioning of the European Union* [2016] OJ C202/47.

WTO, *European Communities — Regime for the Importation, Sale and Distribution of Bananas — Arbitration under Article 22.6 of the DSU*, Decision by the Arbitrators, WT/DS27/ARB/ECU, 24 March 2000.

UNCITRAL, *Working Group III: Investor–State Dispute Settlement Reform — Draft Code of Conduct for Adjudicators in International Investment Disputes* (19 April 2021) UN Doc A/CN.9/WG.III/WP.209.

UNGA Res 70/1, *Transforming our world: the 2030 Agenda for Sustainable Development* (21 October 2015) UN Doc A/RES/70/1.

Anti-Counterfeiting Trade Agreement (opened for signature 1 October 2011, not yet in force).

WTO, *Agreement on Trade-Related Aspects of Intellectual Property Rights* (15 April 1994) Marrakesh Agreement Establishing the World Trade Organization, Annex 1C.

*Rome Statute of the International Criminal Court* (adopted 17 July 1998, entered into force 1 July 2002) 2187 UNTS 3.

World Health Organization, *WHO Global Surveillance and Monitoring System for Substandard and Falsified Medical Products* (WHO 2017).

UNESCO, *Convention for the Safeguarding of the Intangible Cultural Heritage* (adopted 17 October 2003, entered into force 20 April 2006) 2368 UNTS 3.

**Ahmed Essa Al-Sulaiti** is a Lecturer in the College of Law, Hamad Bin Khalifa University. He is also licensed to practice law before the Appellate Court in Qatar. From 2022 to 2025, Ahmed was director of the Office of the State of Qatar to the World Trade Organization (WTO), the World Intellectual Property Organization (WIPO), and the United Nations Conference on Trade and Development (UNCTAD). From 2018 to 2022, he served as a diplomat and legal counsel at Qatar's Permanent Mission to the WTO, WIPO, and UNCTAD in Geneva. He began his career in the oil and gas industry in 2012 before moving to government service.

With over 13 years of experience in international law and diplomacy, Ahmed specializes in trade, intellectual property (IP), and national security. His expertise includes international negotiations, mediation, arbitration, and strategic dispute resolution. He has held several prominent leadership roles, including

President of the International Law Association (ILA)—GCC Branch (2024–present)

Member of the Drafting Committee for WIPO's Diplomatic Conference to conclude and adopt a Design Law Treaty (Riyadh, Saudi Arabia, November 11–22, 2024)

President of the Executive Committee of the Berne Union (2022–2024)

Vice Chair of the Berne Union Assembly (2020–2021)

President of the Beijing Treaty on Audiovisual Performances, with responsibilities for implementation and policy work for the period 2025–2027

Within the WTO, his roles include

Chairperson of the Committee on Anti-Dumping Practices (2021–2022)

Member of the Expert Committee on Subsidies and Countervailing Measures (2024–present)

Member of the Indicative List of Panelists (2023–present)

A recognized expert in Alternative Dispute Resolution (ADR), Ahmed has effectively managed complex international disputes and fostered cooperation among a wide range of stakeholders. He represented the State of Qatar before the WTO Dispute Settlement Body and in legal and diplomatic engagements with the European Union, providing strategic counsel on a broad array of issues.

Ahmed has published peer-reviewed articles in leading international journals, including the *Arab Law Quarterly, the Journal of World Trade*, and the *Global Journal of Comparative Law*. He is also the author of the forthcoming paper *"The Potential Role of Alternative Dispute Resolution (ADR) in Resolving Intellectual Property (IP) Disputes within Qatar: Aligning with Sustainable Development Goal 16,"* which is published in the *Journal of Intellectual Property Law & Practice* (Oxford University Press).

He holds a Juris Doctor (J.D.) degree from Hamad Bin Khalifa University, completed in collaboration with Northwestern University Pritzker School of Law, and a bachelor's degree in International Trade and Management from the University of Tampa.

# From Global Goals to Ground-Level Impact: Translating Climate Negotiation into Local and Regional Action

# 8

Asif Husain-Naviatti

**Abstract**

Ten years after the Paris Agreement, emerging evidence suggests global $CO_2$ emissions growth may be flattening, potentially supported by land-use changes and accelerating decarbonization in major emitting economies. Yet, the climate emergency remains acute: 2024 marked the first full calendar year to exceed 1.5 °C, and current trajectories imply a high likelihood of this being repeated with increasing frequency. The central question is no longer whether 1.5 °C will be breached, but whether an overshoot can be followed by a sustained decline that restores alignment with the target before the end of the century. Such an outcome depends on rapid acceleration of mitigation and avoidance of destabilizing tipping-point cascades.

Against this backdrop, the chapter argues that the persistent gap between international climate ambition and real-world delivery reflects a governance problem of misalignment across scales. It critiques single-level explanations—whether diplomatic, national policy, or behavioral—and advances an integrated view in which international negotiations articulate shared purpose, national institutions provide the machinery for implementation, and regional/local actors connect resources, legitimacy, and practical delivery.

Using the Three-Pillar Model (Sustainable Purpose, Traveling Organization, Connected Resources), the chapter examines how geopolitics, finance, and the design of COP processes shape international discussion, yet demonstrates that the line of sight between rhetoric and delivery remains weak. This limits the capacity of international processes to drive implementation and undermines their relevance in shaping people's lived choices. Meanwhile, a case study of the Yorkshire and Humber Climate Commission (YHCC) in the UK illustrates how

A. Husain-Naviatti (✉)
School of Earth and Environment, University of Leeds, Leeds, West Yorkshire, UK
e-mail: ahn@barbot.co.uk





P. Wollmann et al. (eds.), *The Sustainable Organization – Volume 2*, Future of Business and Finance, https://doi.org/10.1007/978-3-032-23802-3_8

place-based institutions can act as "connective tissue" between global goals and local realities through independent governance, service orientation, political neutrality, evidence-led framing of co-benefits, and inclusive co-production.

The chapter concludes that the decisive challenge of the post-Paris era is operational coherence: accelerating climate action will require multilevel, evidence-driven, and inclusive governance that translates international purpose into national policy and, ultimately, into the lived choices and opportunities that determine societal transformation.

## 8.1 Introduction

According to the Global Carbon Report, there may be signs of a flattening in overall carbon emissions 10 years on from the Paris Agreement. While fossil fuel emissions are actually projected to rise by 1.1%[1] to a record high, despite progression of decarbonization of energy in many countries, this may be offset by a lowering due to changes in land use. There may be other signs of an overall plateauing,[2] especially when high-emitting economies such as China are investing increasingly in decarbonization.[3]

However, the climate emergency has not abated. This merely suggests that the rate of worsening may be decreasing. Concern remains for the state of the world's great carbon sinks on land and water. Meanwhile, the Paris target of 1.5 °C was exceeded for the first time across a whole calendar year in 2024. Preliminary forecasts suggest there is a high probability for this to be repeated in 2025 and for the following 5 years.

It is no longer about whether we reach this threshold before the end of the century,[4] as per the language of the Paris Agreement, as the remaining carbon budget to maintain this level will likely be gone by 2030,[5] but whether the trends will yet permit a downward adjustment following an overshoot before the end of the century, enabling a realization of the target in a technical sense.

For this to happen, two conditions must be met:

1. Implementation of climate action needs to speed up rapidly and rather urgently, such that vastly reduced emissions now lead to recovery in 40 or more years from now from a peak we have yet to reach even in the rosiest of scenarios.
2. A breach of 1.5 °C for any length of time does not lead to systemic collapse that spirals everything out of control and prevents recovery. Scholarly understanding

[1] See: https://globalcarbonbudget.org/fossil-fuel-co2-emissions-hit-record-high-in-2025/.

[2] See: https://www.science.org/content/article/global-carbon-emissions-will-soon-flatten-or-decline.

[3] See: https://www.carbonbrief.org/analysis-chinas-co2-emissions-have-now-been-flat-or-falling-for-18-months/.

[4] See: https://www.nature.com/articles/s41586-025-09802-5.

[5] Ibid.

of tipping points and their long-term impacts[6] is still relatively new and developing but has strong implications for understanding current predictive models in the worst-case stress scenarios. Moreover, the tipping stress point for any particular system may not correlate with a 1.5 °C scenario and could be reached sooner.

The only way we can stave off a failure of condition two is to put our best efforts toward meeting condition one as quickly as possible. This calls for a ramping up of global implementation of climate policies right now. A high premium has been placed on just that in recent Conferences of the Parties (COP) to the UN Framework Convention on Climate Change (UNFCCC). But does the action and outcome meet the rhetoric?

We live in an era of great climate uncertainty and addressing it demands tremendous agility such as we have not yet seen. Few would argue that progress has been slow. Ten years since the adoption of the Paris Agreement, many people are still wondering if there is sufficient evidence to show it is working. Amidst an apparent downturn in global geopolitical cooperation, collapse of the international system seems an increasingly plausible outcome. Even without that, a huge gap remains between international climate discussions and actions on the ground. It is unclear to many exactly how or whether they connect.

## 8.2 The Climate Governance World

At the risk of oversimplification, there are three different ways of looking at the climate governance world:

1. Some see climate change as an **international problem** requiring international solutions, reflecting the true global, trans-boundary, and cross-temporal outcomes of not finding solutions. For them, the bottleneck lies in a geopolitically fragmented world, a complex web of competing global interests, and an imperfect negotiation system that fails to bring countries together around a sufficiently common understanding on climate issues.
2. Others will point out that the real power to effect change exists at the **national level**—nations are sovereign and, ultimately, they are accountable only to themselves, notwithstanding international agreements, frameworks, and conventions. Countries can pull out of international agreements or reinterpret them. The US president signed an executive order in January 2025[7] to pull out of the Paris Agreement again and did not send a delegation to COP30 in Belèm. Compliance on timely submission of Nationally Determined Contributions (NDCs) is poor,[8]

[6] See: https://global-tipping-points.org.

[7] See: https://www.congress.gov/crs-product/R48504.

[8] See: https://theconversation.com/only-15-countries-have-met-the-latest-paris-agreement-deadline-is-any-nation-serious-about-tackling-climate-change-250847#:~:text=Some%20of%20the%20world's%20largest.

and China, India, and Russia—among the highest emitters—did not submit new ones at all in Belèm. This has clear implications for the system's effectiveness as a positive global pressure and influencing mechanism. So many would rather put the emphasis on how to encourage their national governments to take climate action seriously and implement the right policies, notwithstanding competing national interests and domestic agendas.

3. There are still those who believe that only a shift in actual behavior at the **level of people**—which means you, me, businesses, and organizations—will actually make a difference. Then it becomes a question of what motivates that behavior. National policy nudges certainly play a strong role, but ultimately people make their own choices for a variety of different reasons.

They are all right, and this may be recognized in different places to different degrees. However, if we really want to speed up implementation of climate action, we need an approach that recognizes and integrates the close relationships between international, national, and local levels of action.

### 8.2.1 Understanding the Interdependence by Applying the Three-Pillar Model

The Three-Pillar Model (3-P-Model)[9] offers one way of understanding this interdependence:

- *Sustainable Purpose* is articulated most clearly at the international level, where collective scientific understanding generates shared long-term goals.
- *Traveling Organization*—the capacity to translate those goals into systems capable of movement, adaptation, and delivery—depends on national institutions, regulatory frameworks, and political leadership.
- *Connected Resources*—the alignment of stakeholders, networks, finance, evidence, and community legitimacy—are forged most effectively at regional and local levels, where implementation actually occurs.

To be fair, this could never develop serendipitously overnight, and least of all in a big and complex world tackling an even bigger and more complex topic. This is why progress cannot be linear. We should not judge the success of Paris against its first few years, but rather where we go from here.

Different governance levels may place the response lens on different priorities, which may have an impact on the sense of sustainable purpose that pulls everyone together. A lot of climate discussion has focused on the meeting of targets—at the

[9] The Three-Pillar Model was developed between 2019 and 2021 for the so-called VUCA world by a global author community of more than 60 professionals from public and private sectors and published at Springer Nature in up to now five books (eBook and print). First published in: Wollmann, P.; Kühn, F.; Kempf, M. (Eds.): Three Pillars of Organization and Leadership in Disruptive Times—Navigating Your Company Successfully through the twenty-first Century Business World. Cham: Springer Nature, © 2020.

international level, net zero by 2050 and a global average rise in temperatures limited to 2 °C above pre-industrial averages, and striving preferably for a 1.5 °C limit, by the end of the century. The net-zero targets vary at national and subregional levels. Conceptually, the links between net zero by mid-century, global average temperatures by the end of the century, and the outcomes they may or may not have on future lives in different places are difficult for many to comprehend. But when it comes to individual people, international targets do not immediately impact the day-to-day lives that inform choices.

At subregional or local levels, choices may be influenced to some degree by a desire to make progressive ones, but a far stronger pull lies in how we relate to access to services, income disparities, jobs and opportunities, economic opportunities, growth, needs, or simply the convenience of modern amenities. These are much more here-and-now choices. Moreover, climate science is complex, and in the absence of easy solutions, there is a tendency to defer resolution of climate issues to governments. How closely climate issues relate to creating or maintaining a livable world, a thriving economy, improved services and better lives, and how they relate to advancing a modern world is often lost in the mix.

One could talk about how people and their choices, national government policies, and international agreements are interwoven in a complex system of relationships, influence, and empowerment, which operates up and down and side to side. An examination of these dynamics is far beyond the scope of this chapter, where there is space only to allude to these dynamics. But it is important to understand that a complex system of symbiotic relationships does exist between governance levels, and without any pretense of presenting final solutions, I offer here a few hopefully constructive thoughts on bringing this together.

### 8.2.2 The International Impact

International discussions are inherently top-down. Whichever the arena, whether this be the truly global environment of all member states sitting together at the annual COP negotiations, or the smaller and more selective groupings of countries in other formats such as the G7, G20, the BRICS, or even the multilateral discussions on climate that take place regionally, such as the Asia-Pacific Climate Forum, countries make agreements with each other and hold each other to account.

In a truly well-functioning world, each country would recognize the problem through the increasingly sophisticated science presented to the international system by the IPCC,[10] agree that it is in the best common interest, based on that science, to limit and begin to reverse global warming to avert global catastrophe, and proceed to enact policies in their own countries at national level that will lead to that happening. National policies, which generally involve creating an enabling environment for change to happen, would then trickle down to all levels of society by incentivizing progressive behavior, for example, by reducing taxes on electric vehicles to

[10] Intergovernmental Panel on Climate Change.

make them cheaper to run, or by creating the infrastructure needed to remove the nonfinancial obstacles affecting people's progressive choices.

In the real world, geopolitics gets in the way. Climate change speaks to long-term collective benefits. Governments are inherently short-termist and selfishly in the guard of national interests first. The US withdrawal from the Paris Agreement is entirely motivated by putting America and the American people first,[11] with a backdrop of economic and politically ideological concerns. Suggestions that the Kingdom of Saudi Arabia aims to reinterpret landmark COP29 agreements[12] on reducing fossil fuel use are linked with the structure of its oil-based economy. The adjustments needed to effect rapid change may be costly at the national level, and therefore, potentially unaffordable for many developing countries, hence the vital importance of international financing of the transition. COP30 set a new goal to increase international annual financing to $1.3 trillion by 2035.[13] The following graph shows where we actually are on this (Fig. 8.1).[14] This magnitude of resources would need to come from public, private, and multilateral sources, but it is clear that the wealthy industrialized countries, which are therefore also the highest carbon emitters historically and currently, would need to take the lead.

This leads to complicated discussions on equitable distribution of finance to effectively enable global climate action, leading to mismatched expectations, unraveling of trust,[15] a them-and-us mentality, and ultimately entrenchment into gamesmanship motivated by national interests and managing expectations. There has been considerable scholarly interpretation of these dynamics using game theory to predict an outcome for collective action through the lens of the Paris Agreement, with verdicts pointing both to success[16] and failure,[17] including the potential for collapse. I am not aware of any optimistic studies that predict the rapid progress needed, however.

[11] See: https://www.whitehouse.gov/presidential-actions/2025/01/putting-america-first-in-international-environmental-agreements/.

[12] See: https://www.theguardian.com/environment/2024/nov/23/revealed-saudi-arabia-accused-of-modifying-official-cop29-negotiating-text.

[13] See: https://www.ecosostenibile.eu/en/blog-en/cop30-belem-mutirao-decision-fossil-fuels/#:~:text=The%20Belém%20agreement%20sets%20a,finance%2C%20with%20a%20guaranteed%20minimum.

[14] See: https://www.nrdc.org/bio/joe-thwaites/international-climate-finance-goals-where-are-we-where-do-we-need-be-and-how-do-we#target-1.

[15] The failure of developed countries to reach even the earlier pledge of $100 billion annually between its adoption in 2009 and 2022, a level already widely viewed as grossly inadequate, contributed hugely to international distrust on climate action. The COP29 agreement to increase the pledge to $300 billion in 2024 was also viewed as inadequate against needs in the trillions.

[16] https://theconversation.com/trump-has-rejected-the-paris-agreement-again-but-game-theory-shows-how-other-countries-can-still-lead-by-example-246818.

[17] https://blogs.cornell.edu/info2040/2019/10/21/blog-post-1-game-theory-and-paris-agreement/.

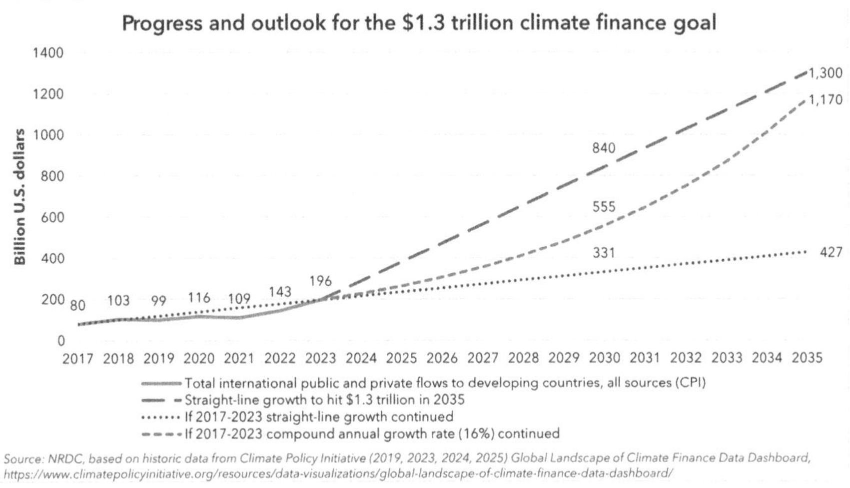


**Fig. 8.1** Progress and outlook for the $1.3 trillion climate finance goal. (Source: Natural Resources Defense Council, NRDC: https://www.nrdc.org/bio/joe-thwaites/international-climate-finance-goals-where-are-we-where-do-we-need-be-and-how-do-we, retrieved on January 12, 2026. Used with permission)

Faltering progress on sufficient international financing encourages many developing countries to hold onto their own dreams for carbonization, as a temporary transitional boost, if not a permanent one, to their failing economies or seemingly insurmountable economic and social problems. In their view, this equalizes the playing field, putting pressure on developed countries to finance the global change, underlining also that the responsibility to put everything right rests with those who contributed the most to the problem in the first place. India's record on this issue is mixed, and with the world's highest population poses a great concern for Paris success if it continues to resist decarbonization.[18] These dynamics merely perpetuate zero-sum views, canceling out—or at least slowing down—the prospects for mutual gain.

But multilateralism does work within the bounds of understanding reasonable expectations, its limits, and not imagining that all the answers are provided by a single system. There are many reasons attributed to the successful negotiations of the Montreal Protocol,[19] which saved the ozone layer, and this has been the subject of many studies. One element, not to be overlooked, is the mission spirit. Notwithstanding the complications and complexities of climate discussion, there is an overall agreement on a common goal, underpinned by a common understanding of the science. The series of IPCC reports[20]—the fabric and foundation of this

[18] See: https://thediplomat.com/2025/12/india-at-cop30-a-mismatch-between-grandstanding-and-climate-action/.

[19] https://theconversation.com/saving-the-ozone-layer-why-the-montreal-protocol-worked-9249.

[20] See: https://www.ipcc.ch/reports/.

understanding—have proved a massive boost to international discussions, given them evidence, informed targets, inserted an underlying constructive trajectory into discussion, and ultimately a shared mission that nuances discussion. They are the closest we have come to making a largely unforeseen future, seen. At least this contributes to incremental advancement and an underlying sense of purpose. It may not stop oil-based economies such as the Kingdom of Saudi Arabia from guarding its economic interests, or even the progressive Europeans from negotiating fresh fossil fuel contracts in the wake of the Russian threat to energy security, but it is a baseline that moves forward.

This goes hand-in-hand with an incremental pattern of negotiations that also inch forward, though it may appear opaque to the man on the street. COP28 language[21] to transition away from fossil fuels, phase down unabated coal power and triple renewable energy capacity by 2030 was hailed as a breakthrough in pointing the finger directly at fossil fuels for the first time and setting up a clear target on renewables. It has not yet been advanced upon, and it did not even make a reiteration at COP30, but it remains an anchored moment that becomes increasingly difficult to deviate from or argue against.

Notably, the G20 Summit held in Delhi[22] in September 2023 targeted tripling of renewable energy capacity by 2030. G20 countries account for around 83% of global emissions, and thus momentum established in a smaller, but highly influential, forum served as a powerful precursor to a more global agreement at the COP. The Indian presidency showed that developing countries can leverage their influence on decarbonization and financing issues in many different ways. Each iteration establishes or reinforces a norm that provides a new baseline for the next. It is not all about the COPs, or COPs alone, and perhaps more attention should be paid to the relationships and contributions to be made by and between different institutions as a way of speeding up negotiations.

Nor do they need to be multilateral ones. The UK provides a positive example of bilateral leadership and influence at a nonpolitical level. The UK's Climate Change Act was passed in 2008 to affect a just transition to a low-carbon and climate-resilient economy and has continued to have the strong support of successive governments. A central enabler of this objective is the establishment of the Climate Change Committee, which assesses progress toward reducing carbon emissions and provides independent and long-term advice, leading to legally binding targets on CO2 reduction by 2050. It does this according to 5-year carbon budgets that set incremental limits, aiming toward net zero by 2050. By 2025, the UK's seventh carbon budget covered emissions up to 2042, projecting a reduction of CO2 emissions by 80% on 1990 levels. Thus, the UK is advancing toward the 2050 net-zero target according to an independently assessed, managed pathway, based on science

[21] See: https://www.weforum.org/stories/2023/11/phase-down-out-fossil-fuel-arguments/#:~:text=The%20deal%20states%20an%20intention,reduction%20in%20fossil%20fuel%20use.

[22] https://www.esgtoday.com/g20-leaders-target-tripling-global-renewable-energy-capacity-by-2030/.

and evidence. The UK actively seeks to engage with other countries around the world to share its experience, as well as learn from good examples set by others. The UK's approach has served as a model or inspiration to many countries, including New Zealand, France, Denmark, Germany, Mexico, and Nigeria, sparking a positive influencing trend in global climate governance.[23]

The soft power of US–China bilateral discussions between John Kerry and Xie Zhenhua,[24] outside the usual political arenas, made a huge contribution to advancing global climate agendas during the Biden era. Agreement between the world's top two emitters—and not traditional allies—proved a powerful and positive influence on COP28 outcomes. Extending this principle of influential nonpolitical action that take place internationally, but outside the confines of negotiated environments, more efforts should be made to build on coalitions of the willing extending to private sector and other nonstate actors, whose influence on climate governance at all levels cannot be ignored.

### 8.2.3 Making Global Local

This highlights another dynamic that the architects of the Paris Agreement foresaw. Transboundary issues—the essence of a country asserting its identity and putting its own interests first—are among the greatest of factors complicating international negotiations that rely at their heart on collective values, the global good, and mitigating against an unforeseen future. Thus, article 12 of the Paris Agreement[25] recognizes the importance of public participation, access to information, and awareness in implementation, which places a high premium on promoting values above national interests. By extension, participation in COPs of nonstate actors that embody these values, though not as parties to the agreement, has long been permitted by UNFCCC.[26]

There is an underlying potential for this to influence negotiations indirectly if not directly, but has it worked? Firstly, lobbying through proximity to negotiations works both ways, and certainly the oil and gas lobby has been identified as contributing to an opposite effect. According to post Belèm analysis, over 1600 oil lobbyists attended COP30,[27] marking the largest share of attendance to date. Was this a factor in a relatively poor outcome falling short of clear commitment on reducing fossil fuels?

---

[23] https://www.theccc.org.uk/2020/10/29/evidence-led-climate-policy-and-robust-climate-governance-are-key-to-delivering-the-paris-agreement/.

[24] See: https://www.bbc.com/news/world-asia-67376471.

[25] See: https://academic.oup.com/oxford-law-pro/book/57184/chapter-abstract/473917642?redirectedFrom=fulltext.

[26] See: https://legalresponse.org/resource/ngos-without-observer-status-engagement-at-cops/.

[27] See: https://globalwitness.org/en/press-releases/fossil-fuel-lobbyists-flood-cop30-climate-talks-in-brazil-with-largest-ever-attendance-share/.

Secondly, the vast majority of nonstate actors attending COPs are confined to zones that limit their access to negotiations and delegations. Given the vastly increasing registered attendance of COPs, there are few opportunities for them to make a meaningful contribution beyond putting on or attending side events. According to Carbon Brief,[28] which analyzes COP attendance, COP30 was attended by over 56,000 people, making it the second largest in history after COP28 that recorded over 80,000. Such attendance has been encouraged in the name of inclusivity, and the UAE declared COP30 to be the most inclusive and accessible ever,[29] with the express intent to grant diverse and marginalized groups a seat at the table. An attendance in the tens of thousands must inevitably compromise such a lofty aim, and there is little evidence to show that this is having a truly beneficial effect. Moreover, the costs and logistics involved in staging such grand events may be pricing smaller nations—including those who depend the most upon positive outcomes—out of any realistic prospect of hosting future COPs. The sheer size of COPs today may be alienating many of the most supportive constituencies who are finding COPs increasingly difficult to comprehend as they continue to operate outside the direct line of sight of the real implementers on the ground—the people, towns, cities, and businesses—who make things actually happen.

This need not necessarily be so, and there is vast scope for improvement. This is a separate topic, albeit worth noting, and well elaborated on in a European Capacity Building Initiative (ecbi) policy discussion note,[30] which makes a number of proposals to put implementation right back in the direct sight of subregional and local-level implementers, using the newly formed climate weeks as a vehicle. It proposes also that the COP negotiations themselves be streamlined to better focus on their original purpose of tracking progress on real implementation and negotiating shared commitments. These proposals are very workable and do not require new institutions, complex treaty reform, or rounds of diplomatic agreement, and they show that there are other ways than negotiation to affect a genuine shift toward implementation. Meanwhile, there is power in non-negotiated outcomes to address barriers to implementation that could better enable national and local actions to increase delivery.

## 8.3 The Local and Regional Impact on National Implementation: The Power of Place-Based Action

If the international system cannot deliver implementation on its own, where can practical momentum be found? Even if COPs can improve line of sight, they cannot deliver action at the point where change is lived. That line increasingly leads to regional and place-based governance structures.

[28] See: https://www.carbonbrief.org/analysis-which-countries-have-sent-the-most-delegates-to-cop30/.

[29] See: https://www.cop28.com/en/accessibility.

[30] See https://ecbi.org/news/ecbi-publishes-note-reimagining-cops-and-climate-weeks.

Societal transformation on climate change will have taken place when people change the way they do things. People can mean everyone from consumers, to producers, to business owners, or individuals living their daily lives and making their everyday choices. People can also loosely refer to larger society structures such as the businesses themselves, larger industry, utilities companies, villages, towns, and cities, and their respective governance systems and institutions. Choices are made at every level. Many of these choices will be made because a national government incentivized it or stimulated an enabling environment through national policies. Tax breaks on electric vehicles are an often-cited first example, and investment in clean energy and infrastructure to reduce costs is another.[31]

Before we go on, it is inherently difficult to speak with universal credibility about change from the bottom up because the world is big. While the UNFCCC convenes the one global climate governance mechanism that gives a voice to all 195 recognized countries, there are multitudes of local and subregional systems within those countries and scattered across the world. There are differences between countries in history, culture, and economic status; in exposure to climate risk; and in their beneficial relationships with industrialization and any immediate potential to gain from carbonization. At the local and subregional levels, there are disparities again. Every rich society experiences inequalities, and every poor society contains pockets of privilege. Reach the local level of any society, and the tapestry of differences, experiences, politics, cultures, and influences of many kinds becomes a rich one indeed. This could be the topic of a whole series of books, let alone one subsection of a single chapter.

With that huge caveat in mind, I turn back to making an impact on society at the local and regional levels in the UK, from the perspective of the **Yorkshire and Humber region**. This is far removed from the acute concerns of African farmers struggling to maintain generations of tradition in water-scarce conditions and a heating climate, or the fears of Pacific Island coastal communities whose lives and livelihoods are threatened by rising seas and inundation. And yet it provides an interesting case study, which itself deserves far more scholarly account than it can afford here.

Yorkshire and the Humber is interesting because it is among the largest of the UK's regions, covering a population of around 5.5 million, equaling many countries in size and scale. It contains two of the country's industrial heartlands—the historic West Riding of Yorkshire and the Humber itself—and indeed parts of the region claim to have birthed the industrial revolution. But there is also a large agricultural economy to the north and huge swathes of natural beauty throughout the region. There are coastal areas suffering from the effects of erosion due to climate change, and the region is increasingly prone to extreme flooding, to which it is adapting. The region is also home to some of the UK's more stark examples of inequality, with

[31] According to IRENA, 91% of new renewable projects in 2024 were cheaper than the cheapest fossil fuel alternatives. See https://www.irena.org/News/pressreleases/2025/Jul/91-Percent-of-New-Renewable-Projects-Now-Cheaper-Than-Fossil-Fuels-Alternatives.

Hull and Barnsely often ranked among the lowest in the UK for life expectancy.[32] One may say that the Yorkshire and Humber region is a microcosm of issues experienced by the UK at large, and therefore—especially given the size of the region and opportunity to influence neighboring regions—success in this region could have a far-reaching impact on realizing the UK's national goals and targets. At the same time, it is a manageable scale, and this is where place-based action comes into its own.

Is it possible to get the whole region—comprising 15 local authorities, 4 combined mayoral authorities, a fair mix of political colors, large businesses, thousands of small businesses, and communities of many shapes and sizes—thinking in unison on climate change? The short answer to this is no, or at least not in those words. But this is what is possible, and I write from a practical and observational perspective with no scholarly pretense.

The Yorkshire and Humber Climate Commission (YHCC) was founded in 2021 as an independent body with a broad remit to bring local authorities, the private sector and third sectors together behind a climate action agenda. By 2021, the Commission had produced a carbon reduction-and-adaptation-focused action plan influenced by the momentum generated by COP26 in Glasgow. By 2024, this had expanded into a full Climate Action Plan for the region,[33] promoting net zero by 2038, against the UK's national target of net zero by 2050. It outlines comprehensively a number of realistic actions to get there across sectors, showing that £250 billions of savings can be realized across the region by 2050 in return for £7 billion annual investment. Importantly, this is based on independent research and a modeling report, the high-level accessible version of which can be found in Our Carbon Story.[34] By 2025, there is an emerging additional perspective of five elected combined mayoral authorities spanning the region, and a further backdrop of nationwide devolution, which has yet to be enacted in parliament at the time of writing. These developments will have profound implications on how local and regional authorities are empowered to act in many arenas, including climate change.

The Commission engages, or will engage, in many sectors of increasing importance to addressing climate change as the years pass, including land and water use across the region, stimulating private investment in climate action and nature-based solutions, promoting the many benefits of health to individuals and the climate, to name a few. However, there are probably five key ingredients to success:

1. *Governance*. The Commission is independent and its consensus decisions are representative of a fair and influential cross-section of society. It comprises around 40 Commissioners derived from local authorities, the utilities including water and energy, industries represented in various forms including the Central Business Institute (CBI), all 12 of the region's universities coordinated by

[32] https://www.ons.gov.uk/peoplepopulationandcommunity/healthandsocialcare/healthandlifeexpectancies/bulletins/lifeexpectancyforlocalareasoftheuk/between2001to2003and2022to2024.

[33] https://yorksandhumberclimate.org.uk/inspire/climate-action-plan/.

[34] https://yorksandhumberclimate.org.uk/inspire/our-carbon-story/.

Yorkshire Universities, and various levels of community interface including individual groups and a Youth Commissioner coordinated by a self-organized Regional Youth Climate Assembly. In true deference to the independent nature of the Commission, Commissioners are actually appointed according to a balance of expertise, rather than as representatives of the entities that give them this expertise.

The Commission relies heavily on the concept of network. Commissioners are expected to devote their time and expertise to promoting the work of the Commission in their respective spheres, which in theory has the potential to carry the Commission's influence throughout their sectoral, business, and governance networks. While supported by a very small commission secretariat, the concept of network and volunteerism carries through to a large number of committed associates from differing backgrounds, who also lend their time, expertise, and networks to the functioning success of the Commission.

The one exception to independent representation lies in the four Vice-Chairs of the Commission, which are representative of the four subregions of Yorkshire, thus embedding the region's formal governance in the otherwise independent framework.

Network building, stakeholder engagement, and embedded influence are from the outset an important part of the Commission's functioning.

2. *Service orientation.* By definition of being a climate commission, there is a strong sense of collective purpose and common mission across its membership. However, the Commission is not mandated, authorized, or resourced to deliver climate benefits directly. How could it, without an annual budget running into billions, and a staff contingent or outreach able to commit impactful climate services—a term that could define or refer to a very wide range of activities—throughout the region. It exists, therefore, to promote a facilitative or enabling environment. But facilitation, enabling, or even coordination is hardly a recipe for inspiring vision and motivating far-sighted action. It is essential then that the commission's "clients" feel the Commission brings something useful, needed, wanted, and impactful to the table, which is—in essence—a service.

Governments at all levels exert a high level of influence on the choices people make—through public policy, behavioral design ("nudges"), economic incentives, and interventions affecting daily behavior. Therefore, the Commission needs to provide a service to the 15 local authorities and Combined Mayoral Authorities that enhances this either qualitatively or even quantitatively. The Commission has knowledge, expertise, and a means to bring local authorities together around key elements of climate policy. This promotes the application of new methods, good practices, the sharing of information between authorities, including successes and failures, and a unified approach.

A good example of this is the Accelerating Adaptive Action Programme. The Yorkshire and Humber region is one of the UK's most flood-prone regions. The River Ouse in York floods its banks frequently, bringing devastation to affected people and high risk to affected businesses. The costs of extreme events are increasing, which has an impact also on insurance costs. The Humber is experiencing rising sea levels—expected to be up to 1.3 meters in the next 100 years[35]—which could potentially cost businesses in the estuary area millions. Moreover, the increasing intensity of storm events has led to more urban flooding.

Clearly, regional governments need to be aware of this and prepare for it on many levels, including planning infrastructure, being aware of the health and food security implications of flooding, and enhancing resilience against increasing events as a whole. This illustrates very keenly how the increasing impact of weather events is not only experienced in developing countries but can be devastating and costly to developed countries too.

YHCC helps local authorities collectively to develop a better understanding of existing adaptation strategies throughout the region, connects other organizations and the private sector toward a more joined-up approach, and builds the capacity of local authorities and other organizations to bring climate adaptation into their decision-making processes and respond effectively. York now has considerably raised its standard of flood risk reduction providing a 1 in 200-year standard of protection, equivalent to Storm Eva that hit the region on Boxing Day 2015. The region's work on flood risk reduction and alerts is arguably world-leading.

3. *Nonpolitical.* The local and regional political outlook exhibits some dynamics that parallel those playing out on the international stage, in that climate can become politicized easily and provoke a wide spectrum of viewpoints. Many remain skeptical about the science behind our understanding of climate change, while others question its urgency. Still others may acknowledge the validity of climate concerns but are more energized by issues they believe need to be tackled with even greater urgency. After all, the cost of living is high and increasing, and there are income and economic disparities across society that impact an individual ability to cope with it. Many people struggle to heat their homes, fear for their jobs, and worry about the long-term viability of the economy. Health costs are rising in a country that rightly puts a high premium on maintaining its national health service for universal benefit, and yet this must be paid for.

These are but a few of the many imperatives that both local and national governments face every day. Amidst such concerns, the discourse around climate issues can easily collapse into idealistic progressive agendas versus the hard realities of making affordable choices, even in developed countries. It is important, then, that the Commission's work is based on practical realities rather than agendas, and hence what is actually needed.

But the true impact of climate action is poorly understood. There is a problem inherent in international rhetoric, which trickles down to countries, but fails to resonate with grassroots concerns. International conferences have become synonymous with the mitigation targets of reducing the global average temperature increase to 2 °C—preferably limiting this to 1.5 °C —above pre-industrial aver-

[35] See: https://consult.environment-agency.gov.uk/humber/strategyreview/user_uploads/humber-2100%2D%2Dstorymap-content-for-website.pdf.

ages. The nuances of international negotiation behind achieving these targets are but a distant concern to local service provision.

International targets are highly relevant to the setting of national targets to reach net zero by 2050, which the UK and many other countries have adopted. This has an impact on national policies that guide the choices that people make. But it is not always clear in the public imagination how global average temperatures are a concern for the ordinary man and woman on the street, or even whether net-zero policies impact them positively.

The Yorkshire and Humber's ambitious net zero by 2038, and individual towns' and cities' own targets are sometimes more ambitious still, such as York's and the internationally renowned steel city of Sheffield's, target of net zero by 2030. Many of these are open to challenge and scrutiny in the absence of locally determined contributions (LDCs). LDCs would typically be focused more on reducing greenhouse gas emissions in locally and regionally relevant areas such as electrification, transport, buildings, and local planning. As such, they could provide valuable complements to nationally determined contributions (NDCs), which are shared internationally through the UNFCCC system, as well as demonstrating the relevance of reducing emissions to people on the ground—a vital, but missing, quantitative bridge in relating the different layers of climate governance.

Indeed, the local-level reality is that addressing climate change means addressing the very socioeconomic concerns that climate skeptics instinctively—and not incorrectly—sense need to be addressed first. Namely, improving health, creating jobs, stimulating growth, making energy more accessible and affordable, and reducing service costs. Encouraging people to rediscover the outdoors, engage more closely with nature, spend more time in open spaces or simply walk more, is beneficial both to mental health—where many of the increasing health costs can be found—and physical health. This, in combination with the creation of a cleaner air environment, can go a long way to reducing respiratory diseases that are also notably on the increase. The result is a massive saving in healthcare costs and a reduced burden of healthcare provision after the event. That there are greater efficiencies in preventative healthcare has long been understood. But has the connection with climate change been equally understood?

Similarly, as the world and technology advances, there are more new jobs and economic opportunities to be created in the so-called green economy and around nature-based activities and solutions, compared with carbon-polluting sectors. The inherent value of nature itself is becoming more and more recognized as important to invest in.

It is entirely possible to address climate change without once mentioning the word climate. There is a far larger common denominator of shared values and concerns between those supporting climate agendas and those who fear such agendas may detract from bigger societal concerns. Indeed, Just Transitions is a central pillar of YHCC activities. It recognizes the disparities that exist in the region and underlines that the benefits of a transition to carbon neutrality must reach all people—especially those more exposed to the adverse socioeconomic effects of climate shocks. The benefits of being more progressive on climate action do not accrue merely to those who can afford to be more idealistic in their approach.

Local government actions have a tremendous influence over built environments across sectors. Were there LDCs, forging the links between emissions mitigating efforts and the wider benefits to communities and cost savings, would be more easily understood.

4. *Based on evidence*. This brings us to evidence. These are good claims, but for them to truly resonate across society, they must be factually understood and backed up by reality. Without this, immediacy wins. If the IPCC provides global evidence of climate realities based on science, the local equivalent is evidence that shows how investment in climate leads to better lives, more jobs, economic growth, accessibility and affordability of energy, better services, and improved health and quality of life.

Strengthening the evidence base is a regional priority. The Yorkshire and Humber region has a Climate Action Plan,[36] published by the YHCC in September 2024. The plan itself covers a number of important issues, including support to the decision-making processes, managing clear priorities such as land and water use, and addressing key elements of the transition such as transforming energy, among many other elements that come together to provide a holistic approach that local authorities, businesses, and organizations can get behind. The Plan explicitly refers to findings from a Carbon Reduction Assessment for the Yorkshire and Humber region,[37] suggesting that investing a little less than 3% of regional GDP every year will meet these challenges while producing even greater savings. It shows that up to £4.5 billion could be saved annually from the regional energy bill alone, and 12,000 jobs created each lasting 20 years. For those that question whether there is sufficient return in climate investment as to attract business interest—which is a key driver of social transformation—a Confederation of British Industry (CBI) report[38] identifies the transition as a £37–57 billion opportunity for British business. Equally, there is credible evidence to suggest that the consequences of not making such investments will lead to additional costs of up to five times more in magnitude.[39] This is compelling evidence.

I cite these examples knowing that this evidence has informed the regional approach behind the Climate Action Plan, and indeed there is additional evidence based on independent research. Independence is a key underpinning to the validity of any credible approach.

However, this may not be enough, and while YHCC has made the gathering of additional evidence a key priority, this remains difficult to stimulate. Research can be

[36] See: https://yorksandhumberclimate.org.uk/inspire/climate-action-plan/.

[37] See: https://yorksandhumberclimate.org.uk/wp-content/uploads/2025/02/YH-Carbon-Reduction-Report-final-edit.pdf.

[38] See. https://www.cbi.org.uk/media/pplbtdca/12820_green_growth_report.pdf.

[39] See: https://www.cambridge.org/core/books/economics-of-climate-change/A1E0BBF2F0ED8E2E4142A9C878052204.

costly, itself requiring greater investment and funding. And yet this is every bit as essential to making progress from the bottom up as scientific evidence in the IPCC reports are from the top down. It is not either-or. The two go hand-in-hand.

5. *Inclusive*. Inclusivity is a prerequisite for success. This can mean many things, but I refer here to an inclusivity of approach. A successful strategy requires legitimacy. Working closely with local authorities is an important element of legitimacy, but it is not the only one. In a democratic society such as the UK, local authorities gain their own sense of legitimacy through elections, and the electorate represents the broadest possible cross-section of society, which is to say everyone aged 18 and above, provided they actually vote in local elections. The YHCC is not elected. It aims to provide a valued service to local authorities through its facilitation and coordination of actors and programs. But why should local authorities listen?

I take the example of the Good Futures project—an upcoming effort designed to equip local authorities with a shared approach to better decision-making on climate issues through the embedding of frameworks into policy and practice, and the requisite training workshops and other capacity-building efforts required to use them effectively. This is an important approach to Just Transitions, which recognizes the pockets of vulnerability across societies, ensuring that decisions taken account for their needs and that the benefits of the green transition are shared by all. It is fully aligned with the YHCC's Climate Action Plan and supports the region's broader strategic priorities for delivering a fair, ambitious, and just climate response.

Note, as mentioned above, however, that as the substance pertains to health, welfare, and many other co-benefits of climate action, the applicability is broad and it is very useable as a tool for those whose objectives are simply and rightly to enhance equity in decision-making. By co-designing the project with local and regional authorities across Yorkshire and the Humber, the Just Transitions team has ensured that its outputs are regionally grounded, actionable, and directly relevant to pressing policy needs, maximizing the potential for significant local and regional impact. Topics covered include digital inclusion and access, co-producing jobs and skills solutions with workers, unions, and business leaders, and the human rights and equalities perspective of access and vulnerability. This approach aligns closely with the YHCC's mission of bridging research, policy, and practice across the region.

More importantly, how did we arrive at this tool? From the very outset of this undertaking, the YHCC undertook to incorporate as broad as possible a constituency of viewpoints, outlooks, perspectives, and opinions, which is the essence of co-designing. The earliest stages of project formulation involved holding workshops representing many different groups and communities representing a wide cross-section of interests. There were four observable benefits of this approach.

1. There is no better way to understand the real concerns of ordinary people than to actually ask them, and no better way to ensure a final product is relevant to them than to actually incorporate those concerns.
2. It is so often the case that people, communities, grassroots organizations, and others come up with solutions that simply do not occur to the politicians, elected representatives, local authority executives and staff, or others who are simply so close to the implementation end they may miss the on-the-ground nuances and realities.
3. By involving people in this way from the outset of the project, and showing that their views were being heard, it generated a sense of co-ownership of the product. This was observable through continued requests to remain connected. People became invested in the outcome.
4. Going back to the principle of networks, drawing these groups into YHCC's overall sense of mission through direct involvement had the benefit of expanding this mission to the contacts and networks of the groups and organizations involved. While not as observable, there is always the potential that the networks surrounding those networks have been similarly touched by YHCC's sense of mission. It becomes an important element of societal transformation.

While I have identified five priorities that, as chair of the YHCC, I feel are particularly potent in our efforts to effect transformation and change across society, these are not the only elements, and this is not an exclusive model. Addressing climate change is such a vast undertaking covering so many overlapping priorities—indeed, everything is somehow a priority—sectors, topics, concerns, controversies, and approaches, it seems impossible to cover everything.

However, in the conclusion of this section, I must observe that the UK is not universally served by place-based Commissions. There are a number of other Commissions in the region, operating at the city level in York, Leeds, and Lincoln, among others, and this is also short of full coverage. These are the legacy of the Place-based Climate Action Network (PCAN), which was an ESRC-funded network running from 2019 to 2024. There are other climate associations and networks outside this framework and operating according to different models. Nonetheless, the size of the Yorkshire and Humber region and the scope of climate issues wrapped up in its diverse landscape render this a particularly interesting case study, and one that has the potential to influence other regions throughout the UK. While remaining in service to local and regional needs, authorities, organizations, and policies, it is also well placed to engage at a national policy level where the interests of local and regional delivery are served, not least of all on all matters pertaining to the English Devolution and Community Empowerment Bill,[40] which may have a significant impact on the powers vested in combined authorities and local councils to improve service delivery in their jurisdictions, including those relating to climate change.

[40] See: https://bills.parliament.uk/bills/4002.

## 8.4 Conclusion: Aligning Global Purpose with Local Reality

The central challenge of the post-Paris era is no longer conceptual but operational. The science is unequivocal, the global mission broadly agreed, and the rulebook complete. What remains unresolved is how to turn this rapidly into implementation—to convert international ambition into national policy, and in turn into the lived choices, behaviors, and opportunities that determine whether climate action succeeds or fails.

This chapter has argued that the tendency to privilege a single governance level—whether international negotiations, national policymaking, or local behavioral change—misreads the nature of climate transformation. Each level contains part of the truth; none is sufficient alone. Climate change is not simply a problem of diplomacy, nor solely one of public policy, nor reducible to personal or organizational behavior. It is a multilevel problem, whose solutions emerge only when these layers work in concert rather than in isolation.

International negotiations can articulate a shared purpose and provide long-range direction, but they cannot determine how policy lands in different national contexts. National governments can legislate, regulate, and finance, yet their decisions only acquire meaning when translated into the daily realities of people, companies, and communities. Local and regional actors, meanwhile, sit closest to where climate impacts are felt and where change is lived, but they often carry responsibilities that exceed their formal powers or resources and are frequently overwhelmed by the enormity of competing challenges.

What often appears as a failure of one layer is, in reality, a failure of alignment between them. The international system presumes domestic machinery ready to operationalize global ambition. National governments presume that local actors can deliver complex transitions with limited fiscal space and uneven capacity. Local communities presume that national or global frameworks will eventually adjust to their needs, pressures, and opportunities. These assumptions leave large gaps through which progress can easily stall, even when the underlying intent is widely shared. The result is a system in which each level is partially right about where the bottleneck lies, yet none alone can unblock it.

The experience of the Yorkshire and Humber Climate Commission (YHCC) illustrates what becomes possible when these governance layers are intentionally bridged. By creating a space where public, private, and third-sector actors can work collectively—grounded in evidence, aligned around shared regional priorities, and supported by practical tools—place-based institutions can act as translators and accelerators of climate action. The YHCC does not replace international agreement or national policy; rather, it supplies what neither can provide: a coherent, grounded, region-wide framework of legitimacy, evidence, networks, and practical delivery, which is the connective tissue through which both can be meaningfully enacted. Its independent governance model strengthens trust; its service orientation enables local authorities to act on common priorities; its evidence base reframes climate action as economic and social opportunity; and its inclusive practice embeds the transition in the lived concerns and values of communities. In doing so, it shows that when governance is rooted in local context, trusted by a diversity of actors, and capable of convening across boundaries, implementation gathers a momentum that formal authority alone cannot produce, linking the global sense of mission through national legislation to the local realities that determine whether that mission succeeds.

This chapter has purposefully contrasted the international, national, and regional perspectives not to elevate one above the others, but to show how easily each can be misunderstood when taken in isolation. Climate governance functions best when purpose is clear, when systems can adapt and travel across scales, and when resources—financial, institutional, social, and informational—are connected in ways that reflect how real societies operate. These qualities may appear in different forms at each level, but they reinforce one another when allowed to interact.

The lesson is not that every region should replicate Yorkshire and Humber's model exactly, but that place-based governance is indispensable to the implementation challenge facing the world. International agreements can define purpose, but they cannot create legitimacy on the ground. National governments can set policy direction, but they cannot build the networks of trust required for societal transformation. Local communities can innovate, but they cannot do so at scale without aligned frameworks, investment, and enabling conditions.

As the world enters a decade in which climate impacts intensify and overshoot becomes increasingly likely, coherence across levels of governance will matter as much as ambition within them. The challenge ahead is not merely to strengthen any one part of the system, but to build the relationships through which each level enables the others and allows them to reinforce one another.

If global goals are to become ground-level realities, this integrative work—less visible than negotiations and less celebrated than technological breakthroughs—will be indispensable. The work of implementation is fundamentally relational. It requires agility, distributed leadership, and a willingness to see climate action not merely as compliance with international goals but as a pathway to healthier, more resilient, and more equitable societies. The question is no longer whether we can act quickly enough, but whether we can act coherently enough. Achieving this coherence demands multi-level, evidence-driven, and inclusive governance meaningfully connecting international purpose, national organization, and local resources.

**Asif Husain-Naviatti** is chair of the Yorkshire and Humber Climate Commission, a senior visiting Researcher at the University of Leeds, and a writer on international climate governance. With nearly three decades of experience at the United Nations and the World Bank, he specializes in sustainable development and multilateral governance in politicized contexts. He has served as secretary of the Advisory Commission on UNRWA and as senior adviser to UN Secretary-General Kofi Annan on AIDS. His climate change work spans sustainable forestry in Latin America, coastal zone mitigation in the Asia-Pacific region, and policy research on international climate governance. An International LEAD Fellow on Sustainable Development, he also holds directorships in sustainable forestry and contributes to academic and policy journals. In his spare time, he is an avid pianist and family man.

# From Awareness to Action: The Urgent Role of Education and Communication in Advancing the UN Sustainable Development Goals (SDGs)

9

Christina D'Asaro

**Abstract**

This chapter aims to provide a basic framework of the current global landscape related to sustainability and the UN SDGs. It argues there is an urgent need for a globally coordinated, multi-level educational and communication effort—aligned across sectors, responsive to local contexts, and designed to move from understanding to measurable impact—to accelerate progress in both regards.

The primary goal of sustainability is to meet our current needs without compromising the ability of future generations to meet theirs. This requires balancing the needs of people, the planet, and prosperity. To achieve a sustainable, better future for all, we must come together in shared purpose, guided by a North Star, committed to reaching the same destination. Education and Communication are powerful forces that turn intention into momentum. They are the tools that allow us to build a shared vocabulary, a collective sense of responsibility, and a common vision of what's possible—empowering individuals, communities, and institutions with the knowledge, support, and motivation required to take meaningful action. When strategically designed and implemented effectively, they mobilize the collective will required to shape a better future.

## 9.1 Introduction

Introduced in 2015, the UN's 17 Sustainable Development Goals (SDGs) outline a comprehensive Agenda for Sustainable Development by 2030, addressing global challenges like climate change, poverty, inequality, and others to ensure a future in which we can not only survive, but thrive (Fig. 9.1).

C. D'Asaro (✉)
Brooklyn, NY, USA


P. Wollmann et al. (eds.), *The Sustainable Organization – Volume 2*, Future of Business and Finance, https://doi.org/10.1007/978-3-032-23802-3_9

**Fig. 9.1** UN SDG (source: https://www.un.org/sustainabledevelopment/), used with friendly permission of UN. The content of this publication has not been approved by the United Nations and does not reflect the views

Ten years after the signing of the Paris Agreement, the 2025 SDG Report[1] and reporting by the World Meteorological Organization (WMO) confirm that 2024 was the hottest year on record (with a temperature ~1.55 °C above the pre-industrial level), and the past decade, 2015–2024, has been the warmest on record.

> There's still time to avoid the worst of climate catastrophe. But leaders must act—now, declared UN Secretary-General António Guterres following its release.[2]

The world is perilously close to becoming unsustainable, with an increasingly unpredictable climate, and extreme weather events and conditions that will render parts of it (particularly for the most vulnerable), uninhabitable.

The rapid advancement and adoption of Artificial Intelligence (AI) allows for potential sustainability solutions that can be implemented at scale—from optimizing energy systems to monitoring environmental change. Yet, it also poses new challenges due to the significant energy consumption required to develop and run these systems, its potential to amplify misinformation and disinformation, further the digital divide, and upend the social and economic world order.

Regarding the UN SDGs, the 2025 Report notes that 35% of SDG targets are on track or making moderate progress. Nearly half are moving too slowly, and,

[1] https://unstats.un.org/sdgs/report/2025/

[2] https://wmo.int/news/media-centre/wmo-confirms-2024-warmest-year-record-about-155degc-above-pre-industrial-level?utm_source=chatgpt.com

alarmingly, 18% are in reverse.[3] Despite the challenges, however, there are meaningful signs of progress and reasons for cautious optimism. There are numerous activities and projects beyond the officially flagged and internationally and nationally reported measures in place, currently underway. Breakthrough is still possible. With bold action and investment, the SDGs can still be achieved (or largely achieved).[4]

It cannot be ruled out that current public perception is based too much on rollbacks and lack of success, and not on the broad momentum and multitude of successful measures happening at the global and local levels.

There have been several key legal and governance advances between 2023 and 2025, including the following.

### 9.1.1 Historic High-Seas Treaty: The First Binding Treaty for Biodiversity Beyond National Jurisdiction

On January 17, 2026, the High Seas Treaty, formally the Agreement on the Conservation and Sustainable Use of Marine Biological Diversity of Areas Beyond National Jurisdiction (BBNJ) entered into force, protecting biodiversity in international waters and closing a foundational governance gap.

**Importance** This UN Treaty establishes, for the first time, a legal framework to protect biodiversity in international waters—a significant historic milestone for both global ocean protection and multilateral cooperation. Parties will now have formal tools to create Marine Protected Areas (MPAs) on the high seas, regulate potentially harmful activities, and coordinate science and capacity building.[5]

As Singapore's Ambassador for International Law, Rena Lee, who served as President of the Treaty negotiations and was pivotal in guiding its successful adoption in 2023, noted: "The entry into force of the BBNJ Agreement today stands as a monument to multilateralism and to the years of dedication, dialogue and cooperative action by countless committed people around the world. Global challenges, such as the climate and biodiversity crises, affect all of us. As such, global cooperation is not a choice—it is a necessity. This Treaty embodies hope, resolve, and a shared commitment to a brighter future for the ocean and our planet."[5]

---

[3] https://unstats.un.org/sdgs/report/2025/

[4] https://sdg.iisd.org/news/amid-real-but-unequal-progress-breakthrough-still-possible-un-stats-report/

[5] https://highseasalliance.org/2026/01/16/historic-high-seas-treaty-enters-into-force-launching-a-new-era-of-global-ocean-governance/

### 9.1.2 Global Plastics Treaty Negotiations Are Advancing: A Legally Binding Instrument on Plastic Pollution Is in Progress

Negotiations under the UN Environment Programme's Intergovernmental Negotiating Committee (INC) continue toward a global, legally binding plastics treaty. Multiple INC sessions have produced drafts and negotiations in 2024–2025 focused on a myriad of issues, from product design and waste management to possible production controls. The progress is meaningful, despite consensus on production reduction and other measures being challenged.[6]

**Importance** A global plastics treaty would create binding obligations across the plastics life cycle (production → use → waste) and help reduce plastic in the ocean—a major transboundary risk.

### 9.1.3 Local Efforts

At the Local2030 Coalition's event at the High-Level Political Forum on Sustainable Development (HLPF) 2025 on using local action to speed up the SDGs, the City of Yokohama showed how local innovation can drive global sustainability progress. Through initiatives like the Y-SDGs Certification System, which now recognizes over 800 local organizations for advancing ESG goals, and creative community programs such as SDGs Food Lockers that reduce food waste, Yokohama has demonstrated measurable results in the circular economy and engaging citizens.[7] The city's international partnerships, including a collaboration with Mandaue City in the Philippines on waste reduction and carbon cuts, reinforced that achieving the SDGs depends on empowering local actors to lead globally relevant, community-driven solutions.

Other positive progress includes the development of industry-specific pacts (e.g., Ocean Tourism Pact), voluntary frameworks, and increased corporate commitment, offering new governance norms and private-sector accountability mechanisms for ocean-dependent industries.[8] While voluntary, pacts such as these often accelerate the adoption of responsible practices and create legal/contractual pathways for finance and standards—especially where formal treaty law moves slowly.

Nevertheless, with only 5 years until the deadline for Agenda 2030, there is an urgent need—now, more than ever—for global, coordinated, multi-level action on sustainability and advancing the SDGs.

And that starts with Education and Communication.

---

[6] https://www.unep.org/inc-plastic-pollution

[7] https://businessyokohama.com/blog/2025/07/21/hlpf2025-2/

[8] https://bluetourisminitiative.org/news/ocean-tourism-pact-launch-at-the-blue-economy-and-finance-forum-in-monaco/

## 9.2 Education and Communication: The Foundation for Mobilization

Education and communication form the foundation for mobilizing a global community around pressing challenges. When information is conveyed clearly and compellingly, it helps everyone (from governments and institutions to private sector partners and individuals) quickly understand the severity of a situation and why it matters to them.

Effective communication translates complex goals into relatable, actionable steps, creating a shared sense of urgency and purpose. This is critical, as it aligns various disparate groups, accelerates decision-making, fosters collaboration, and inspires individuals to take ownership.

The result is momentum—the ability to transform awareness into action within a compressed time frame, which is often critical for addressing global issues that demand immediate, coordinated responses.

When communications clearly articulate *why* something matters (risk, collective or individual consequences), people are more likely to feel urgency (*Proceedings of the National Academy of Sciences* (PNAS) by Bolsen et al.).

### 9.2.1 Case Study: "Communication and Collective Actions: A Survey Experiment on Motivating Energy Conservation in the U.S."

Authors: Toby Bolsen, James N. Druckman, Fay Lomax Cook[9]

**Key Findings** Messages that emphasize both *individual responsibility* and *collective environmental benefit* are more effective at motivating people to take conservation actions, even when individual benefit is minimal or indirect.[9]

**Takeaways** Framing (education + values + purpose) helps people understand the gravity of an issue and motivates them to act. This is particularly effective when communicating a global goal where each person's contribution makes a difference.

**Education** From an education perspective, much more can be done to integrate sustainability and the UN SDGs into the global ecosystem, from the youngest to the eldest. Programs can, and should, be built into kindergarten curricula through universities and beyond. Innovative ways of educating different populations, such as

[9]Toby Bolsen, James N. Druckman and Fay Lomax Cook (2014). Communication and Collective Actions: A Survey Experiment on Motivating Energy Conservation in the U.S. Journal of Experimental Political Science, 1, pp 24–38 doi:10.1017/xps.2014.2, see https://faculty.wcas.northwestern.edu/jnd260/pub/Bolsen%20Druckman%20Cook%20JEPS%202014.pdf

indigenous groups, who often live in remote areas, have languages passed down orally for generations, the illiterate, and the unconnected, can be reached by partnering with organizations such as Audiopedia,[10] as well as through the United Nations Institute for Training and Research, with its vast expertise in capacity building in both areas.

### 9.2.2 Case Study: "Audiopedia Foundation"

Audiopedia stands out as a prime example of how digital platforms can directly contribute to meaningful connectivity and, by extension, women's empowerment. Recognized by the SDG Digital Acceleration Agenda, Audiopedia is a digital platform that offers accessible, relevant content in local languages, specifically designed for illiterate or semiliterate women. Aligning with Meaningful Connectivity Goals: Audiopedia's approach aligns perfectly with the goals of meaningful connectivity by addressing several critical barriers:

- Accessibility: Through the use of audio content, Audiopedia makes information accessible to those who cannot read, overcoming one of the significant hurdles in the path to digital inclusion.
- Relevance: The content provided by Audiopedia is meticulously curated to meet the specific needs of its target audience. It covers a wide range of topics, from health and nutrition to rights and education, ensuring that the information is not only accessible but also profoundly relevant to the women it aims to serve.
- Local Languages: By offering content in local languages, Audiopedia breaks down language barriers, ensuring that the information is understandable and culturally resonant for the listener.
- Empowerment Through Knowledge: The ultimate aim of Audiopedia is empowerment. By providing women with knowledge that is directly applicable to their lives, Audiopedia enables them to make informed decisions, improve their and their families' health, and pursue opportunities for economic advancement.
- Impact and Scalability: The impact of Audiopedia is evident in the testimonials of its users, who report significant improvements in their quality of life and a greater sense of autonomy. Moreover, the digital nature of the platform allows for scalability, offering a model that can be replicated and adapted across different contexts and regions.

Audiopedia's success[11] underscores the potential of digital solutions to contribute meaningfully to connectivity and empowerment, serving as a beacon for future initiatives aiming to bridge the digital divide.

---

[10] https://www.audiopedia.foundation

[11] See for example recognition and awards like:

- In 2022 selected by the United Nations Development Program (UNDP) for their Digital-X Catalogue
- In 2021reached the Grand Finale of Indias largest Social Innovation Challenge, the TATA

This case study compels action, urging multilateral and multi-stakeholder processes to support and scale identified solutions. It cautions that without deliberate, collaborative efforts to amplify the impact of proven initiatives, the global community risks falling short of the SDG5 objectives.

This premise applies not only to SDG5, but rather to achieving all the SDGs and a collective, sustainable future.

In line with this chapter's overarching premise that a global, coordinated, multi-level effort is required to achieve Agenda 2030, the document calls for a united front through strategic partnerships, policy reforms, and resource mobilization to "ensure that meaningful connectivity reaches every woman, paving the way for a more equitable, sustainable future."

More on Education can be found in other chapters of this book; the focus of this chapter will pertain to Communications.

## 9.3 Words Matter: Why Clear, Concise, Compelling Communication Is Critical

### 9.3.1 Introduction

Johan Karlsson, Managing Director of Better Shelter, a social enterprise that develops and provides innovative housing solutions for persons displaced by armed conflicts and natural disasters, and one of the top five sustainability entrepreneurs in Sweden, recently mused on LinkedIn[12] under the headline of: *What If NGOs Wrote Your Ads (a reflection)* (Fig. 9.2).

The power of this example is its ability to illustrate some of the key differences between the way the public and private sectors communicate in a way that people can immediately recognize, relate to, and understand.

When it comes to (external) communications, the private sector approach is typically direct, actionable, and urgent.

Public sector communications, on the other hand, traditionally take a more academic approach. Often written for a highly educated audience, messaging includes more complex concepts and terminology, shaped by political considerations, geopolitical dynamics, multiple stakeholders, and an obligation to remain neutral.

---

Social Enterprise Challenge

- In 2020 selected for the first cohort of the UNLOCK Accelerator program by Wikimedia Germany
- In 2020 one of the winners of the Smart Development Hack by the German Federal Ministry for Economic Cooperation and Development and the European Commission
- In 2020 selected as a game-changing health innovation by Health in Your Hands (HIYH)
- In 2019 recognized by the World Summit on the Information Society (WSIS)
- In 2019 won the digital volunteering award of state government of Rhineland-Palatinate, Germany
- In 2018 selected for the first cohort of the Hatch Colab Accelerator in Geneva

[12] https://www.linkedin.com/posts/johan-karlsson-01822b5_what-if-activity-7374739753788686336-W3bx/?utm_source=social_share_send&utm_medium=member_desktop_web&rcm=ACoAAAHx58cBSn-DnZE2t0GOMcMm2QG7_kADyPw

[Sport Shoe Industry] [Furniture Industry] [Automotive Industry]

**Fig. 9.2** Pictured above are three examples elaborated by the author (logos deleted) provided in the post, which ended with: "Just a light-hearted exploration, not tied to any brand or organization. But words matter." Source: Johan Karlsson on LinkedIn (see footnote 12). Used with permission

Regardless of the sector, developing clear, compelling communications that resonate with the target audience is critical to ensure key messages are heard and ultimately motivate people to take the desired action (See Sect. 9.4 for more about target audience messaging).

As they develop messaging, communicators must contend with the increasingly dangerous threat of misinformation and disinformation—both of which erode trust in institutions, expertise, and shared facts.

"Misinformation" is "false or inaccurate information."[13] Often amplified through social media, AI-generated summaries, and data that is incomplete or without context, it can weaken public confidence—particularly when institutions are slow to respond, or worse—fail to respond at all.

"Disinformation" is "false information that is intended to mislead; deliberate misinformation."[14] Created purposely to mislead, destabilize, or manipulate public opinion, it exploits emotional narratives, delegitimizes multilateral institutions, distorts global agreements, and undermines consensus on issues like climate change, public health, and others.

The implications of both are serious—particularly for the UN (which is expected to be *neutral* in an environment that is *anything but*), governments, and public-interest institutions. Each operates in a *contested information space* where:

- Trust is uneven across regions
- Narratives are politicized (by design)
- Disinformation is often state-aligned or ideologically driven
- Authority alone no longer persuades, and silence or delayed explanation is often seen as complicity or failure
- The erosion of trust leads to a weakened mandate, resulting in reduced compliance and funding

[13] See Oxford's English dictionaries.

[14] Ibid.

Addressing this threat requires that the UN, governments, and international organizations:

- Shift from reactivity (correcting information) to proactivity:
  - Invest in and prioritize transparent communication
  - Anticipate predictable false narratives
  - Strengthen media and information literacy as a critical part of public education
  - Establish clear protocols that allow for rapid response
  - Set strong standards for the ethical and transparent use of AI
  - Localize messages through trusted regional and community voices for credibility

The private sector also has a crucial role to play—particularly technology platforms, media companies, and AI developers. They should:

- Reduce algorithmic amplification of content that is demonstrably false
- Improve transparency around content generation and distribution
- Supporting independent verification and journalism
- Embed safeguards against misuse of generative AI

Combating misinformation and disinformation is a shared responsibility—one that protects trust, supports a shared reality grounded in reliable information, and is vital for cooperation, which is ultimately critical to achieving the SDGs and a sustainable future.

### 9.3.2 Charting the Course: Why a Communications Strategy Is Crucial

Developing communications without a strategy is like setting sail with no destination. The wind may carry you forward, but where? And, at what cost? For centuries, sailors looked to the North Star as their guide. Navigating through stormy seas, with ever-changing tides, fleeting gusts, and temperamental weather, the Star was constant. It remained a beacon for the entire crew, confirming they were headed in the right direction with a shared destination and purpose.

### 9.3.3 A Communications Strategy Is an Organization's North Star

It provides clarity in a world that is increasingly noisy, chaotic, and changing at an unprecedented scale and speed. Trends roll in and out. Audiences shift. Platforms and technology evolve. A strategy grounds communications in something sound—clarity of purpose. "The underlying principles of strategy are enduring, regardless of technology or the pace of change."—Michael Porter[15]

A strong communications strategy:

- Provides a structured framework to define the organization's mission, values, and priorities to stakeholders and specific objectives it looks to achieve

[15] https://quotefancy.com/quote/1371191/Michael-Porter-The-underlying-principles-of-strategy-are-enduring-regardless-of

- Addresses the *why, who, what, and how* of communication:
  - Why the organization is engaging
  - Who it needs to reach
  - What messages will resonate, and how those messages will be delivered and evaluated (Cornelissen, 2020; Zerfass et al., 2018)
- Ensures consistency across channels, fosters trust, and aligns communication activity with broader organizational goals

It aligns leadership, staff, and partners to ensure everyone is working together, headed toward the same destination. No matter how talented the crew, if each member is following their own direction, the ship will go in circles. The North Star is a shared point of reference. It ensures every adjustment throughout the journey contributes to moving the entire organization forward.

Strong communication strategies also extend internally, unifying employees and partners around shared direction, which research has shown improves engagement, organizational commitment, and overall performance (Men & Bowen, 2017; Tsai & Men, 2017).

Storms are inevitable. Many times, the path will become unclear, and conditions can become severe. There may be disagreements as to how to best navigate through the tumult. It is during these times that the North Star becomes essential.

It helps organizations reorient, steady the ship, and ultimately reach their destination—stronger and more resilient.

### 9.3.4 Risks of Drifting Without a North Star

Too often, organizations (in both the public and private sectors) leap directly into "doing communications." Well-intentioned, they issue press releases, create social media campaigns, or send emails without defining where they are headed.

Initially, it can look like progress. After all, the sails are full, and the ship is moving. But moving without a clear direction and destination (strategy) is just drifting. And the risks are real:

1. *Wasted Resources*:

Without a solid strategy, time, energy, and budgets are often put into disparate efforts, leading to little return on investment (ROI), or, worse, a complete lack of ability to measure ROI, since clear objectives and ways to measure success are lacking.

2. *Inconsistency*:

Different teams and/or communication materials say different things, confusing audiences and diluting trust.

3. *Reputational Damage*:

Without a unifying strategy, even well-intentioned communications can cause reputational damage, causing significant, if not irreparable, damage to the entire organization.

Before an organization drafts a press release, launches a campaign, or publishes a single post, it should define its North Star, answering, "*What is the unwavering purpose that will guide us, no matter what?*"

Antoine de Saint-Exupéry, the aviator and writer, emphasized the importance of inspiring a shared vision: "If you want to build a ship, don't drum up people together to collect wood and don't assign them tasks and work, but rather teach them to long for the endless immensity of the sea."

A clear message that resonates emotionally and inspires everyone to stay focused on their shared destination/goal, undeterred by challenges, and committed to doing whatever it takes to get there will always be more powerful than one that is purely intellectually based.

Charting the course in advance, having that North Star, ensures that when public opinion or circumstances change, or unexpected challenges arise, the organization can come together and adjust accordingly—never losing sight of its shared destination.

### 9.3.5 A First Takeaway: A Communications Strategy Is Not a Luxury

It is a critical discipline and business driver that transforms ad hoc activity into purposeful engagement, ensuring that every interaction supports an organization's long-term mission and credibility and aligns with broader organizational objectives.

The most successful organizations are those that know where they are going, keep focused on the North Star, know their stakeholders extremely well and how to move them, and inspire others to follow.

## 9.4 A Strategy Is Only as Good as Its Execution

### 9.4.1 Introduction

In Communications, this means translating strategy into a clear, actionable plan, which includes:

Identifying your key audiences, truly understanding them, and developing messages that resonate with them. It requires a deep understanding of the media they consume, and when, where, and how your message will reach them efficiently and effectively.

Finally, it requires designing campaigns that are not only effective in the short-term, but also build trust, loyalty, and growth long-term. This is how organizations build lasting impact.

### 9.4.2 A Crucial Task: Knowing the Recipients of Your Communication Well and How to Move Them

One of the most underestimated skills in modern communication is the ability to *truly* understand the people you're trying to reach. Not just who they are in terms of demographics or title, but what *moves* them. Without that insight, your message will get lost in the abyss. As a communicator, when you deeply understand your audience, you can connect with them. By putting yourself in their position and being able to understand, identify, and relate to their point of view, you can tap into shared experiences, pressures, and motivations that really resonate. Connecting with your audience authentically, getting their attention, and keeping it—over time—is everything.

Every audience has its own nuances. Political stakeholders operate under almost constant scrutiny and changing priorities, and are typically moved by clarity, consequences, and alignment with their goals. Journalists filter the world through what is "newsworthy" and valuable, and generally respond to honesty, precision, and a strong angle. Influencers seek purpose and resonance. They champion messages that align with their values and those of their audience and followers. The public, on the other hand, typically responds to simplicity, humanity/authenticity, and the feeling that "this matters to me."

*People are moved when they feel understood.* When your message reflects a shared experience, a challenge they've gone through, a hope they recognize, or a frustration they've felt, it creates a spark. They see something that makes them stop, because it speaks to something inside them.

Communication is powerful when executed well. It motivates people to stop, reflect, (re)consider, and act. It translates into understanding, which leads to momentum, which drives impact.

Identifying and understanding your audience(s), developing messages that resonate, determining the most effective ways to reach them with those messages (e.g., digitally, through events, direct mail, podcasts), gaining their attention, and then galvanizing them to move toward shared goals, creates momentum. Effective Communication builds momentum that sustains action toward the UN SDGs—turning shared purpose into lasting impact.

### 9.4.3 Measuring Effectiveness

Recognizing the power of data over subjective assessment, Grace Murray Hopper, a US Navy rear admiral and computer science pioneer who helped lay the groundwork for modern computing, said, "One accurate measurement is worth a thousand expert opinions."[16]

---

[16] https://wearetechwomen.com/inspirational-quotes-grace-hopper-computer-scientist-u-s-navy-rear-admiral/

This principle holds for Communications, where metrics can offer objective, actionable insights that are more reliable than guesswork. Measuring things like audience sentiment, reach and delivery, brand awareness, engagement levels, etc., provides valuable insight into what is working and where improvements can be made. This results in efficiencies over time, better use of resources, and ultimately greater impact. Every Communications plan should include a measurement and evaluation plan. Without it, success is undefined, and impact is impossible to prove.

### 9.4.4 Measurement and Evaluation

- *Ensures Accountability*—Shows leadership and stakeholders that communications deliver value and support organizational goals
- *Defines Success*—Outlines specific, measurable targets (e.g., "increase engagement by 15% over six months") vs. "raise awareness"
- *Allows for Course Correction*—Data shows what's working and what's not in real-time, allowing the team to pivot quickly and optimize results
- *Builds Institutional Learning*—Leads to lessons and insights that improve future strategies and build institutional knowledge
- *Builds Credibility*—Positions communications as a strategic function, backed by evidence vs. anecdotes
- *Illustrates Impact*—Links communications activities to outcomes that matter (e.g., behavior change, stakeholder engagement, reputation, revenue, and/or mission impact)

## 9.5 A Best Practice Use Case: Case Study—The Global Indigenous Youth Summit on Climate Change (GIYSCC)

### 9.5.1 Overview

The United Nations Institute for Training and Research (UNITAR) organized the Global Indigenous Youth Summit on Climate Change (GIYSCC) on the 9th of August 2023 (the United Nations International Day of the World's Indigenous Peoples),[17] along with the Science Diplomacy Center, Springer Nature, the International Network for Government Science Advice Africa, and dozens of other esteemed partners (like Science Diplomacy Center, SpringerNature, Ingsa Africa, University of Lagos, Nigerian Young Academy, UArtic, International Institute of Science Diplomacy and Sustainability, Program on Negotiation Harvard Law School, etc. …[18]).

[17] Details and pictures see: UNITAR https://unitar.org/about/news-stories/news/unitar-and-partners-host-global-indigenous-youth-summit-climate-change

[18] Details and pictures see: UNITAR (https://unitar.org/about/news-stories/news/unitar-and-partners-host-global-indigenous-youth-summit-climate-change

The summit was a global dialogue, developed by, for, and among Indigenous youth, which took place across three 8-h time zones, in 24 h.

Illustrating UNITAR's continued commitment to reaching the furthest first and leaving no one behind, the Summit:

- Provided a "one of a kind" platform to accelerate global engagement with Indigenous youth communities, cultures, languages, ethnicities, countries, and continents across the world
- Encouraged lifelong learning among Indigenous youth to contribute with knowledge and skills in the evolution of the United Nations Framework Convention on Climate Change (UNFCCC) and related global agreements that have regular Conferences of the Parties
- Built stronger collaboration across generations to advance UN development efforts at local, national, and global levels

### 9.5.2 Goals

The Summit had Multiple, Interlinked Goals:

1. Accelerate global engagement with Indigenous youth worldwide, across cultures, languages, and ethnicities, in addressing climate change
2. Share perspectives, observations, concerns, and strategies from Indigenous youth regarding local climate change impacts and resilience
3. Stimulate lifelong learning among Indigenous youth, including in relation to global climate diplomacy (e.g., UNFCCC) and sustainable development
4. Enhance intergenerational connections under the UN Declaration on the Rights of Indigenous Peoples (UNDRIP) and related frameworks
5. Inspire Indigenous youth leaders to think long-term (throughout their lives) with sustainable development and climate action in mind

### 9.5.3 Format and Approach

- The agenda was co-designed with Indigenous youth, including decisions on topics and modes of participation
- Included multiple formats—panel dialogues, short documentaries, artwork, music, cultural dance, and photo galleries, giving participants the flexibility to express themselves in culturally relevant forms
- Emphasized language inclusion—sessions included many languages, prerecorded as needed, with communication hubs and remote and low-bandwidth considerations

### 9.5.4 Public Echo

A *Nature commentary* ("Indigenous Youth Must Be at the Forefront of Climate Diplomacy") was published in advance by the Indigenous youth leaders of GIYSCC. Not only did the Summit convene youth, but it also produced policy-oriented texts aimed at influencing the international conversation.

The full press release read:

> In celebration of the International Day of the World's Indigenous Peoples on August 9th, UNITAR, in collaboration with the Science Diplomacy Center, Springer Nature, and the International Network for Government Science Advice Africa, as well as dozens of other esteemed partners, organized the Global Indigenous Youth Summit on Climate Change. This unique event created a platform for Indigenous youth worldwide to collaboratively address climate challenges and explore sustainable solutions.
>
> With sessions spanning three 8-hour time zones, the summit facilitated an inclusive dialogue among Indigenous youth, connecting voices across cultures and languages. Throughout the summit, Indigenous youth shared crucial insights into climate change impacts and resilience strategies. The event spanned from 00:00 to 24:00 GMT on the International Day of the World's Indigenous Peoples. The summit saw engagement from over 1300 registrants representing 112+ nations and communicating in 88+ languages across a continuous Zoom event and parallel YouTube livestream.
>
> The Global Indigenous Youth Summit on Climate Change exemplifies the impact of youth voices in steering global conversations toward actionable change. UNITAR and its partners remain committed to fostering progress and honoring the knowledge of Indigenous communities towards the implementation of the Sustainable Development Goals.

### 9.5.5 Evaluation of the Use Case, Lessons Learned, and Recommendations

This Use Case is a strong example of how co-creating safe and inclusive summits/forums for discussion and open exchange, across the globe, incorporating various cultures, languages, generations, and perspectives, is critical to accelerating the SDGs. Equally important is the investment of time and resources into communicating about these forums—with partners who galvanized communities online and, in the field, accounting for various languages, time zones, levels of digital availability/connectivity, as well as other factors to ensure the summit was as accessible and inclusive as possible, for maximum impact.

When audiences are invited to co-create and collaborate from the beginning, they are naturally more invested in the entire process and contributing to a successful outcome. They are more engaged, and more likely to remain engaged and committed to the goals of the initiative(s) and the organizations themselves, long-term. This is how progress and impact are compounded over time, with exponential gains.

**Recommendations** Encourage co-creation of events with various audiences, partners, and stakeholders from the onset. Make them multigenerational, global, inclusive, and accessible. Create a safe space for sharing ideas, cultures, and solutions. Incorporate ways to keep momentum and progress moving forward, including follow-up conversations, summits, and forums, and consider creating smaller groups/ action committees focused on the implementation of specific solutions, as suggested and agreed to during or following the discussions.

## 9.6 Similar Use Cases to Learn

Similar Use Cases include the collaboration between the United Co-Creation Lab + UNITAR, which included a side event during the United Nations General Assembly in September of 2024 held in collaboration with the Ministry of Foreign Affairs of Denmark—Permanent Mission of Denmark to the UN in New York and the Ministry of Foreign Affairs and Human Mobility of the Republic of Ecuador to explore how co-creation between youth and global leaders can foster sustainable development and inclusive societies. This event, "Co-Creation for a Sustainable Future: Empowering Youth in Global Dialogue," marked the official launch of the United Co-Creation Lab,[19] an initiative dedicated to bridging the generational gap and bringing young voices into decision-making processes.

United Co-Creation Lab is dedicated to fostering intergenerational solutions to global challenges by facilitating ethical co-creation processes between children and young people (ages 10–20) and political leaders. Rooted in human rights principles, the lab develops research-based methods and practices that ensure equality across age, culture, and experience. These methods empower youth to actively participate in democratic development processes, enhancing their skills and promoting creative global citizenship. Through collaborative initiatives, the lab bridges generational gaps, enabling the co-creation of innovative solutions to pressing global issues.

Ms. Charlotte Sahl-Madsen, President of United Co-Creation Lab, underscored the important role co-creation has in including young people as equal partners in shaping the future.[19]

## 9.7 A Brief Reference to the Three-Pillar Model

The Three-Pillar Model is based on three key building blocks (pillars) for organization design and leadership in the VUCA world. The starting point of the developing global author community[20] in 2019 was to design principles for a world with almost no principles left. The model was since then frequently applied—and its

[19] https://unitar.org/about/news-stories/news/unitar-and-united-co-creation-lab-youth-and-leaders-join-forces-co-create-sustainable-future

[20] The very diverse global author community consists of practitioners, leaders and academics from all continents and more than 20 countries. They represent knowledge stemming from their roles within more than 20 large global public institutions (incl. UN Agencies) and global enterprises and more than 30 different local public organizations or national companies.

applicability and usefulness proven—in various global use cases in the private and public sector (so far >80 described in five books[21]).

1. The first Pillar: *Sustainable Purpose*

Sustainable Purpose is the raison d'être of an organization, having a guiding objective means bringing orientation and certainty to the people for their joint endeavor and success. This is especially important in times of crisis and transformation. It is also the key source for the narrative of transformation. It is the North Star mentioned earlier in the chapter.

2. The second Pillar: *Travelling Organization*

Travelling Organization means for an organization accepting to be in a permanent state of flux and transformation, while interacting with the external requirements of its journey with rapid learning and adaptivity. Ideally, this is the mindset of loving the unknown and being well able to manage it. The Travelling Organization stands for all stakeholders to be actively or passively affected by the Sustainability and UN SDG Coverage transformation—this means nearly everybody in the world in all fields.

One has an intuitive idea of a Travelling Organization from the metaphor of an expedition to Africa in former times. The participants of the Travelling Organization must be curious, open to learn, courageous, keen to experiment, and they deal well with uncertainty, stress, unforeseen incidents in a totally new world. This mindset of the Travelling Organization is the core of a Learning Organization. It is impossible for the Travelling Organization not to learn. Learning is a question of survival.

3. The third Pillar: *Connectivity*

Connectivity is all about the ability to interconnect on the journey with all needed parties and resources inside and outside the silos, therefore creating high efficacy and consistency and during this process building also shared learning networks. If the whole world is affected by a transformation, it has to connect with everybody.

[21] Wollmann, P.; Pemler, D.; Ndrevataj, M. (Eds.): The Sustainable Organization—How Organizations Address the 17 UN SDGs Using the 3-P-Model. Cham Springer Nature, © 2025. Wollmann, P.; Stricker, L.; Pemler, D. (Eds.): Reading an Unknown Organization—Leaders in New Roles Thriving in Challenging Environments with the 3-P-Model Mindset. Cham Springer Nature, © 2024.Wollmann, P.: Püringer, R. (Eds.): Transforming Public and Private Sector Organizations—Implementing Sustainable Purpose, Travelling Organization and Connectivity for Resilience. Cham: Springer Nature, © 2022.Wollmann, P.; Kühn, F.; Kempf, M.: Püringer, R. (Eds.): Organization and Leadership in Disruptive Times—Design and Implementation of the 3-P-Model. Cham: Springer Nature, © 2021.Wollmann, P.; Kühn, F.; Kempf, M. (Eds.): Three Pillars of Organization and Leadership in Disruptive Times—Navigating Your Company Successfully through the 21st Century Business World. Cham: Springer Nature, © 2020.

## 9.8 Overall Conclusion and Takeaways

To fully achieve the SDGs and a sustainable future, coordinated, multi-level, and intergenerational educational and communication efforts are essential. While our challenges are profound, so is our capacity to meet them. Simply put, there is no time to waste. We have proven use cases, the knowledge, and the ability to implement solutions that we know can fast-track progress toward the SDGs and a sustainable future. AI, if used responsibly, ethically, and strategically, paired with human insights and intelligence, has the potential to expedite solutions to even the most complex issues exponentially—enabling us to regain ground and propel us toward achieving the SDGs, and a sustainable future for all.

**Takeaways: What Every Organization Can and Should Do**
- Develop a communications strategy
- Get to know and deeply connect with their audience(s)
- Co-create—intergenerationally, with partners, across sectors
- Develop messaging and initiatives that make people feel something emotionally and motivate and inspire them to act
- Provide people with a clear call-to-action and a shared destination (North Star)
- Proactively and vigorously address misinformation and disinformation
- Measure effectiveness of communications and optimize for maximum impact
- Invest in (continuous) communication efforts to build momentum that sustains action and drives (lasting) impact

**When: Start Today**
The window for action is closing rapidly. Success will require an unwavering commitment and the dedication to motivate and inspire action at every level—through both Education and Communication. Only through genuine and focused collaboration across sectors, industries, and disciplines at the local, national, and global levels can we effectively mobilize the scale of change needed before that window closes for good. Success is our shared destination—and the time to act is now.

**Christina D'Asaro** is a strategic global communications leader with 20+ years of experience supporting business growth and global impact across integrated agencies, Fortune 500 companies, and the United Nations.

# Part IV

# Advancing Sustainable Development in Urban and Rural Areas

# Challenges for Sustainable Corporate Management in the Food Supply Chain

# 10

Reinhard Grandke

**Abstract**

This article analyzes and describes the special features and challenges of the food supply chain. The integration of a sustainability management system at individual stages, but above all throughout the entire food supply chain, can make a significant contribution to achieving the SDGs (UN Sustainable Development Goals, s. https://sdgs.un.org/goals).

One of the special features of the food supply chain is that it ultimately serves eight billion consumers whose daily food supply must be secured. Another special feature is the agricultural stage, which differs fundamentally from the upstream and downstream stages in terms of the organization and size of the farms. The special features of the agricultural stage are described in detail. Requirements for the management of companies in the food supply chain are derived from the analysis of the upstream and downstream stages and the agricultural stage. Concrete proposals are developed and described for the strategy process, organization, cultural development, research and development, and leadership. Finally, these are combined with the Three-Pillar Model to derive models for the transformation of companies.

## 10.1 Introduction

Sustainability is not a temporary trend, but a central guideline and responsibility for society, politics, and business in order to secure the future of our planet.

A key contribution to achieving sustainability goals is to make sustainability a defining, self-evident part of modern corporate management, thereby not only

---

R. Grandke (✉)
Dreieich, Hessen, Germany
e-mail: reinhard.grandke@upgrade-partner.de





P. Wollmann et al. (eds.), *The Sustainable Organization – Volume 2*, Future of Business and Finance, https://doi.org/10.1007/978-3-032-23802-3_10

contributing to the social and global goal of sustainability but also aligning companies and business models for success and sustainability.

In recent years, there have been several initiatives and agreements that make it mandatory to integrate sustainability in corporate strategies and define the frameworks for the future.

A major milestone was the adoption by the United Nations in 2015 of Agenda 2030,[1] a joint roadmap for "transforming the world for the better." At the heart of Agenda 2030 is a catalog of 17 sustainable development goals that take into account all three dimensions of sustainability—economic, environmental, and social. The Sustainable Development Goals (SDGs) call on politics, business, and society to pool their potential to address the challenges of our time (Fig. 10.1).

The Paris Climate Agreement,[2] which was adopted on December 12, 2015, at the World Climate Conference, was also in line with this agreement, 195 countries committed themselves to curbing and transforming the global economy in a climate-friendly way.

In the wake of these fundamental decisions and agreements, numerous initiatives have been launched at EU level, but also in individual countries.

**Fig. 10.1** UN SDG (source: https://www.un.org/sustainabledevelopment/), used with friendly permission of UN. The content of this publication has not been approved by the United Nations and does not reflect the views

[1] https://www.2030agenda.de/de/jahr/2015

[2] https://unfccc.int/process-and-meetings/the-paris-agreement

The European Green Deal[3] was a paradigm that, in its fundamental nature, initiated an almost disruptive change in economic policy at the EU level.

Legislation such as the Supply Chain Due Diligence Act[4] and the obligation for large companies to report on CSRD (Corporate Sustainability Reporting Directive) as part of their annual financial statements provide a legal framework for developments that commit companies to sustainable business practices. The focus is not only on the company itself, but on the entire supply chain in which the companies operate.

Following the European Parliament elections in 2025, the ambitious goals of the Green Deal have been scaled back, but the trend toward sustainability remains, and the introduction of binding sustainability strategies remains one of the key challenges for companies.

## 10.2 The Food Supply Chain

As shown in the simplified diagram in Fig. 10.2, the food supply chain is a regionally, nationally, and internationally networked value chain that meets the basic need for nutrition. The strategic orientation of the value chain varies depending on the continent, country, and region. However, the basic structure of the food supply chain, as shown in Fig. 10.2, is identical.

The food supply chain has a number of special features that fundamentally distinguish it from other supply chains. One key distinguishing feature is the number and organization of the players involved in the individual stages. What makes the food supply chain unique is that it ultimately serves over eight billion people as consumers.[5]

The upstream supplier stage is dominated by a small number of often internationally active companies, such as manufacturers of agricultural machinery, milking machines, and operating resources such as seeds, fertilizers, and pesticides.

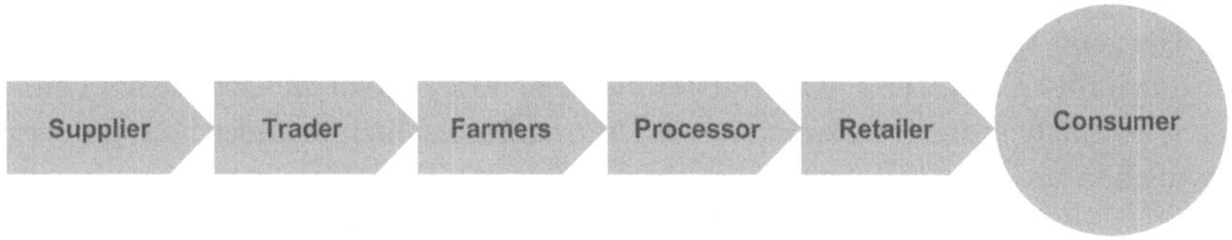


**Fig. 10.2** Simplified schematic representation of the food supply chain, drafted by the author. Used with permission, all rights reserved

[3] https://commission.europa.eu/strategy-and-policy/priorities-2019-2024/european-green-deal_en

[4] https://www.csr-in-deutschland.de/EN/Business-Human-Rights/Supply-Chain-Act/supply-chain-act.html

[5] See https://www.worldometers.info/world-population/

The trade stage prior to the agricultural stage is also dominated by a small number of players, but these are often regional in nature. Cooperatively organized trading houses are also key players in this stage. In the trade stage, taking Germany as an example, around 80% of the market is also dominated by four players (Aldi, Lidl, REWE, Edeka) or their subsidiaries.[6]

The processing stage is also characterized by concentration processes, with 80% of the quantities delivered in Germany, for example, being purchased by a small number of companies.

The agricultural stage is different. At this stage, there are many more players than at the other stages of the value chain. In a country like Germany, there are around 255,010 farms[7] in operation, which cultivate an average of approximately 65 ha of land.

This particular constellation gives rise to the special features of the food supply chain. While the players outside the agricultural stage are usually organized as national or global companies and have corresponding organizational structures, the agricultural stage is characterized by a different structure, often centered on the farm manager and his family.

The final stage of the food supply chain is also unique. This is where the consumer comes in. The food supply chain differs from other supply chains in terms of the number of consumers. Worldwide, there are over eight billion people at the end of the food supply chain who want and need to be fed every day from the production processes.

## 10.3 Food Supply Chain and Sustainability

The food supply chain plays an essential and significant role in many of the SDGs, as it ensures food security as a basis for life, but also represents a key factor influencing the economic development and prosperity of countries.

The agricultural sector is particularly important, as it plays a central role in food security, but also in shaping habitats, agriculture, resource consumption, and food production.

The food supply chain and its stakeholders have great potential for achieving sustainability goals. The FAO estimates that over 30% of total greenhouse gas emissions originate in the food supply chain.[8] Agriculture, which includes both primary production and land use change, is a large-emitter.[9]

---

[6] See: https://www.bundeskartellamt.de/SharedDocs/Interviews/DE/2016/Fuldarer_Zeitung__Die_Big_Four_haben_85_Prozent_Marktanteil.html

[7] German Federal Statistical Office, 2023.

[8] https://www.fao.org/statistics/highlights-archive/highlights-detail/greenhouse-gas-emissions-from-agrifood-systems.-global%2D%2Dregional-and-country-trends%2D%2D2000-2022/en

[9] Details see: FAOSTAT ANALYTICAL BRIEF 50, Greenhouse gas emissions from agrifood systems—Global, regional and country trends, 2000–2020 https://openknowledge.fao.org/server/api/core/bitstreams/121cc613-3d0f-431c-b083-cc2031dd8826/content

Management must not focus exclusively on its own level. Rather, it is important to develop management approaches that operate across organizational levels and boundaries and integrate them into strategies and organizational design. Sustainability management at the individual level, as well as contributing to sustainability management across the entire food supply chain, are central pillars of modern corporate organization and management.

Since companies operate at the regional, national, and international levels, this results in a high degree of complexity with multidimensional requirements. A special feature here is that the "agriculture" level of the food supply chain usually operates regionally (agriculture), while the other players operate at least nationally, if not globally.

If we take Germany as an example, the requirements are diverse. The regulatory framework is strict due to national and European requirements. In addition, German consumers are sensitive to quality and manufacturing processes, which is reinforced by the influence of powerful NGOs. At the same time, production processes are geared not only toward the domestic market but also toward exports, as shown in Fig. 10.3, and must therefore compete internationally.

This poses various challenges for companies operating in these markets. To meet these challenges, companies are increasingly focusing on holistic management of

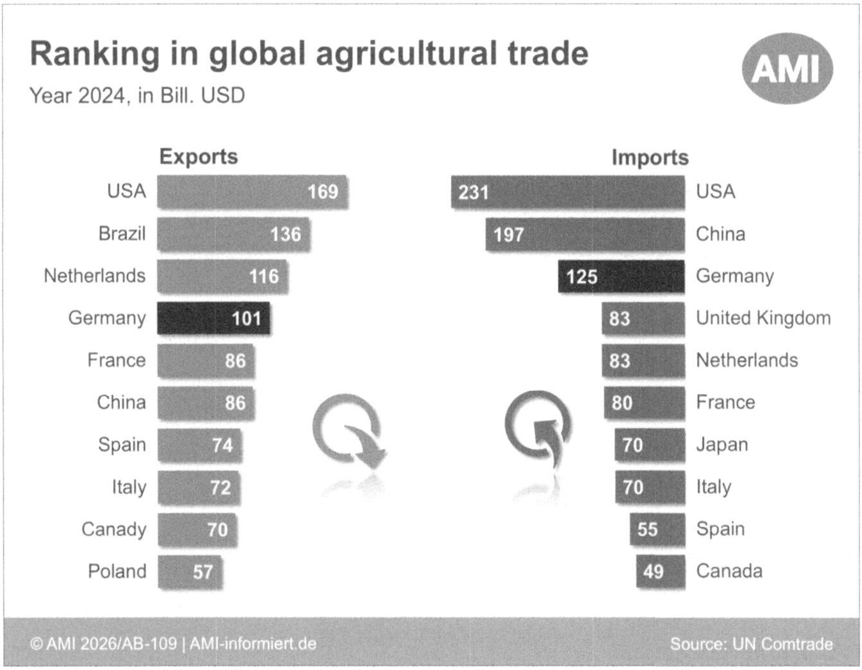


**Fig. 10.3** Ranking in global agricultural trade (source: AMI 2026/AB-109). Used with friendly permission of AMI, all rights reserved at AMI

the production process in order to comply with regulatory requirements, their own sustainability goals, and customer requirements.

One example of this is Arla, a large European dairy group that is pursuing an ambitious sustainability program. The main goal is to reduce $CO_2$ emissions.

The potential for savings at the company level (Scope 1 and 2) is limited, accounting for only 4% of total emissions. The greatest leverage lies with farm businesses, which account for 81% of emissions from milk production (Scope 3).[10]

Arla has therefore specifically extended its sustainability strategy to include farms and actively supports farmers in the management process. Around 200 data points are collected on the farms and included in the calculations.

These include not only information on milk production, but also data on overall farm management to develop coordinated measures to reduce emissions. This reduces the carbon footprint along the entire food supply chain and improves the sustainability of the end product, milk.

The research, definition of standards, procurement of measurement systems, interpretation of measurement data, and development of recommendations for action are developed by Arla and implemented in collaboration with farm managers on the farms.

Farms participating in the program benefit from a higher milk price, with a total of around 500 million euros per year being distributed to participating farmers.

## 10.4 Challenges for Sustainability Management in the Food Supply Chain

The increasing importance of sustainability management requires the players involved in the food supply chain to realign their strategies and, as a result, to establish new or adapted corporate structures.

These adjustments go beyond ordinary strategic developments and transformations, as there are often conflicts between sustainability goals and purely economically oriented corporate goals.

The changes within companies increasingly integrate ecological and social targets as components of corporate success. They must therefore be incorporated into strategic objectives and corporate culture, including that of suppliers and customers.

The challenges for sustainability management in the food supply chain include:

- *Aligning corporate strategy with sustainability goals*
  Sustainability is becoming increasingly important for players in the food supply chain. This issue is linked to an expansion of the strategic perspective beyond one's own value creation stage (see example of Arla). This requires a new,

[10] Source: Arla Sustainability Report 2023, p. 34 (https://www.arla.com/493575/globalassets/arla-global/company%2D%2D-overview/investor/annual-reports/2023/arla_annual-report-2023_uk_v2.pdf

expanded strategy process that is geared not only to the competitive advantages of the individual company, but also to those of the entire food supply chain.

- *Developing a common strategic understanding with other players in the food supply chain*
  There are often differing requirements for strategy within the food chain. Consumers demand greater consideration of environmental aspects in production but are not always willing to pay higher prices. This poses a particular challenge for upstream suppliers, who are required to produce more sustainable products despite conflicting objectives. In order to develop a common strategic understanding, it is necessary to develop scientifically sound criteria and integrate them into producer practices.
  At the same time, it is also important to promote a realistic understanding of sustainability at the individual stages of the food supply chain among buyers and consumers.
- *Deepening knowledge of the production processes of other actors in the food supply chain*
  Sustainability requires comprehensive knowledge of production processes along the entire value chain. On the one hand, this is necessary in order to understand the relevance of individual sustainability criteria at other stages and, on the other hand, to identify common approaches for action.
  For example, consumers are increasingly demanding process criteria such as husbandry, feeding, and animal health, which are being integrated into product labeling (e.g., pasture-raised milk, husbandry levels 1–4).
  Only through a common understanding among retailers, processors, producers, and consumers can credible and comprehensible sustainability indicators (KPIs) be developed and used.
- *Implementation of sustainability KPIs in the company's management systems*
  In order for sustainability criteria to become equal control variables, the recording systems and evaluation algorithms must be adapted to existing business management systems.
  The challenge lies in integrating the recording, evaluation, interpretation, and implementation of measures into company routines.
- *Implementation of sustainability criteria in the company's reporting obligations (Supply Chain Act, CSRD)*
  Sustainability reporting is becoming increasingly important. To this end, suitable indicators must be developed that both meet regulatory requirements and withstand critical scrutiny by consumers and NGOs.
- *Development of a communication concept on sustainability for the company*
  In order to integrate sustainability management in the company, regular reporting on sustainability KPIs should also be implemented, analogous to financial reporting.
  It must be ensured that this is not only project-related but is established as an integral part of company routines.
- *Integration of the sustainability strategy into product presentations, including positioning, marketing, and pricing*

Sustainability is increasingly becoming a competitive factor. Customers and consumers actively seek out sustainability information and incorporate it into their purchasing decisions.
It is therefore necessary to include the costs of data collection, marketing, and positioning in pricing in order to secure or increase value creation potential.

- *Development of pools of measures that include effective levers for achieving sustainability goals*
  While there are many standardized and proven measures in classic corporate management, this has not yet been sufficiently the case in the context of sustainability.
  Therefore, pools of measures, organizational preparations, and risk-oriented management with preventive and emergency measures must be developed.
- *Linking internal company KPIs with the KPIs of the food supply chain required by customers and consumers*
  Translating the sustainability criteria demanded by customers and buyers into requirements for upstream suppliers is a major challenge. A uniform definition and interpretation of KPIs is a basic prerequisite for efficient sustainability management along the entire food supply chain.
- *Implementing sustainability in corporate culture. Developing a common understanding with employees and stakeholders*
  Integrating sustainability in corporate culture is a continuous learning process. It goes beyond understanding one's own stage. All players in the food supply chain must develop a common understanding and must know and be aware of which people with which education, culture, and background are active in the food supply chain.
  The agricultural stage plays a special role here: The other players should understand how farms operate, while farmers should gain insights into the up- and down-stream stages.
- *Developing solutions that enable the agricultural stage to meet the requirements set by suppliers and customers*
  Since the CSRD reporting requirements primarily affect upstream and downstream players, agricultural businesses often do not have sufficient human and financial resources.
  The other players in the food supply chain must therefore provide support services, develop suitable tools, and offer training to enable coordinated implementation.
- *Design and implementation of an environmental screening for sustainability (regulations, market requirements, development of standards, etc.)*
  With the increased focus on the food supply chain, the strategic focus of the players is also shifting. In addition to monitoring their own stage, active networking with other players is necessary in order to identify market developments, regulatory changes, and new standards at an early stage.
  In addition, a monitoring system should be established to track consumers and NGOs in order to systematically identify trends and social expectations.
- *Anchoring sustainability management in the corporate organization*

Given the scope of sustainability management for companies, it is necessary to integrate management in the organization to establish data collection and evaluation in an equivalent and effective manner. To this end, it is necessary for the project reference or affiliation as a staff unit to be integrated into the "normal" structures.

- *Creation of exchange platforms with other players in the food supply chain*
  Data and its collection are the challenge for the food supply chain. Data systems and interfaces are of utmost importance here. Even within companies or within value chain stages, standards and interfaces are often insufficiently defined. The food supply chain requires systems that enable data exchange beyond its own stage.
- *Integration of sustainability requirements into contractual relationships or new forms of cooperation with suppliers and customers*

In order to effectively integrate the above challenges, structures and contractual relationships must be redefined and adapted.

## 10.5 Special Features of the Agriculture Stage in the Food Supply Chain

As already explained, the agriculture stage in the food supply chain differs fundamentally from the other stages and their actors. This poses challenges for actors in the agriculture stage, but also for players throughout the entire food supply chain. At the same time, this stage offers great potential for promoting environmental and biodiversity effects.

- *Heterogeneous farm structures*
  Agricultural businesses are highly diverse in terms of structure. There are numerous small, medium-sized, and large businesses that are managed in very different ways—for example, in arable farming, livestock farming, energy production, or direct marketing. This heterogeneity makes it difficult to standardize sustainability measures.
- *Lack of reporting requirements*
  To date, there has been no legal obligation for agricultural businesses to report on sustainability in their annual financial statements or sustainability reports. As a result, there is often a lack of standardized procedures for collecting and disclosing sustainability-related key figures.
- *Non-standardized data collection*
  There are no uniform, standardized processes for collecting operational sustainability data. Although the increasing use of digital technologies and sensors on machines is already generating extensive data sets, these are often not interconnected.
  There are currently numerous initiatives aimed at integrating this data and developing interoperable interfaces. However, it should be noted that the integration of

the various systems into a farm management system has not yet been fully completed, and no leading system has yet established itself on the market.

- *Limited resources and structures*
  Agricultural businesses often have limited human and financial resources and lack established operational structures to systematically develop and implement sustainability management systems.
- *Lack of training and qualifications*
  To date, there is no comprehensive, specialized education or training program focused on the introduction and organization of operational sustainability management. As a result, there is often a lack of methodological and organizational knowledge.
- *Low level of specialization compared to downstream stages*
  Unlike processing or trading companies, agricultural businesses are often less specialized. Many farms combine several branches of production, such as arable farming and milk production, while downstream players such as dairies or mills usually specialize in a single area of production. This diversity poses additional challenges for the standardization and comparability of sustainability indicators.

## 10.6 Approaches to Sustainability Management in the Food Supply Chain

The challenges identified in Sects. 10.4 and 10.5 give rise to various approaches for establishing effective sustainability management among individual actors, but also along the entire food supply chain.

Some key approaches are described below.

### 10.6.1 Strategy Process

The strategy process in the food supply chain needs to be expanded to include two key dimensions. The first concerns the integration of sustainability goals into the company's own strategy. The second involves the development of a comprehensive strategy process that goes beyond the company itself and is oriented toward the entire food value chain.

The individual players must integrate sustainability KPIs into their corporate strategy. This offers an opportunity to link classic strategy elements with CSRD instruments. The double materiality analysis carried out as part of sustainability reporting plays a central role here. By involving internal and external stakeholders, relevant indicators and fields of action can be identified.

This modified materiality analysis should be carried out in advance of the strategy process in order to create a sound basis for defining objectives. In a joint, moderated process (e.g., involving mediation-based procedures), diverging interests can be balanced and a common commitment to strategic sustainability goals can be

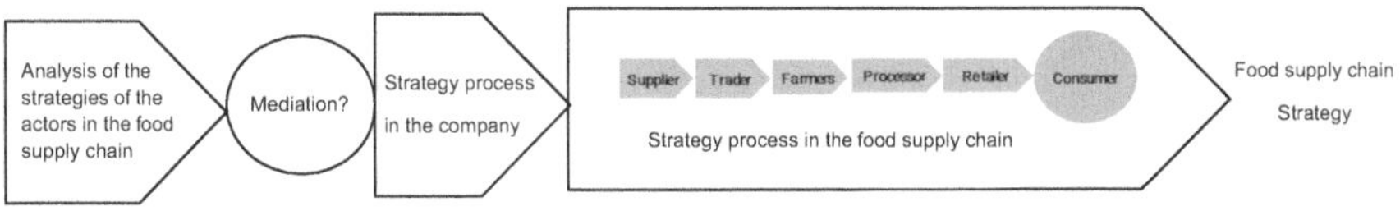


**Fig. 10.4** Extended strategy development process for the food supply chain, drafted by the author. Used with permission, all rights reserved

achieved. Separate strategy development processes are required for product groups with particularly differentiating sustainability characteristics.

The results are then incorporated into the company's internal strategy and, in the next step, integrated into the overall strategic process of the food supply chain. This is where the development and coordination of a comprehensive supply chain strategy take place.

This process is shown schematically in Fig. 10.4.

As explained in Sect. 10.5, the agricultural stage of the food supply chain in particular is structured fundamentally differently in organizational terms. Comparable strategic planning processes often do not exist there. In order to involve these players, suitable organizational forms must be developed, for example in the form of cooperatives or producer associations. In addition, new cooperation and contract models must be designed to enable greater integration into the overall strategy of the food supply chain.

### 10.6.2 Organization

Sustainability management must be integrated into the organization of the company. Companies differ fundamentally in terms of their organization. The food supply chain includes corporations, medium-sized companies, and agricultural enterprises.

In large corporations, sustainability management is usually already integrated in various areas of the organization due to regulatory requirements. In many medium-sized companies, however, the introduction of continuous sustainability management has been postponed until 2028 or later due to the postponement of CSRD reporting requirements and the EU's omnibus procedure. Nevertheless, there are already numerous initiatives at the company level that address sustainability aspects, such as process optimization, waste reduction, and the use of more environmentally friendly raw materials. However, these activities often take the form of projects and are not yet systematically integrated into the organization. Nevertheless, there is demand for data from corporations in the context of reporting requirements, and this demand must be met.

Table 10.1 lists typical organizational solutions for integrating sustainability management into companies.

No standard has yet emerged for how sustainability management is integrated in organizations.

**Table 10.1** Organizational approaches to sustainability management in companies, drafted by the author

| Driving force within the company | Organizational position in the company |
|---|---|
| Interest of individual promoters (often shareholders or C-level executives) | Project or staff unit |
| Regulatory requirements | Legal, investor relations, controlling |
| Customer requirements | Sales or production |
| Stakeholder requirements | Investor relations, communications |
| Marketing | Marketing |



However, the requirements profile is clear. Sustainability management must be integrated in the company on an equal footing with existing measurement and management systems. This argues in favor of integrating data collection in the controlling department. The measures should then be delivered to management in the same form as the financial key figures. Management then decides which measures are to be implemented.

Good and effective project management will be necessary, especially at the beginning, as the implementation of a data collection and evaluation system must take place in project form.

In the long term, it is advisable to set up a "food supply chain" functional area to coordinate management, controlling, and communication along the entire value chain. These results should then be integrated into "normal routines."

In agricultural companies, efforts are also being made at the corporate level to improve sustainability. Savings in fertilizers, pesticides, or diesel fuel contribute to sustainability, but are often driven more by economic factors. Some companies are involved in food supply chains and meet the sustainability requirements of suppliers or customers. Examples of this include organic farming, milk production (see the Arla example), and poultry meat production. In these integrated chains, however, the criteria and measuring instruments are usually implemented by one player on the supplier or customer side. Much of the internal data generated by the machines is used to optimize production. Programs that link and evaluate data and implement it in action plans on the farms are still in development.

However, it is foreseeable that developments will proceed "bottom up," as can be seen in Fig. 10.5.

### 10.6.3 Corporate- and Food Supply Chain Culture

A sustainability-oriented corporate culture means that values, behaviors, and decisions are permanently aligned with ecological, social, and economic goals—at all stages of the food supply chain. Sustainability must not only be integrated in

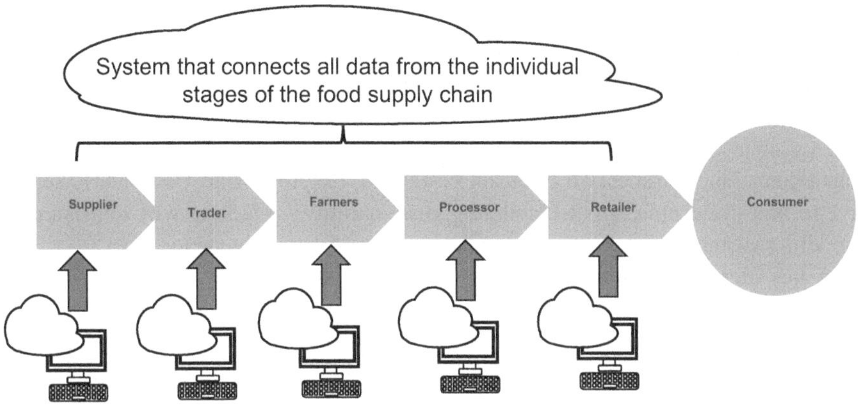


**Fig. 10.5** From systems at individual stages to a comprehensive digital system for the complete food supply chain. Drafted by the author. Used with permission, all rights reserve

guidelines but also lived out in the daily actions of all stakeholders—from upstream suppliers to consumers.

Key elements include transparency, cooperation, a willingness to learn, and openness toward all stakeholders. It is important to build trust and establish a culture across all stages. This must also be proactively promoted through measures.

The culture can also include a new understanding of stakeholders. In the food supply chain in particular, there are traditional dogmatic positions, often toward NGOs. In the culture of the food supply chain, goals that were previously mutually exclusive are now common goals that require a shared understanding and set old positions into question. This means that new partnerships must be formed, or at least an understanding of each other's positions must be developed, in order to successfully shape the food supply chain together. The first steps have been taken and are becoming clear when you look at the cooperation between NGOs such as animal welfare associations, the WWF, and organic associations with food retailers and/or discounters.

However, continuous work is necessary to ensure this. On the one hand, exchange platforms must be created to share information and explain background information.

Furthermore, a training program must be developed that explains sustainability in the food supply chain and promotes a common understanding.

Consideration should be given to whether, for example, joint trainee programs could be developed to give young people the opportunity to learn from players throughout the food supply chain and experience different cultures in practice.

Another tool could be "job rotation," not only within one's own company, but across the entire chain.

These measures help build culture, here as well, continuity is necessary to build lasting trust and create a shared culture.

### 10.6.4 Leadership

Forward-looking leadership in the food supply chain credibly combines economic success, social responsibility, and environmental sustainability. Managers set and live by clear sustainability goals, integrate them into the corporate strategy, and make them a measurable part of business success. Employees are particularly sensitive to and discerning about "modeling sustainability," which is why consistently modeling sustainability is also a success factor in "sustainable leadership."

A key element is authentically exemplifying sustainability, as employees perceive this behavior very closely. Sustainable leadership also means looking beyond one's own level and promoting cooperation on equal terms among all stakeholders. Long-term partnerships, open communication, and a common KPI basis are key success factors for effective leadership in the food supply chain.

### 10.6.5 Research and Development

Research into sustainability in the food supply chain focuses on reducing environmental impact, using resources more efficiently, and embedding social responsibility throughout the entire food supply chain. However, the field of management research should not be neglected either. Sustainability management is making the requirements and processes in companies more complex. Cooperation throughout the food supply chain is a complex challenge for which there are currently insufficient methods and procedures.

The application of supply chain legislation and the requirements for CSRD reporting require intensive research. In particular, the development of meaningful KPIs for sustainability management is of central importance. In addition, recording systems, the development of sensors, the measurement of KPIs, and the meaningful and effective linking of these are important research questions.

One focus will be on digitalization and data. The emphasis will be on data-driven decisions, real-time tracking, blockchain for traceability, and AI-based solution proposals.

## 10.7 Conclusion, Takeaways and Outlook

In the future, sustainability management systems will become the standard for companies and the value chains in which they operate. For the management of the companies involved, this means that tasks will become more complex, as they will relate not only to traditional targets but also to environmental and social ones.

Not only will there be a greater number of control variables to manage, but conflicts between the target variables will also have to be resolved.

Furthermore, traditional supplier/customer relationships are changed, and new partnerships and organizational models are emerging along the value chain, which must be incorporated into strategy and organization.

Data management will become increasingly important. The definition of standards and interfaces will be a key success factor. Overall, the development of sustainability management will accelerate the digitalization of industries.

However, sustainability is not just a corporate goal. Sustainability also means transferring behavior from the "private" to the "professional" sphere and vice versa. This makes it even more important to align the culture of companies and supply chains with this goal.

The food supply chain is certainly the most complex and extensive value chain. But right there the eight billion consumers are confronted with sustainability daily. The food supply chain brings business processes into society and must therefore also take social requirements into account in a special way. As a pioneer for many other supply chains and a development laboratory for methods and behavior, it has a role that can shape and accelerate the comprehensive introduction of sustainability management systems worldwide.

However, the food supply chain does not only have a special role as a pioneer in the introduction and implementation of sustainability management systems. It combines well with the Three-Pillar Model[11] in companies, as the dynamics with which the food supply chain is changing must also be reflected in companies. The pillars and the commitment of organizations to lasting change are a decisive competitive advantage.

Integrating sustainable benefits into the company's mindset is a basic prerequisite for thinking and acting across organizational and cultural boundaries within the supply chain.

When you consider the challenges facing companies, as outlined in Sect. 10.6, companies must embark on a journey, as both internal strategic understanding and an outward-looking organizational self-concept must be developed for employees, stakeholders, and the organization. The discussions about nutrition, the introduction and weighting of sustainability are diverse, and there are no standard solutions as yet. Openness to experience and learning from others are key success factors in generating corporate success from these approaches.

The third pillar of the "Three-Pillar Model,"[12] Connectivity, is a crucial one. Both effectiveness and efficiency are based on connectivity in a food supply chain system. This means Connectivity between individuals, between people, companies,

---

[11] The so-called Three-Pillar Thinking Model for organization design and leadership consists of the pillars *Sustainable Purpose* (the raison d'être of an organization, bringing orientation and certainty to the people for their joint endeavor and success), *Travelling Organization* (an organization accepting to be in a permanent state of flux and transformation) and *Connectivity* (the ability to interconnect on the journey with all needed). Details see: Wollmann, P.; Kühn, F.; Kempf, M. (Eds.): Three Pillars of Organization and Leadership in Disruptive Times—Navigating Your Company Successfully through the 21st Century Business World. Cham: Springer Nature, © 2020. See also Chap. 2 of this book or chapters 1 and 2 of the previous book: Peter Wollmann/Doris Pemler/Mersida Ndrevataj (eds.): The Sustainable Organization—How Organizations Address the 17 UN SDGs Using the 3-P-Model, Vol. 1, Cham: Springer Nature 2025.

[12] Ibid.

products, and processes. Without Connectivity, the complex food supply chain in particular cannot be managed effectively.

For optimized management of the food supply chain, it is necessary to develop innovative management systems and continue to adapt them during implementation. This is one of the success factors for the future.

**Reinhard Grandke** is an experienced consultant and sparring partner for companies in the areas of strategy, project management, and organizational development. He has extensive experience in corporate management (CEO) and as a member of supervisory and advisory boards with a focus on the agribusiness and food industry. He has been teaching at Justus Liebig University for 30 years as a lecturer, private lecturer, and professor of project, environmental, and sustainability management. He is an educated farmer and studied agricultural sciences. He earned his doctorate and habilitation on topics related to the implementation of innovations in organizational strategies.

# Sustainability as an Organizational Learning Practice: The Case Study of a Rural Area in Germany

11

Bianka Renn

**Abstract**

This article examines the institutional learning processes and governance adaptations that unfolded in the German Schleiden region (Eifel) after the 2021 flood, which acted as a catalyst for sustainability-oriented transformation. The theoretical framework links evolutionary resilience—the capacity to use crises as opportunities for institutional renewal—with polycentric governance, which explains adaptability through decentralized networks and institutional diversity. The 2021 flood event exposed the limits of hierarchical administrative structures and underscored the importance of locally embedded, interconnected actors for operational flexibility and effective crisis management. As a structural response and expression of institutional learning, the Gesellschaft für Wirtschaft, Tourismus und Veranstaltungen mbH, Schleiden (*GfW Schleiden*; in English: Company for Economy, Tourism, and Events, LLC) was established as a municipal development organization functioning as a learning-oriented node within the region's polycentric governance system. A complementary actor is the *DLG Eifel eG* (in English: Service Cooperative Eifel), a privately organized service cooperative that operates as a non-territorial, thematic arena for cross-sectoral knowledge and practice exchange. Together, GfW Schleiden and DLG Eifel eG (Service Cooperative Eifel) constitute complementary spaces of interorganizational learning that jointly support the emergence of an adaptive "travelling region." The findings show that resilience and sustainability are operationalized through connectivity, functional redundancy, and institutional openness, illustrating the gradual formation of a learning region whose developmental logic aligns with the concept of the Learning Organization and the so called Three-Pillar Model of the

B. Renn (✉)
Geschäftsführung, Gesellschaft für Wirtschaft, Tourismus und Veransatltungen mbH, Schleiden, Germany
e-mail: bianka.renn@schleiden-eifel.com





P. Wollmann et al. (eds.), *The Sustainable Organization – Volume 2*, Future of Business and Finance, https://doi.org/10.1007/978-3-032-23802-3_11

Sustainable Organization (known from Volume I), and whose postcrisis trajectories evolve over the long term and in nonlinear ways.

## 11.1 Introduction

The ability not only to respond to crises but also to derive institutional learning processes from them is increasingly regarded as a core element of sustainable regional development within European spatial development policy.[1] In the *Long-term Vision for Rural Areas*, resilience is understood as the ability of rural regions to diversify their economic structures, sustain social cohesion and quality of life, and operate through effective multi-level governance with empowered local actors. At the same time, resilience is framed as a forward-looking capacity that enables rural areas to anticipate and actively manage long-term ecological, digital, and demographic transformations. Within this understanding, the European Commission emphasizes that resilience ultimately depends on the capacity of local actors to learn institutionally from crises and to develop adaptive governance structures that ensure long-term sustainability.[2] Within this vision, multi-level governance and local action groups are identified as key mechanisms for implementing resilient rural development.

Against this background, current institutional developments in rural–urban areas can be understood as part of a broader European transformation logic: sustainability emerges where governance structures are designed to be capable of learning, networked, and adaptive. In parallel, recent organizational research—building on the concept of the Travelling Organization—conceptualizes sustainable institutions as dynamic, mobile systems that evolve by continuously transferring knowledge, adapting practices, and reconfiguring relationships across spatial, sectoral, and institutional boundaries.[3] This perspective becomes empirically visible in the case of Schleiden, where the 2021 flood acted as a catalyst for institutional learning and governance adaptation, leading to the establishment of networked organizational structures aimed at strengthening long-term regional resilience and sustainability.

The Gesellschaft für Wirtschaft, Tourismus und Veranstaltungen mbH, Schleiden (abbreviated from now on as: *GfW Schleiden*) can, in this context, be conceptually located as an institutional innovation following a crisis experience—an empirical case demonstrating the transformative potential of governance renewal within the

[1] European Commission: *European Spatial Development Perspective—Towards Balanced and Sustainable Development of the Territory of the European Union*, Luxembourg 1999.

[2] European Commission: *A long-term vision for the EU's rural areas—Towards stronger, connected, resilient and prosperous rural areas by 2040* (COM(2021) 345 final), Brussels 2021.

[3] The Travelling Organization is the crucial pillar of the so-called Three-Pillar Model developed in Wollmann, P.; Kühn, F.; Kempf, M. (Eds.): Three Pillars of Organization and Leadership in Disruptive Times—Navigating Your Company Successfully through the 21st Century Business World. Cham: Springer Nature, © 2020. More details see in footnotes 18 and 19 and in Chap. 2 of the book.

rural–urban fabric, consistent with both the European understanding of resilience and the notion of sustainability as an ongoing, movement-oriented learning process.

## 11.2 Theoretical Framework for Urban–Rural Sustainability

### 11.2.1 Concept and Paradigms of Urban–Rural Sustainability

The long-established dichotomy between "urban" and "rural" in classical regional and social geography is now considered outdated. Contemporary spatial research approaches understand cities and the countryside as functionally interconnected systems whose development is shaped by mutual dependencies—an urban–rural continuum encompassing social, economic, ecological, and institutional dimensions.[4]

According to the OECD, urban–rural linkages refer to the flows of people, goods, capital, information and ecosystem services between urban and rural areas, which collectively shape regional resilience and well-being.[5] UN-Habitat similarly describes them as dynamic two-way interactions through which both urban and rural areas contribute to sustainable development across the entire territory.[6]

These interconnections determine the resilience and sustainability of regional systems by enabling the continuous exchange of resources, knowledge, and services. The denser and more reciprocal these relationships are, the greater a region's ability to respond to and adapt following economic, ecological, or social shocks. In this regard, the OECD emphasizes that regions with strong, diversified, and reciprocal urban–rural linkages demonstrate higher adaptability and faster recovery in postcrisis contexts.[7]

Urban–rural linkages are not limited to interregional relations but also operate intensively within regions—for instance, between urban centers and their surrounding municipalities. However, the relevance of dense and reciprocal intra-regional interdependencies remains comparatively underexamined, even though such localized networks often prove particularly resilient during crises, throughout their management, and in the development impulses that follow. This perspective resonates with the argument that rural regions are best understood as relational spaces, shaped by multi-scalar flows and interdependencies rather than as isolated territorial units.[8] Their coordination requires polycentric governance structures that are based on cooperation, institutional diversity, and shared responsibility.

[4] OECD: *Rural well-being: Geography of opportunities*, OECD Rural Studies, Paris: OECD Publishing 2020, p. 162.

[5] OECD: *Rural well-being: Geography of opportunities*, Paris: OECD 2020, p. 41.

[6] UN-Habitat: *Urban–rural linkages: Guiding principles. Framework for action to advance integrated territorial development*, Nairobi 2019, p. 8.

[7] OECD: *Rural well-being: Geography of opportunities*, Paris 2020, pp. 44–45.

[8] Michael Woods: Rural, London/New York: Routledge 2021.

In a region such as Schleiden and the Eifel National Park area, where municipalities, economic actors, cooperatives, and civil society organizations are closely interconnected, the strength of these intra-regional exchange and cooperation networks forms the foundation of sustainable development. Here, resilience emerges not only through external connections to larger centers but especially through the mutual empowerment of local actors within a shared, polycentric structure.

Thus, the approach of urban–rural sustainability is expanded here by an intra-regional dimension: sustainability arises not only from external linkages between urban and rural areas but equally from the internal quality of cooperation, knowledge flows, and institutional learning.

### 11.2.2 Governance and Resilience in the Urban–Rural Context

#### 11.2.2.1 Resilience as a Governance Task: From Stabilization to Learning Orientation

In spatial science debates, resilience is understood as the processual dimension of sustainability—particularly relevant for regions such as Schleiden, which operate within the interrelation of urban, rural, and climate risks.[9] While sustainability aims at maintaining the balance among ecological, social, and economic systems,[10] resilience refers to their capacity to respond, adapt, and evolve in the face of disturbances. In this sense, resilience becomes a prerequisite for sustainable development, especially in functionally interconnected urban–rural regions characterized by complex dependencies and uncertainties.[11]

In the urban–rural context, the concept of resilience has evolved from the idea of merely restoring a pre-crisis state to that of evolutionary resilience.[12] It refers to the adaptive and transformative capacity of regions to use crises as opportunities for institutional and organizational learning. Davoudi describes resilience as a continuous process of reorientation and reflection, in which systems do not simply stabilize aftershocks but proactively reconfigure and adjust to changing conditions: Drawing on evolutionary approaches, resilience is understood not as a simple return to a

[9] Thomas Elmqvist et al.: Urban Planet: Knowledge towards Sustainable Cities, Cambridge: Cambridge University Press 2019; Carl Folke et al.: “Resilience Thinking: Integrating Resilience, Adaptability and Transformability”, in: Ecology and Society, Vol. 15, No. 4 (2010), Art. 20; Charles L. Redman: “Should Sustainability and Resilience Be Combined or Remain Distinct Pursuits?”, in: Ecology and Society, Vol. 19, No. 2 (2014), Art. 37.

[10] World Commission on Environment and Development (WCED): Our Common Future, Oxford: Oxford University Press 1987; United Nations: Transforming Our World: The 2030 Agenda for Sustainable Development, New York 2015.

[11] OECD: Rural Well-being: Geography of Opportunities, Paris: OECD Publishing 2020.

[12] Simin Davoudi: “Resilience: A Bridging Concept or a Dead End?”, in: Planning Theory & Practice, Vol. 13, No. 2 (2012), pp. 299–307.

pre-crisis state, but as a process that emphasizes adaptation, learning and the reorganization of institutions and practices in response to change.[13]

From this perspective, resilience becomes a core governance responsibility. It requires institutional openness, cooperative learning processes, and a willingness to question established structures. Regions that interpret resilience as a learning process thereby create the foundation for understanding sustainability not as a fixed state but as a development-oriented practice.

In Schleiden, the 2021 flood became a catalyst for such learning and adaptation processes. The GfW Schleiden—as a municipally owned company tasked with economic development, tourism, and public events—can be understood as an institutional response to this crisis. Its orientation reflects the recognition that these responsibilities can only be fulfilled effectively within networked and learning-oriented structures. In this sense, GfW Schleiden embodies key aspects of evolutionary resilience: resilience as the conscious advancement of institutions and practices that view change not as a threat but as a condition for sustainable development.[14]

#### 11.2.2.2 Polycentricity and Adaptive Capacity as Structural Principles of Regional Resilience

Building on the concept of evolutionary resilience as a learning- and transformation-oriented understanding of resilience, the question arises as to which institutional architectures are capable of supporting such learning processes. In this respect, Davoudi provides a theoretical foundation. Regional resilience is not conceptualized as a static economic indicator, but in terms of the adaptive capacity of institutional networks. This capacity is reflected in the ability to anticipate crises, to exchange knowledge across sectoral boundaries, and to act in ways that go beyond mere reactive responses. The analytical focus shifts to procedural and relational mechanisms, namely how effectively a region organizes collective learning and institutional adaptation. Polycentric and networked governance structures—linking municipalities, economic actors, civil society and research institutions—tend to enhance adaptive capacity, as resilience ultimately depends on the ability of institutions to continuously reconfigure themselves.[15]

Institutional diversity and redundancy are critical determinants of resilience due to research on adaptive systems. Polycentric systems create alternative courses of action that enhance stability in times of crisis and foster knowledge exchange across different institutional levels. Such structures enable continuous institutional learning and strengthen the capacity not only to absorb shocks but to use them productively for transformation. From this perspective, redundancy is not a sign of

[13] Simin Davoudi: "Resilience: A Bridging Concept or a Dead End?", in: Planning Theory & Practice, Vol. 13, No. 2 (2012), p. 330.

[14] Davoudi 2012, pp. 302–305.

[15] Davoudi, 2012, pp. 299–307.

inefficiency but an expression of a learning- and adaptation-oriented system in which overlap provides robustness and safeguards against systemic failure.[16]

In the context of Schleiden after the 2021 flood, these theoretical approaches provide a fitting analytical framework. The disaster revealed that highly hierarchical administrative systems may be limited in their responsiveness and learning capacity, whereas locally networked actors often display greater operational flexibility and stronger capacities to mobilize resources and knowledge.

Within this context, the GfW Schleiden is positioned as a pivotal node in a polycentric network that connects municipalities, businesses, civil society, and cooperative structures. Its mission aligns with the principles of polycentric adaptive capacity: diversity, cooperation, and shared responsibility as governance foundations that strengthen sustainable regional resilience without treating it as a static achievement. Through its cooperative and learning-oriented structure, GfW Schleiden contributes to embedding resilience in practice—conceived as a system of shared responsibility, mutual adaptability, and institutional openness. The effectiveness of this orientation must, however, be assessed over time and especially under conditions of crisis.

#### 11.2.2.3 Interim Conclusion: Sustainability as a Learning-Oriented and Mobile Governance Practice

The theoretical approaches of Urban–Rural Sustainability, Evolutionary Resilience, and Polycentric Governance can be understood as complementary perspectives on sustainable regional development. Collectively, they emphasize that sustainability should not be conceived as a static end state but as a continuous process of institutional learning and adaptation.

Within the urban–rural continuum, sustainable structures emerge where actors across multiple levels—municipal, economic, civil society, and scientific—collaborate in cooperative and reflective ways. Resilience functions here as the operational dimension and driving engine of sustainability: it describes the ability to use crises as opportunities for learning and for further developing institutional arrangements.[17]

Polycentrically organized governance systems foster such learning processes by institutionalizing diversity, redundancy, and communication across decision-making levels. In these mechanisms lies the foundation of adaptive capacity—the ability to respond to uncertainty not through rigidity, but through creative adaptation. This processual and relational perspective connects with the concept of the Travelling Organization, first introduced in the context of the Three-Pillar Model[18] and detailed

[16] Low, B., Ostrom, E., Dietz, T., & Beam, J. (2003): Redundancy and Diversity: Do they Influence Optimal Management? In: Berkes, F./Colding, J./Folke, C.: Navigating Social-Ecological Systems. Cambridge University Press, S. 83–114.

[17] Simin Davoudi: "Resilience: A Bridging Concept or a Dead End?", in: *Planning Theory & Practice*, Vol. 13, No. 2 (2012), pp. 299–307.

[18] The so-called Three-Pillar Thinking Model for organization design and leadership consists of the pillars *Sustainable Purpose* (the raison d'être of an organization, bringing orientation and certainty to the people for their joint endeavor and success), *Travelling Organization* (an organization accepting to be in a permanent state of flux and transformation) and *Connectivity* (the ability to

as a value-based system for creating and sustaining social and economic value in the context of first volume of this book series: The Sustainable Organization Volume 1.[19]

From this perspective, sustainable organizations are not static entities but learning and moving systems that transfer and transform knowledge, values, and practices across institutional, sectoral, and spatial boundaries. The Pillar Model of Sustainability is thus understood not only as a structural but also as a dynamic steering model: sustainability is "on the move"—it emerges through an organization's ability to continuously reposition itself within changing contexts, to learn continuously, and to renew cooperation.[20]

Against this backdrop, the present contribution conceptualizes sustainable regional development as a learning- and movement-oriented governance practice: a dynamic interplay of actor networks, institutional diversity, and ongoing reflection that enables transformation and continually regenerates sustainability. This understanding forms the theoretical framework for the following case study of the GfW Schleiden, which exemplifies how organizational learning and adaptation processes can be institutionally embedded in an urban–rural context.

## 11.3 Methodology and Case Study Design

This study applies a qualitative–interpretive research approach aimed at understanding institutional dynamics in the context of sustainable regional development. It conceptually links sustainability as a normative orientation with resilience as an operational capacity: sustainability provides the normative long-term direction, while resilience captures the adaptive ability to respond to disturbances, engage in institutional learning, and reconfigure organizational structures.[21]

The research follows an embedded single case study design, focusing on the region of Schleiden as an empirical setting of institutional learning.[22] The case was selected for two interrelated reasons: first, the 2021 flood as an external shock that triggered extensive adaptation processes; and second, the establishment of the GfW Schleiden as a municipal company integrating economic development, tourism, and event-related functions. This constellation enables an in-depth analysis of the interplay between crisis experience, governance structures, and sustainable regional development.

---

interconnect on the journey with all needed). Details see: Wollmann, P.; Kühn, F.; Kempf, M. (Eds.): Three Pillars of Organization and Leadership in Disruptive Times—Navigating Your Company Successfully through the 21st Century Business World. Cham: Springer Nature, © 2020.

[19] Peter Wollmann/Doris Pemler/Mersida Ndrevataj (eds.): The Sustainable Organization—How Organizations Address the 17 UN SDGs Using the 3-P-Model, Vol. 1, Cham: Springer Nature 2025.

[20] Ibid.

[21] Carl Folke et al.: "Resilience Thinking: Integrating Resilience, Adaptability and Transformability", in: *Ecology and Society*, Vol. 15, No. 4 (2010), Art. 20; Simin Davoudi: "Resilience: A Bridging Concept or a Dead End?", in: *Planning Theory & Practice*, Vol. 13, No. 2 (2012), pp. 299–307.

[22] Robert K. Yin: *Case Study Research and Applications: Design and Methods*, 6th ed., Los Angeles et al.: SAGE 2018.

The author's insider position—as a participant in key decision-making processes—provides contextual depth and insight into institutional logics under crisis conditions. In line with reflexive case study methodology, this proximity is treated analytically by consciously balancing engagement with critical distance.[23]

Empirical material comprises publicly accessible and institutionally verifiable documents, including city council resolutions (Allris system), the articles of association of GfW Schleiden, and selected documents from DLG Eifel eG. These sources are complemented by the author's systematic observations during the organizational build-up phase since 2022. Data analysis followed a theory-guided pattern-matching approach, informed by the concepts of evolutionary resilience and polycentric governance.

The objective of the study is not statistical generalization, but theory-based contextual understanding. It seeks to deepen insight into how sustainable development is practically enacted within a real governance system, with particular attention to decision logics, learning dynamics, and institutional adaptability.

## 11.4 Case Study

### 11.4.1 Spatial and Institutional Context

#### 11.4.1.1 Geographic and Administrative Setting

The city of Schleiden is located in the southwest of North Rhine-Westphalia, within the district of Euskirchen, and forms part of the Eifel region, which extends across three countries and two German federal states: North Rhine-Westphalia and Rhineland-Palatinate in Germany, the German-speaking Community of Belgium and Luxembourg. This cross-border position makes the Eifel a transnational rural area in which political, administrative, and cultural responsibilities overlap.[24]

The region is part of the Regional Economic Development Programme of North Rhine-Westphalia, which designates Schleiden as a structurally weak rural area with particular development needs.[25] Administratively, Schleiden belongs to the Cologne governmental district and functions as a medium-sized regional center, providing key services for several surrounding municipalities.[26,27]

[23] Mats Alvesson/Kaj Sköldberg: *Reflexive Methodology: New Vistas for Qualitative Research*, 3rd ed., London et al.: SAGE 2018.

[24] EUREGIO Maas-Rhein: *Cross-border Cooperation in the Eifel–Ardennes Region*, Maastricht 2023, online: https://www.euregio-mr.eu

[25] Ministerium für Wirtschaft, Industrie, Klimaschutz und Energie des Landes Nordrhein-Westfalen (MWIKE NRW): *Regionales Wirtschaftsförderungsprogramm NRW (RWP)*, Düsseldorf 2023.

[26] Information und Technik Nordrhein-Westfalen (IT.NRW): *Kommunalprofil Stadt Schleiden* [Municipal profile of the City of Schleiden], Düsseldorf 2024, online: https://www.it.nrw

[27] Landesbetrieb Wald und Holz Nordrhein-Westfalen: *Nationalpark Eifel—Aufgaben und Management* [Eifel National Park—Tasks and management], Düsseldorf 2023.

Schleiden hosts the administrative headquarters of the Eifel National Park, a state institution established in 2004 that serves both as a protected area and as a development space for sustainable tourism, environmental education, and ecological awareness.[27] In addition, the Nature Park Nordeifel[28] e. V. and several LEADER regions contribute to sustainable regional development by promoting cross-border cooperation and initiating EU-funded projects.[29]

A key actor in this regional governance landscape is the DLG Eifel eG (Service Cooperative Eifel), a service-oriented cooperative that operates as a privately organized regional cooperation network.[30] Within this multilayered institutional framework, the GfW Schleiden is closely connected with public, economic, and civil society actors—an embodiment of polycentric governance.[31]

This institutional complexity shapes the preconditions for sustainable development in Schleiden and the wider region. Sustainability here does not emerge through hierarchical control, but through cooperation, coordination, and institutional learning across administrative and sectoral boundaries. In this context, GfW Schleiden acts as an institutional node within a polycentric network of urban–rural interrelations. In doing so, it translates the theoretical correlation between institutional thickness and regional resilience into the practice of post-disaster governance—a learning—and adaptation-oriented form of steering that evolved in response to the 2021 flood.

#### 11.4.1.2 Structural Characteristics and Development Trends

The Eifel region belongs to Germany's rural and peripheral areas and is characterized by low population density, small-scale economic structures, and the topographical features of a low mountain range. For the city of Schleiden and the southern district of Euskirchen, these conditions create specific development dynamics that combine opportunities with structural challenges.[32]

With a population of around 13,000 inhabitants, Schleiden has a relatively low population density of approximately 110 residents per square kilometer.[33] As in many rural areas of North Rhine-Westphalia, demographic change is evident in an

---

[28] North Eifel Nature Park.

[29] Naturpark Nordeifel e. V.: *Über uns* [About us], Schleiden 2023, online: https://www.naturpark-nordeifel.de; European Commission: *LEADER Programme for Rural Development*, Brussels 2023, online: https://agriculture.ec.europa.eu

[30] DLG Eifel eG: *Unsere Mitglieder und Projekte* [Our members and projects], Schleiden 2023, online: https://www.dlg-eifel.de

[31] Elinor Ostrom: "Polycentric systems for coping with collective action and global environmental change", in: Global Environmental Change, Vol. 20, No. 4 (2010), S. 550–557.

[32] Bundesinstitut für Bau-, Stadt- und Raumforschung (BBSR): Raumbeobachtung Deutschland: Indikatoren zur nachhaltigen *Entwicklung ländlicher Räume*, Bonn 2022.

[33] Information und Technik Nordrhein-Westfalen (IT.NRW): *Kommunalprofil Schleiden, Stadt—Kreis Euskirchen*, Düsseldorf 2024, online: https://statistik.nrw/sites/default/files/municipalprofiles/l05366036.pdf

ageing population and a gradual decline in the working-age population.[34] These trends affect the labor market, the provision of public services, and the local capacity for innovation—key dimensions of sustainable regional development.[35]

The regional economy is predominantly characterized by small and medium-sized enterprises, with a strong orientation towards tourism, crafts, and public services. Large-scale industrial enterprises are largely absent. At the same time, digitalization and sustainability-oriented transformation are gaining importance within local businesses.

Overall, the Eifel region is classified by the State of North Rhine-Westphalia as structurally weak.[36] This structural condition increases its vulnerability to economic and demographic shifts, yet also opens up potential for sustainable, place-based development strategies.

As a medium-sized regional center, Schleiden provides essential services for surrounding municipalities. The public sector and tourism play a particularly significant role, reinforced by the city's function as the administrative seat of the Eifel National Park and as a destination offering numerous regionally and nationally relevant visitor attractions.[37]

Despite recent progress, the region continues to face infrastructural, digital, and workforce-related challenges. At the same time, there is growing awareness of sustainability as a guiding principle for regional development—one that combines ecological responsibility with economic and social viability. From this awareness emerge key conditions for sustainable regional development: institutional cooperation, innovation-oriented adaptability, and the integration of ecological, social, and economic objectives within a learning-oriented governance structure.

#### 11.4.1.3 Institutional Networks and Governance Structure

Sustainable regional development in the Eifel region is based on a dense network of institutional actors whose responsibilities and tasks frequently overlap. Contrary to the common assumption that such overlaps lead to fragmentation, they can in fact have productive effects, as they promote cooperation, knowledge exchange, and shared learning processes. Particularly in polycentrically organized regions, these functional redundancies enhance both the adaptability and the agency of the regional system.[38]

At the state level, the Regional Plan Cologne provides the overarching framework that spatially and functionally integrates the Eifel into North Rhine-Westphalia's

---

[34] Ministerium für Wirtschaft, Industrie, Klimaschutz und Energie (MWIKE NRW): *Regionales Wirtschaftsförderungsprogramm*, 2023.

[35] BBSR 2022.

[36] MWIKE NRW 2023.

[37] Landesbetrieb Wald und Holz Nordrhein-Westfalen: *Nationalpark Eifel: Jahresbericht 2023*, Düsseldorf 2023.

[38] Low, B., Ostrom, E., Dietz, T., & Beam, J. (2003): *Redundancy and Diversity: Do they Influence Optimal Management?* In: Berkes, F. et al.: *Navigating Social-Ecological Systems*, Cambridge University Press, p. 83–114.

regional development policy.[39] Within this framework, the districts of Euskirchen and Düren, as well as the City Region of Aachen, hold key responsibilities in economic development, tourism, and spatial planning.

In the field of tourism, multiple levels of cooperation exist. The Eifel Tourismus (ET) GmbH functions as a transregional organization for the entire Eifel, while Nordeifel Tourismus GmbH operates as a municipally supported company within the district of Euskirchen.[40] Programs such as LEADER and the Nature Park Nordeifel e. V. complement these structures by fostering participatory, project-based approaches to regional development.[41]

An important component—and in many respects a key building block—of the regional governance landscape is the DLG Eifel eG (Service Cooperative Eifel). As a cooperative network, it is rooted not in a formal institutional mandate but in the intrinsic motivation and voluntary commitment of its member businesses and associated institutions. Its activities are structured around five strategic pillars: health, human resource management, digitalization, diversity and equal opportunity, and knowledge and competence. Through this thematic breadth, the cooperative connects businesses, municipalities, educational institutions, and civil society actors, fostering practice-oriented collaboration that supports sustainable economic development.[42]

The GfW Schleiden is closely embedded in this network. As a municipally owned company responsible for economic development, tourism development, cultural development, and event management, it functions as an institutional hub within the region's polycentric governance structure. In cooperation with partners such as DLG Eifel eG, it develops projects and formats that strengthen knowledge transfer, collaboration, and regional value creation.

The institutional landscape of the Eifel region thus illustrates that sustainable regional development does not emerge through hierarchical control, but through functional overlaps, joint learning processes, and a culture of shared responsibility. Schleiden and the wider Eifel region exemplify a polycentric governance structure in which the theoretical concepts of networking, diversity, and resilience take practical shape.

#### 11.4.1.4 Implications and Conclusion

The Eifel region, and within it the city of Schleiden, exemplifies both the institutional complexity and the learning capacity of urban–rural spaces. Its sustainable

---

[39] Bezirksregierung Köln: *Regionalplan Köln*, Cologne 2023, online: https://www.bezreg-koeln.nrw.de

[40] Eifel Tourismus (ET) GmbH: Über uns—Die Organisation [About us—the organisation], n.d., retrieved 2025 from https://www.eifel.info; Nordeifel Tourismus GmbH: Tourismus in der Nordeifel [Tourism in the Northern Eifel], Kall 2025, online: https://www.nordeifel-tourismus.de

[41] European Commission: *LEADER—Community-led Local Development*, Brussels 2023, online: https://agriculture.ec.europa.eu/common-agricultural-policy/rural-development/leader_en; Naturpark Nordeifel e. V.: *Naturpark Nordeifel*, Schleiden 2023, online: https://www.naturpark-nordeifel.de

[42] DLG Eifel eG: *Über uns* [About us], Schleiden 2023, online: https://www.dlg-eifel.de

development does not rely on hierarchical control but on connectivity—the ability to interlink actors, levels, and sectors in ways that enable the sharing of knowledge, responsibility, and resources. These functional overlaps prove productive, as they generate the redundancies that, in line with the concept of polycentric governance, enhance the region's flexibility and resilience.[43]

Within this interconnected network, the GfW Schleiden plays an active role grounded in its self-conception as a service-oriented and cooperative institution. Rather than acting as an administrative instrument, the GfW positions itself as part of a broader system and operates—together with other local and regional stakeholders—as a node in a polycentric governance structure. This stance reflects a strategic choice to conceptualize economic and social development as a shared responsibility between the municipality, local businesses, and civic partners.

Its collaboration with DLG Eifel eG, educational institutions, and both municipal and civil society actors illustrates the close interrelation between resilience and sustainable development.[44] Sustainability is thus not perceived as a fixed condition, but as a learning-oriented practice in which cooperation, knowledge exchange, and institutional openness constitute key success factors.

Viewed through the lens of the Pillar Model of the Sustainable Organization, the Eifel region—and equally the DLG Eifel eG as a non-geographical but thematically integrated space of cooperation—can be interpreted as a travelling region: a constellation in continuous transformation, where sustainability emerges from collective learning processes across organizational and territorial boundaries.[45] Although not tied to a specific municipal territory, the DLG provides an interorganizational arena in which knowledge, practices, and experiences circulate beyond local administrative limits. This demonstrates that sustainable development does not evolve from actors thinking within their own spatial confines, but from their ability to embed themselves in wider cooperative networks and mobilize diverse resources and perspectives across sectors and regions.

In this perspective, the GfW Schleiden and the DLG Eifel eG can be understood as complementary arenas of interorganizational learning that jointly shape the travelling region. While the GfW operates within a defined municipal framework and anchors development processes locally, it simultaneously extends these processes outward by connecting with regional partners. The DLG, in turn, opens a non-territorial cooperative space in which businesses, municipalities, and institutions collaborate across administrative boundaries. Together, they constitute mutually

[43] Elinor Ostrom: *Beyond Markets and States: Polycentric Governance of Complex Economic Systems*. In: *American Economic Review*, 100(3) (2010), 641–672.

[44] Simin Davoudi: "Resilience: A Bridging Concept or a Dead End?", in: *Planning Theory & Practice*, Vol. 13, No. 2 (2012), pp. 299–333; Carl Folke et al.: "Resilience Thinking: Integrating Resilience, Adaptability and Transformability", in: *Ecology and Society*, Vol. 15, No. 4 (2010), Art. 20.

[45] See: Wollmann, P.: *A Recap of the Three-Pillar Model and its Importance for the Book's Objectives* in: Wollmann P., Pemler D.; Ndrevataj, M (eds.): *The Sustainable Organization—How Organizations Address the 17 UN SDGs Using the 3-P-Model*, Cham: Springer Nature, 2025

reinforcing interfaces through which knowledge, practices, and competencies circulate.

Both organizations thus contribute to a governance arrangement that transcends spatial delimitations: the GfW by strengthening intra-regional cohesion and operational coordination, and the DLG by fostering cross-regional exchange and sectoral diversity. Their shared commitment to cooperation, learning, and adaptive development positions them not as isolated actors, but as interdependent components of a broader polycentric learning system—each enhancing the other's capacity to act.

Thus, the theoretical arc of this case study comes full circle: sustainable regional development in Schleiden and the wider Eifel region embodies a travelling form of governance—a dynamic, adaptive system that draws its resilience and innovative strength from cooperation, diversity, mutual embeddedness, and shared responsibility.

## 11.4.2 Institutional Learning and Adaptive Governance After the 2021 Flood

### 11.4.2.1 Contextualizing the 2021 Flood as a Trigger for Institutional Learning

In July 2021, the low-pressure system *Bernd* brought extraordinary heavy rainfall to large parts of western Germany. Within 2 days, more than 150 mm of rain fell across the district of Euskirchen and adjacent regions.[46] In the City of Schleiden, whose settlement structure is shaped by the narrow valleys of the Olef and Urft rivers, the flooding caused severe destruction. Residential areas, roads, and utility infrastructure were damaged or destroyed, and communication and energy supplies were partially disrupted. According to the investigative committee of the State Parliament of North Rhine-Westphalia, nine people lost their lives within the city area.[47] Official flood and heavy rainfall hazard maps issued by the State of North Rhine-Westphalia show that most affected zones largely corresponded to previously identified risk areas, although in several cases the flooding extended significantly beyond them.[48]

The flood acted as an external shock for Schleiden and the wider Eifel region, exposing the limitations of existing administrative structures. The scale and complexity of the situation overwhelmed hierarchically organized crisis management mechanisms at both state and municipal levels, which were not designed for rapid

[46] Deutscher Wetterdienst: Analyse der Starkregen- und Hochwasserlage Juli 2021, Offenbach am Main 2021/2022, online: https://www.dwd.de

[47] Landtag Nordrhein-Westfalen: *Bericht des Parlamentarischen Untersuchungsausschusses IV (Drucksache 18/9500)*, Düsseldorf 2024.

[48] Ministerium für Umwelt, Naturschutz und Verkehr des Landes Nordrhein-Westfalen: *Hochwasser- und Starkregengefahrenkarten für Nordrhein-Westfalen*, n.d., retrieved 2025 from https://www.flussgebiete.nrw.de

adaptation or resource mobilization.[49] At the same time, local and informal networks emerged among public administration, businesses, relief organizations, and civil society that responded more quickly and flexibly than formal structures. These spontaneous collaborations—through volunteer initiatives, ad hoc coordination, and civic engagement—proved to be crucial resilience elements for social and organizational stabilization.

In the sense of evolutionary resilience, the flood can be interpreted as a catalyst for transformation. The disaster not only initiated reconstruction processes but also triggered institutional reflection and reorganization. It opened a window of opportunity for rethinking local governance arrangements, roles, and communication pathways.[50] During this period, the need became evident to foster stronger connections between administration, the local economy, and the community.

This crisis thus marked the beginning of an institutional learning process, in which local actors experienced trust, cooperation, and mutual coordination as central resources. Out of initially improvised collaboration emerged new forms of coordination, which gradually evolved into more permanent institutional structures. This development illustrates the core idea of evolutionary resilience: resilience is less about bouncing back and more about bouncing forward—rethinking, reorganizing and reshaping institutions, practices, and mindsets.[51]

The 2021 flood demonstrated that resilience does not stem from stability, but from adaptability, learning capacity, and connectedness—qualities that now form the foundation of future-oriented and sustainable regional development.

At the same time, the developments, observable in the aftermath of the flood, reveal both the opportunities and the challenges inherent in such a transformation. The reconstruction of physical infra-structure and the parallel evolution of governance structures unfold not as linear processes, but as part of a long-term trajectory characteristic of a travelling region—marked by continuous adaptation, reconfiguration, and learning. Whether the region will fully succeed in harnessing this momentum, embedding the emerging practices, and sustaining the collaborative dynamics that surfaced during the crisis remains an open question. It is a process in motion, and only time will show to what extent the window of opportunity has been translated into lasting institutional and regional change.

#### 11.4.2.2 Governance and Institutional Responsiveness

The management of the 2021 flood disaster posed an exceptional test for the governance structures in Schleiden and the Eifel region. The crisis required coordination across multiple administrative levels—from municipal and district authorities to state and federal institutions. Evaluations of the disaster response in North Rhine-Westphalia show that institutional procedures were largely hierarchical and based

[49] Arjen Boin/Paul 't Hart/Eric Stern/Bengt Sundelius: *The Politics of Crisis Management: Public Leadership under Pressure*, 2nd ed., Cambridge: Cambridge University Press 2016.

[50] Boin et al. 2016, p. 103.

[51] Simin Davoudi: "Resilience: A Bridging Concept or a Dead End?", in: *Planning Theory & Practice*, Vol. 13, No. 2 (2012), p. 302.

on linear information flows.[52] During the acute phase, this structure proved only partially adaptive: communication channels between governance levels were disrupted, responsibilities were ambiguously distributed, and operational coordination had to be reorganized ad hoc in several instances.

At the same time, local actors—particularly municipal administrations, volunteer organizations, and businesses—demonstrated a strong capacity to form informal cooperation networks that remained functional beyond formal jurisdictions. These networks bridged institutional boundaries and enabled rapid decision-making based on mutual coordination and local knowledge. This empirical observation corresponds closely with what polycentric governance theory identifies as a defining feature of resilient systems: the ability to organize decision-making and learning processes in a decentralized manner and to use institutional redundancy productively.[53]

In the Eifel region, resilience during crisis response thus did not primarily arise from central control, but from local self-organization. Citizens, local businesses, and associations-initiated relief efforts, organized logistics and material flows, and maintained communication structures. These decentralized response mechanisms complemented formal emergency systems and contributed significantly to the stabilization of the overall situation. They can be interpreted as an expression of emergent governance—a dynamic interplay between formal and informal mechanisms that enhances the adaptability of complex regional systems.[54]

As the immediate crisis subsided, these experiences were increasingly reflected upon and translated into institutional learning processes. At the municipal level, a growing awareness emerged of the need to consolidate cooperation with local actors on a more permanent basis. The establishment of the GfW Schleiden in 2022 can be understood as a structural response to this realization. The organization assumed responsibilities for economic and tourism development as well as for coordinating municipal events. By choosing the legal form of a limited liability company (GmbH), the municipality intentionally departed from the constraints of classical hierarchical cameralism. This organizational shift provides the GfW with the operational flexibility and budgetary agility necessary to react to dynamic regional challenges. Its structure and mode of operation embody a learning-oriented and adaptive governance approach that integrates flexibility, cooperation, and reflexivity into local administrative practice.[55]

---

[52] Landtag Nordrhein-Westfalen: *Abschlussbericht des Parlamentarischen Untersuchungsausschusses IV (Drucksache 18/9500)*, Düsseldorf 2024; Ministerium des Innern des Landes Nordrhein-Westfalen (IM NRW): *Bericht zur Aufarbeitung der Hochwasserkatastrophe (Bericht an den Innenausschuss)*, Düsseldorf 2022.

[53] Elinor Ostrom: *Beyond Markets and States: Polycentric Governance of Complex Economic Systems*. In: *American Economic Review*, 100(3) (2010), 641–672.

[54] Carl Folke et al.: "Resilience Thinking: Integrating Resilience, Adaptability and Transformability", in: *Ecology and Society*, Vol. 15, No. 4 (2010), Art. 20.

[55] Stadt Schleiden/Gesellschaft für Wirtschaft, Tourismus und Veranstaltungen mbH Schleiden: *Aufgabenprofil der GfW Schleiden*, n.d., retrieved 2025 from https://www.schleiden.de

The analysis of the response phase yields two central insights. First, hierarchical governance structures tend to reach their limits during acute crises when flexibility and local agency are required. Second, polycentrically networked systems—characterized by shared knowledge, responsibility, and resources—exhibit a higher degree of adaptability and resilience. These insights form the foundation for the institutional learning processes that followed, marking the transition from crisis management toward more sustainable and adaptive governance arrangements.

#### 11.4.2.3 Learning Process and Institutional Transformation

The experiences from the acute phase of the 2021 flood in Schleiden were gradually transformed into organizational and institutional learning. At the operational level, it became evident that improvised cooperation, short communication pathways, and shared responsibility significantly increased the region's capacity to act. These insights led to concrete adaptations in administrative routines, including improved coordination between departments, the early involvement of local stakeholders in decision-making, and the systematic use of existing networks for project implementation and knowledge exchange. In line with the concept of evolutionary resilience, this transition marks the shift from short-term crisis response toward ongoing learning and adaptation processes.[56]

As an immediate organizational outcome, the city administration of Schleiden established a Task Force for Exceptional Events (auf Deutsch: Stab für außergewöhnliche Ereignisse, SAE), which serves as an administrative interface with operational emergency services and complements the crisis units at district, state, and federal levels.[57] This unit institutionalizes crisis response mechanisms and enables coordinated action in cases of flooding, energy disruption, or extreme weather. It formalizes the principle of rapid internal coordination that had proven effective during the flood and integrates it into the city's administrative structure.

In parallel, the city joined the Intermunicipal Flood Protection Cooperation Erft (hwsErft) and contributes to catchment area planning for the Urft and Olef rivers, coordinated by the Erftverband.[58] These initiatives aim to coordinate preventive measures, risk analyses, and technical protection systems across municipal boundaries. They represent a structural response to the lessons learned during the flood, namely the need to integrate water management, spatial planning, and disaster prevention. In doing so, they reflect the development of a learning and adaptive regional system.[59]

---

[56] Simin Davoudi: "Resilience: A Bridging Concept or a Dead End?", in: *Planning Theory & Practice*, Vol. 13, No. 2 (2012), p. 302.

[57] Stadt Schleiden: *Stab für außergewöhnliche Ereignisse (SAE)*, Schleiden 2023, online: https://www.schleiden.de. Or https://www.schleiden.de/rathaus/buergerservice/brand-und-katastrophenschutz/sae-stab-fuer-aussergewoehnliche-ereignisse/sae-organisationsuebersicht.pdf?cid=1eby

[58] Erftverband: Hochwasserschutz an Urft und Olef—Hydrologische und hydraulische Grundlagenarbeiten, Bergheim 2023; Kreis Euskirchen: Interkommunale Hochwasserschutzkooperation Erft (hwsErft)—Projektbeschreibung und Bilanz, Euskirchen 2024.

[59] Carl Folke et al.: "Resilience Thinking: Integrating Resilience, Adaptability and Transformability", in: Ecology and Society, Vol. 15, No. 4 (2010), Art. 20.

This learning trajectory culminated institutionally in the establishment of the GfW Schleiden, registered on 27 December 2022 (HRB 9242, District Court of Düren) and operational since 1 January 2023. The GfW Schleiden consolidates economic and tourism development as well as event management as a municipally owned company and operates, among other functions, the Tourist Information Office in Gemünd.[60] This structural decision exemplifies adaptive governance: responsibilities are organized in a way that facilitates connectivity, project capacity, and continuous knowledge transfer between the municipality, businesses, educational institutions, and civil society.

From a methodological perspective, this process can be interpreted as an event-induced reconfiguration: an evolution from informal, network-based crisis practices toward formalized cooperation and implementation structures.[61] As result, the focus shifted from restoring isolated functions to institutionalizing learning-oriented arrangements—a core element of regional adaptability and transformative capacity.

#### 11.4.2.4 Conceptual Positioning: From Evolutionary Resilience to the Learning Region

The developments in Schleiden and the Eifel region following the 2021 flood illustrate a transition from evolutionary resilience toward the concept of a learning region. While evolutionary resilience emphasizes adaptation and institutional renewal in response to crisis events, the learning region perspective extends this logic by embedding learning and reflexivity into ongoing regional governance processes. In this view, resilience becomes not merely a reaction to disruption but a continuous mode of organizing, in which institutions evolve through iterative cycles of reflection, cooperation, and experimentation.[62]

In Schleiden, this shift becomes visible through the institutionalization of learning mechanisms. The establishment of the Task Force for Exceptional Events (SAE), participation in intermunicipal flood protection cooperation, and the creation of the GfW Schleiden collectively demonstrate a movement from ad hoc coordination toward structured and adaptive governance arrangements. These configurations function as cross-sectoral learning infrastructures that bridge the gaps between public administration, technical water management, and the local economy. They enable the transformation of crisis experiences into durable

[60] Stadt Schleiden/Gesellschaft für Wirtschaft, Tourismus und Veranstaltungen mbH Schleiden: *Über uns—GfW Schleiden (seit 01.01.2023)*, n.d., retrieved 2025 from https://www.schleiden-eifel.de

[61] Robert K. Yin: *Case Study Research and Applications: Design and Methods*, 6th ed., Los Angeles et al.: SAGE 2018.

[62] Simin Davoudi: "Resilience: A Bridging Concept or a Dead End?", in: *Planning Theory & Practice*, Vol. 13, No. 2 (2012), p. 302; Carl Folke et al.: "Resilience Thinking: Integrating Resilience, Adaptability and Transformability", in: *Ecology and Society*, Vol. 15, No. 4 (2010), Art. 20.

organizational and institutional capacities, ensuring that knowledge is not siloed but shared across various governance levels.[63]

The learning region thus emerges as a polycentric system that thrives on connectivity, redundancy, and mutual trust. Knowledge is not centralized but circulates among actors, enabling co-creation, shared responsibility, and continuous adjustment. In this sense, Schleiden and the Eifel region exemplify a regional governance model in which resilience and sustainability converge through institutional learning—a defining feature of regions that are both adaptive and forward-looking.[64]

## 11.5 Conclusion

The case of Schleiden demonstrates that sustainable regional transformation can emerge from crisis when institutions are capable of learning, adapting, and reconfiguring themselves. The 2021 flood functioned as a critical juncture, exposing the limitations of hierarchical governance while simultaneously enabling new forms of cross-sectorial collaboration, coordination, and collective problem-solving.[65]

Through the institutionalization of these emergent practices—most notably in the establishment of the Task Force for Exceptional Events and the GfW Schleiden—the city has begun to operationalize the principles of adaptive and polycentric governance. This trajectory reflects an evolutionary resilience pathway, in which crises are transformed into opportunities for reflection, structural innovation, and institutional renewal.[66]

Conceptually, Schleiden and the wider Eifel region illustrate how resilience—as the operational dimension of sustainability—can evolve into a learning region: a governance system that sustains itself through connectivity, diversity, and continuous learning. This transformation underlines that the foundation of sustainable development lies not in stability, but in the capacity for adaptation, cooperation, and shared meaning-making. In this sense, the regional evolution described here directly embodies the concept of the "travelling organization" as defined by Wollmann et al.

---

[63] Stadt Schleiden: *Stab für außergewöhnliche Ereignisse (SAE)*, Schleiden 2023; Erftverband: *Hochwasserschutz an Urft und Olef—Hydrologische und hydraulische Grundlagenarbeiten*, Bergheim 2023; Kreis Euskirchen: *Interkommunale Hochwasserschutzkooperation Erft (hwsErft)*, Euskirchen 2024; Stadt Schleiden/Gesellschaft für Wirtschaft, Tourismus und Veranstaltungen mbH Schleiden: *Über uns—GfW Schleiden (seit 01.01.2023)*, n.d., retrieved 2025 from https://www.schleiden-eifel.de

[64] Elinor Ostrom: *Beyond Markets and States: Polycentric Governance of Complex Economic Systems*. In: *American Economic Review*, 100(3) (2010), 641–672.

[65] Arjen Boin/Paul 't Hart/Eric Stern/Bengt Sundelius: *The Politics of Crisis Management: Public Leadership under Pressure*, 2nd ed., Cambridge: Cambridge University Press 2016.

[66] Simin Davoudi: "Resilience: A Bridging Concept or a Dead End?", in: *Planning Theory & Practice*, Vol. 13, No. 2 (2012), pp. 299–307; Carl Folke et al.: "Resilience Thinking: Integrating Resilience, Adaptability and Transformability", in: *Ecology and Society*, Vol. 15, No. 4 (2010), Art. 20.

(2025).[67] The Eifel region has transitioned into a "travelling region"—a dynamic system that does not seek a static equilibrium but maintains its viability through the continuous, reflexive movement of its institutional arrangements and the proactive reconfiguration of its governance networks. Networks. These qualities are likely to define the resilient and sustainable governance arrangements of the future.

**Bianka Renn** has long-standing experience in international higher education, brand management, and strategic governance in academic and territorial contexts. For almost two decades, she has been engaged in teaching, curriculum development, and applied research at international higher education institutions, predominantly in Italy.

Her professional focus lies in strategic management, brand and organizational development, organizational resilience, sustainable transformation, crisis and change management. Alongside her academic activities, she is involved in practice-oriented projects at the interface of business, education, and regional development.

Since 2018, she has been active in the economic development of rural regions. She serves as Managing Director of the Gesellschaft für Wirtschaft, Tourismus und Veranstaltungen mbH, Schleiden (GfW Schleiden), a municipally owned development company focused on resilient spatial and sustainable regional development and is an honorary board member of DLG Eifel eG (Service Cooperative Eifel)—a regional cooperative working on learning-oriented governance, institutional change, and cross-sector collaboration.

[67] Wollmann, P.; Pemler, D.; Ndrevataj, M. (eds.): The Sustainable Organization—How Organizations Address the 17 UN SDGs Using the 3-P-Model (Volume I), Cham: Springer Nature 2025 and in Chaps. 1 and 2 of this book.

# Addressing Sustainability and UN SDG Coverage in Rural Areas of India: A Case Study of Bunker Roy's Endeavors to Develop Rural Regions and Its People

12

Dagmar von Tschurtschenthaler

**Abstract**

Every movement begins with a special person who dares to see the world differently. For Bunker Roy, that moment came when he left a life of privilege to live and work alongside the poor in rural India. What he discovered became a revolution in how we think about progress, knowledge, and empowerment, especially in rural areas. Through The Barefoot College (Social Work and Research Centre), he proved that true innovation doesn't come from power or privilege, it begins where courage meets community. It is an encouraging use case. This is what we all celebrate: people who turn bold ideas into lasting change and remind us that the future is built by those brave enough to begin.

## 12.1 Introduction

After the description of the Use Case of *Friends for Hope e.V., Munich,*[1] in the previous book[2] it might make sense to take this time the broader perspective of a use case of rural development in India in general and introduce in more detail one of the most impressive actors in this area, Bunker Roy who was already mentioned in the first volume. The use case presented in this chapter was developed in conversations

[1] See: https://www.friends-for-hope.de

[2] Thirolf, T.; von Tschurtschenthaler, D.: Friends for Hope—A Special SDG Coverage Project in India, in: Wollmann, P., Pemler, D., Ndrevataj, M.: The Sustainable Organization—How Organizations Address the 17 UN SDGs Using the 3-P-Model; Cham 2025, SpringerNature (https://link.springer.com/book/10.1007/978-3-031-89549-4).

D. von Tschurtschenthaler (✉)
Munich, Bavaria, Germany



P. Wollmann et al. (eds.), *The Sustainable Organization – Volume 2*, Future of Business and Finance, https://doi.org/10.1007/978-3-032-23802-3_12

with him in September 2025, the respective documentation by the author represents a kind of legacy with important messages.

### 12.1.1 Introduction of Bunker Roy: His Vita

*Sanjit "Bunker" Roy* is an Indian social activist and educator who founded the Barefoot College. He was born in 1945 in Burnpur, Asansol, and had a privileged, elite upbringing. He went to the Doon School from 1956 to 1961 and attended St. Stephen's College, Delhi from 1962 to 1967.

But a visit to Bihar during the famine in 1966 affected him deeply and made him question the privilege he enjoyed. This led him to reject a prospective career in the private sector to work in rural India.

Bunker began working on water issues in the drought-prone region, digging wells for the Catholic Relief Services and Rewards from 1967 to 1971. After moving to Tilonia, in 1972 Bunker set up the Social Work and Research Centre, and their work expanded from water and irrigation to include broader issues such as empowerment and livelihoods. The Social Work and Research Centre is built on Gandhian principles of practicality, local indigenous knowledge and self-sufficiency.[3]

He is a founder of what is now called Barefoot College. After conducting a survey of water sources in 100 drought-prone villages, Roy established the Social Work and Research Centre in 1972. Its mission soon changed from a focus on water and irrigation to empowerment and sustainability. The programs focused on installing water hand pumps near villages and training the local population to maintain them without dependence on outside mechanics, providing training as paramedics for local medical treatment, and on solar power to decrease dependence and time spent on kerosene lighting.

He was recognized in 2010 in *Time* for the programs of the college which have trained more than three million people in skills including solar engineers such as Solar Mamas, teachers, midwives, weavers, architects, and doctors. He was selected as one of Time 100's 100 most influential personalities in 2010 for his work in educating illiterate and semiliterate rural Indians. Roy was awarded the Padma Shri by Giani Zail Singh in 1986. Since the beginning in 1972 Barefoot College was running a big amount of Night Schools for children who had to work during the days. He married ex-IAS Aruna Roy in 1970.

Though his health has slowed him down, his spirit remains unstoppable. When I met him recently, despite his age and illness, I found him as passionate as ever—asking about my work for Tilonia education.

Bunker received numerous awards, such as (selection):

- 1985: "Jamnalal Bajaj Award" for Application of Science and Technology for Rural Development[4]

[3] For details see for example: https://vajiramandravi.com/current-affairs/gandhian-principles/

[4] https://en.wikipedia.org/wiki/Jamnalal_Bajaj_Award

- 2003: Won The 2003 "St Andrews Prize for the Environment"[5]
- 2003: One of 20 people to be selected as "Social Entrepreneurs of the Year" by Schwab Foundation for Social Entrepreneurship[6]
- 2009: Received a "Robert Hill Award" for his contribution to promotion of photo-voltaics (solar energy)[7]
- In 1998, he was awarded the Indira Gandhi Paryavaran Puraskar (Indira Gandhi Environment Award), by the Ministry of Environment and Forests, Government of India[8]
- In 2003, the Barefoot College won an Ashden Award for its work bringing solar power to rural villages[9]
- In 2013, Bunker Roy received a "Clinton Global Citizen Award," along with young activist Malala Yousufzai[10] and Major Bloomberg

### 12.1.2 Bunker Roy: Basics of His Work Vision

There is no urban solution to a rural problem—Bunker Roy

The impact of Barefoot College goes well beyond the thousands of rural women from across the world who have been trained as solar engineers to include innovation in water, energy, livelihoods, education, and sustainability.

Bunker said the first lesson he learned was never dependent on professionals from outside, urban professionals from outside. Always develop the capacity and competence of people from within the organization first because they are the ones who will be there to stay forever and ever. That means one has to develop the grass-root leadership to carry the organizations. Or in other words: the success of a community as well as of an organization like an enterprise depend on its key strategic capabilities which cannot be delegated to third parties. Applying this key insight to rural communities makes a significant difference in thinking styles and acting.

The second thing Bunker learnt was that there was significant a difference between literacy ("the capability of identifying, reading, writing, understanding, and using the written word to communicate effectively and participate in society") and education ("the overall development of an individual through gaining knowledge and skills, habits, and values")[11] which is crucial for acting in the rural areas (Fig. 12.1).

---

[5] https://www.st-andrews.ac.uk/st-andrews-prize/previous-winners/

[6] https://www.thehindu.com/todays-paper/tp-national/tp-tamilnadu/swiss-award-for-bunker-roy/article27870788.ece

[7] https://www.thehindu.com/todays-paper/tp-national/tp-otherstates/Global-honour-for-barefoot-wonder-Bunker-Roy/article16523579.ece

[8] See https://en.wikipedia.org/wiki/Barefoot_College, section awards and achievements.

[9] Ibid.

[10] Ibid.

[11] Cited from https://www.21kschool.com/in/blog/difference-between-education-and-literacy/

**Fig. 12.1** Bunker Roy. All rights reserved by Bunker, used with permission

### 12.1.3 Bunker on the UN SDGs and the Three-Pillar Model

What have we achieved—and what do we urgently need to do differently now? Bunker Roy

The 17 UN SDGs (see Fig. 12.2) form the master plan for the development of the world by 2030 to make it a significantly better place in many perspectives. The history of the ambitious underlying transformation program was described in more detail in chapters 1 and 3 of the previous book[12] and again in Chap. 1 of this book.

Bunker has a clear opinion on the UN SDGs and their progress so far:

1. *Progress? Yes. But far too slowly in our opinion from the perspective of our work in rural India.*

*Some areas are showing positive developments—for example, access to education, the fight against child mortality, or the expansion of renewable energies.*

*But many goals are seriously at risk:*

   (a) *Poverty is increasing again in many regions—partly due to wars, partly due to climate extremes.*
   (b) *Inequality is growing—between states, but also within societies.*
   (c) *Climate targets are quickly being missed across the board—despite promises and technologies.*

*That means: the 2030 Agenda isn't a sure-fire success. And nice goals and conferences alone won't change the world at the end. Bunker Roy underlines that from his perspective the respective purpose doesn't change—only the path evolves. Sustainability also consists of individual phases and continues to evolve.*

2. *From paper to practice: It's not a lack of knowledge—it's a lack of will.*

[12] Wollmann, P.; Pemler, D.; Ndrevataj, M. (Eds.): The Sustainable Organization—How Organizations Address the 17 UN SDGs Using the 3-P-Model. Cham Springer Nature, © 2025; (https://link.springer.com/book/10.1007/978-3-031-89549-4).

**Fig. 12.2** UN SDG (source: https://www.un.org/sustainabledevelopment/ ), used with friendly permission of UN. The content of this publication has not been approved by the United Nations and does not reflect the views

*The SDGs are ambitious. But they often remain stuck in strategy papers—instead of translating into concrete, locally anchored action.*

3. *Hope? Absolutely. But not naive.*

*There are courageous initiatives, innovative alliances, and committed people around the world who are bringing the SDGs to life:*

*In municipalities, companies, schools, research, and civil society.*

The author of the chapter would like to add that for India's rural areas SDG6 "Clean Water and Sanitation" and SDG 4 "Education for All" are central in the UN SDGs.

For the needed journey to significantly develop rural India and finally overcome poverty and health, education and connected issues the Three-Pillar Model (3-P Model)[13] (see in more detail Chap. 2) in terms of Sustainable Purpose (a), Travelling Organization (b), and Connectivity (c)—at least implicitly—also plays a key role for the activities in rural areas. The Sustainable Purpose Bunker Civil's work to develop rural areas in India relates to nearly all UN SDGs—with a focus on SDG 1 (No Poverty), SDG 4 "Education for All," and SDG6 "Clean Water and Sanitation." The metaphor of a Travelling Organization means in rural India going very flexibly on an adaptive journey partly through unknown territories always including all parties and stakeholders.

[13] Wollmann et al. (Eds.): Three Pillars of Organization and Leadership in Disruptive Times—Navigating Your Company Successfully through the 21st Century Business World. Cham: Springer Nature, © 2020.

The 3-P Model is designed for organization and leadership in disruptive times. Its focus is on helping the reader and activists to implement the model and providing a wide variety of application cases for these VUCA times (Volatility, Uncertainty, Complexity and Ambiguity), including global crises like the COVID-19 pandemic. The previous books[14] cover a broad range of organizations: private and public sector, civil society, local and global governmental institutions, global organizations such as UN, etc.

The understanding of the metaphoric concept of a Travelling Organization in the rural areas of India has to be connected to the tradition and values of Mahatma Gandhi. This means that it is particularly important that all implementations on the ground are deeply rooted in the consciousness of the local people and are conceived from their perspective. The local village communities, teachers, parents, and children must always be involved to form a successful movement with the necessary mutual respect, engagement and participation, flexibility and fast adaptability. And the third pillar, Connectivity, is crucial for the needed collaboration and cooperation with all involved parties which are needed for joint success.

### 12.1.4 Framing of the Conversation on the Use Case and the Questions Asked

The key questions in the conversation with Bunker are on the Agenda 2030—Where do we really stand? In 2015, the United Nations made a historic promise with the 2030 Agenda. Today in 2025, we are now more than halfway through the planned timeframe. But what is the true state of implementation of the Sustainable Development Goals (SDGs)? What have we achieved—and what do we urgently need to do differently now?

We still have 5 years until 2030. We don't need any further explanations. Sustainability requires more than goals—it requires attitude, courage, and action. And what do the answer on these questions mean for rural development in India in detail?

## 12.2 The Starting Point: The Trigger for Commitment

India is the world's largest country in terms of population, which is why the success of the world's sustainability transformation and the coverage of the UN SDGs will be largely decided there—and especially in the not yet well-developed rural areas. The need for action and the challenges are correspondingly huge.

So, it is interesting to have a closer look at Bunker's original intention when he started his projects in India's rural areas out? What was his initial focus?

[14] See full list under: https://link.springer.com/search?query=&content-type=Book&dateFrom=&dateTo=&contributor=%22Peter+Wollmann%22&sortBy=relevance

In the conversation, Bunker stressed he was not motivated by any UN resolution on sustainable development and the goals that were predicted. Instead, the state of poverty and availability of water and the information collected from 500 villages in Silora Block Ajmer district was a full importance and relevance to study the impact of grassroot programs on the quality of lives to that extend what the Barefoot College has achieved is worthy of mention and replication as a model for civil society in the south.

The author can well understand what triggered Bunker's commitment; Friends for Hope was founded in a similar way. When coordinating activities to cover sustainability and the UN SDGs, it could be important to pay more attention to what has already been created independently of the UN SDG regime and to consider how it can be integrated without dominating it. And this proceeding is especially important in a player like India with its special importance for the world as a whole.

## 12.3 The General Proceeding in Rural India

In the conversation the proceeding in detail was touched. Did Bunker define a bundle of specific projects as a top-down priority from the outset within a comprehensive concept in place from the beginning or did he rather take an urgent need as a bottom-up starting point for an initial project and develop the comprehensive holistic concept over time and as the project pursued progressed?

The answer is that there was and is no question that the rural poor from these 500 villages were consulted and priorities established. Work is the first urgent need to get money from government programs to be able to bring food to the table. Second access to portable drinking water through traditional methods like rainwater harvesting construction of tanks for percolation with a view to recharging traditional water sources.

## 12.4 The Key Outcomes

Asking Bunker what the key outcomes of his work were in his perception so far leads to a clear answer: It is on the one hand the quality and availability of portable water leading to self-sufficiency for drinking and agriculture purposes. Second is the use of technology for turning fluoride water from 3000 PPM to 500 PPM using RO plants on a large-scale, especially around the salt lake of Sambhar in Rajasthan, there is no choice but to use RO technology on the large-scale providing water to over 25,000 people. Let's have a closer look at the background:

- In large parts of India, groundwater naturally contains high levels of fluoride and poses a health risk. In rural areas of India, over 90% of groundwater sources are naturally contaminated with varying amounts of fluoride.[15]

[15] See also https://www.eawag.ch/de/info/portal/aktuelles/news/wo-sich-in-indien-zu-viel-fluorid-im-grundwasser-befindet/;https://fluoridealert.org/de/news/how-fluorosis-is-crippling-india-and-needs-urgent-attention/

- Reverse osmosis (RO) is a physical separation process used to concentrate dissolved substances from liquids. It is based on reversing the principle of osmosis, which occurs in cellular metabolism. Reverse osmosis is frequently used for water desalination.

At the beginning Barefoot College's work concentrated on water, clean drinking water. And therefore, on installing hand pumps. This was the main way to support the communities.

In 1985 the ground water level started seriously decreasing, the dip came with a severe drought in 1986–1997. Hand pumps were not installed after that.

On the questions:

- If the Barefoot College did initially only install hand pumps, or did they also build rainwater tanks?
- When did their focus shift?
- Does BC (Barefoot College) still build hand pumps now?

The answers are:

- In 2012, the Barefoot College became the first NGO partner with UNESCO's Global Partnership for Girls' and Women's Education.[16]
- More details see below.

**Why Rainwater Harvesting?**

The Barefoot College did not want to rely longer on wells and unpredictable groundwater levels to provide potable water to hundreds of thousands of people in need.

Of all the water solutions that the College has tried and tested, rainwater harvesting has been the most sustainable and effective. It is a sustainable solution to the extreme shortage of clean drinking water sources and at the same time a low-cost method with maximum benefits. Rainwater harvesting has been one of the technologies that the College has implemented in partnership with local communities as part of its holistic community-based approach to development.

It provides clean, potable drinking water for villages, irrigates fields and sustains livestock—all important criteria for rural communities that depend on agriculture and animal husbandry for their livelihoods.

Rains are often undependable and ground water has unhealthy levels of arsenic or fluoride, and in many areas is too saline to drink. Rainwater—as a direct source of potable water when filtered—can be collected from the rooftops of schools and stored in underground storage tanks. Communities contribute to constructing the system.

Rooftop rainwater harvesting is a traditional approach used for hundreds of years in India to collect rainwater and store it to provide clean water for daily needs in these arid regions. Roofs of schools and community buildings are adapted to catch infrequent rainwater and channel it to underground tanks. Traditional knowledge,

[16] *"India's Barefoot College and UNESCO join forces for Girls' and Women's Empowerment". United Nations Educational, Scientific and Cultural Organization. Retrieved 23 November 2012.*

local materials and community skills are used in the construction of these tanks by the local community.

**Long-Term Impact** Women may walk long distances to collect water for their families. Attendance of children in school is increased, and the dropout rate is reduced particularly for girls. A safe source of clean drinking water improves maternity, childhood and community health and reduces illness (Fig. 12.3).

**And a Word on Education** As Bunker Roy says: purpose doesn't change—only the path evolves.

**Night Schools**

Since the 1980s, the Barefoot College has run night schools. These are designed for children in small, remote villages who work during the day. At times, the organization operated up to 108 schools in North India, including 58 in Rajasthan alone (Fig. 12.4).

In recent years, the focus has shifted. Government schools in the villages have made many night schools redundant. The number of girls attending school has increased significantly. Their educational attainment has also risen; many girls are now in secondary education, learning English, and have IT skills. At the same time,

**Fig. 12.3** Rainwater harvesting tank, Residential Bridge School, Singla. Photo credit D.v. Tschurtschenthaler, used with permission

**Fig. 12.4** Night school Chhir. Photo credit Sascha Montag, used with permission

it is unusual for girls from remote villages and various ethnic communities to move freely, even without Internet access, and visit an Internet café in the nearest larger town.

### Digital Village Learning Centers

The Barefoot College has responded to this development by establishing Digital Village Learning Centers.

The Centers address the shortage of resources that are dedicated to empowering and educating females in rural communities.

Barefoot College created these Digital Village Learning Centers to address specific challenges faced by rural girls and women:

- Women are not allowed to attend regular college, so after leaving school they must rely on correspondence courses to complete their education.
- The Digital Village Learning Center offers young married women an own space, where they can spend at least 2 h each day on the Internet, working on online courses and gaining valuable knowledge and skills.
- In this area, custom dictates that young women are not given access either to mobile phones or Internet facilities at home. The Digital Village Learning Center provides wi-fi and e-devices so that they can tap into the resources and information they need, not just to prepare for examinations but also for their own general interests.

**Fig. 12.5** Village Learning center Tilonia. Girls preparing their exams. Photo credit D.v. Tschurtschenthaler, used with permission

The Digital Village Learning Center is a vision going beyond that of a mere building. As a safe space for women, it encompasses the idea of comfortable, friendly learning in a rich environment offering a library and Internet access to the world well beyond the village, as well as the prospect of independence through vocational training (Fig. 12.5).

**Residential Bridge School Singla**

Besides that, a Residential Bridge School had been founded in the village Singla when Barefoot College noticed that the night schools left out of their purview children from areas where despite the presence of local schools many weren't enrolled in them owing to financial disadvantages as well as the nomadic nature of parent's livelihoods (migrating workers). In this scenario a residential bridge school emerged as an effective alternative to night schools (Fig. 12.6).

Since its inception in 2013 the Residential Bridge School has walked down a path marked by constant change, innovation and growth. With every passing year, the Bridge School moves closer to the goal of creating a platform for holistic, contextually suited and relevant platform, including basic English and IT lessons. Today the school has the capacity to host approx. 100 students.

Barefoot College was also aware of the challenges resulting from the IT boom of the early 2000s, that transformed India and the world. While this has benefited global companies, it has created enormous challenges for rural India.

**Fig. 12.6** Residential Bridge School, Singla. Photo credit D.v. Tschurtschenthaler, used with permission

The widening digital divide hinders rural youth's ability to engage in a job market that prioritizes IT skills and English proficiency. Rural educators struggle with limited resources and often lack the necessary training to equip students with these in-demand skills. High teacher turnover, driven by better prospects in urban centers, disrupts educational continuity and leaves rural students unprepared for a digitized global economy.

But there are other post-boom sustainability challenges as well. Migration to urban centers has strained the social fabric of rural communities and created demographic challenges related to aging. The technology gap continues to widen as rural communities lack access to digital tools and high-speed Internet.

## 12.5 Fundamental Insights on Social Transformation in Rural Areas

When being asked what his fundamental insights and lessons learned after many years of your involvement are, Bunker stressed three topics in the conversation:

- First: that unless you carry the rural poor people, the beneficiaries, the women and children and talk about what matters most in their lives there is no shortcut to sustainability. This means that a clear individually tailored bottom-up approach is necessary. If the UN SDGs are interpreted and applied as a top-down approach to implement solutions they will not be successful, top-down approaches to offering standardized solutions in general have not worked anywhere in the world and it is unfortunate that many institutions trying to realize the UN SDGs have still not learnt that this is the wrong approach.

- Second: once the people are mobilized and they are aware of the issues confronting them through dialogue and discussion there is no alternative to talking about this on a national scale with other like-minded civil society groups and affected should be made to bring them on a common platform and share their knowledge skills and wisdom with powers that be.
- And third: important for the framing in general—in Bunker's opinion, the most important SDGs for India are those covering "livelihoods, addressing poverty and hunger, women and children, education for all and technology."

## 12.6 Key Conclusions and Takeaways

Bunker stressed in the conversation four key central takeaways:

- First: the need to identify the vast knowledge, skills, and wisdom held by people living in poverty worldwide and to share at the community level this wealth of knowledge freely. The Solar Mamas of Tilonia[17] are a good example of what is meant by sharing.
- Second: the hope is that every community living in poverty around the world learn from each other.
- Third: the conviction that cities shouldn't be the only places where opportunity lives. When work, wages, and welfare are clustered in a few urban centers, migration becomes survival, not choice. Sustainable livelihoods begin when jobs are created closer to home—in small towns and rural economies. When local industries thrive, communities stay rooted, families stay together, and cities breathe easier. Sustainable growth isn't about pulling people into cities. It's about pushing opportunity outward ensuring dignity, security, and stability everywhere.
- Fourth: the importance of learning by doing and understanding. It is crucial to understand that all achievements are built on trying, failing, learning, and trying again. The message is clear: success isn't the goal—learning is. And through that learning, success finds its way. Based on this the purpose of the endeavor doesn't change—only the path evolves.

The Barefoot College (Social Work and Research Centre), a powerful initiative, originally approached with a simple idea: teach rural women how to build solar panels. Today, those same women are travelling across countries, training others in solar energy, and transforming lives—not just in their village but around the world. The spirit is: "Keep trying. Whether it's a success or a failure, it always teaches you something. Keep doing things. Keep experimenting. Even if it fails." That's the spirit of Tilonia. A community constantly evolving, learning, and daring to do something different.

---

[17] See https://barefoot.college/impact/solar/

**Dagmar von Tschurtschenthaler** was even during her studies in history and politics very interested in India, which she repeatedly visited privately. In 1990, she became acquainted with Barefoot College, which was concerned with the problems of the marginalized rural population and immediately inspired her. After a family time with four daughters, she started freelancing for Studiosus Munich in 1993, designing itineraries for North India and working as a hotel handling agent. Her work in India increased her interest in the country and awakened her desire to work there for children without future prospects. Barefoot College has always worked according to the principles of responsible leadership. This philosophy also shaped her work from the very beginning. Every year, she spent longer periods of time in India to get to know the project even better and to get involved. During this time, she also worked as a scout for ZDF in India (1996–2006) for various films. Scientific editing and management on location (India) and editing of various APA travel guides, India. In the meantime, through Friends-for-Hope e.V. she concentrates entirely on supporting Barefoot College, which she visits several times a year and which still inspires her even after 35 years.

# Social Innovation in Tourism in Venice: A Three-Pillar Model Analysis of Purpose, Adaptability, and Connectivity

13

Olga Tzatzadaki

**Abstract**

This chapter examines Social Innovation in Tourism (SIT) as a response to over-tourism in Venice, adopting the Three-Pillar Model—Sustainable Purpose, Travelling Organization, and Connectivity—to analyze how local cultural heritage initiatives support community resilience and counter depopulation. Moving beyond technology-centered tourism innovation, SIT is conceptualized as a value-driven, place-based process mobilizing tangible and intangible heritage in line with UN SDGs 8, 10, 11, and 17.

Using a mixed methodology combining literature review and semi-structured interviews with ten stakeholders (2024), the findings highlight governance fragmentation, limited institutional support, and coordination gaps as key barriers, while confirming SIT's potential as a lever for inclusive and sustainable territorial regeneration.

## 13.1 Introduction: Tourism Innovation and Trends

### 13.1.1 Tourism Innovation Beyond Technology: Social, Cultural, and Territorial Challenges

Tourism innovation has become increasingly central to scholarly debate and policy discussions, reflecting the sector's growing economic, cultural, and social

O. Tzatzadaki (✉)
University Iuav of Venice, Venice, Veneto, Italy
e-mail: otzatzadaki@iuav.it


P. Wollmann et al. (eds.), *The Sustainable Organization – Volume 2*, Future of Business and Finance, https://doi.org/10.1007/978-3-032-23802-3_13

relevance.[1] However, literature has been predominantly shaped by a technological perspective, focusing on digital tools, information systems, and tech-driven services that enhance tourism management and competitiveness.[2] While these contributions are valuable, they risk narrowing the conceptualization of innovation by neglecting its social, organizational, and cultural dimensions, which are essential for long-term sustainability and community well-being.[3]

The relevance of innovation becomes evident when considering the uneven distribution of tourism's effects across territories. Urban destinations affected by overtourism face substantial pressures that undermine residents' quality of life, such as housing displacement, loss of public space, and cultural commodification.[4] These processes often generate a growing sense of alienation among local inhabitants, who may feel like "strangers in their own home."[5] Such dynamics are not only economic but also social and cultural, leading to the erosion of community identity and weakening social cohesion.[6] In contrast, rural and inland areas frequently experience undertourism, depopulation, and economic decline, resulting in unemployment, out-migration, and loss of cultural continuity. These contrasting realities highlight the need for a purpose-driven approach that can guide innovation toward outcomes that support local resilience and the preservation of cultural heritage.[7]

In this context, innovation serves as a "lighthouse" that anchors tourism in community values and long-term goals. Rather than focusing solely on visitor numbers or economic growth, purpose-led innovation emphasizes the preservation of local identity, the enhancement of social cohesion, and the promotion of inclusive

---

[1] Sørensen, F.: *The geographies of social networks and innovation in tourism*. Tourism Geographies, 9 (1), 22–48, 2007.

Hjalager, A. M.: *A review of innovation research in tourism*. Tourism Management, 31 (1), 1–12, 2010.

Işık, C.; Günlü Küçükaltan, E.; Taş, S.; Akoğul, E.; Uyrun, A.; Hajiyeva, T.; Turan, B.; Dırbo, A. H.; Bayraktaroğlu, E.: *Tourism and innovation: A literature review*. Journal of Ekonomi, 1 (2), 98–154, 2019.

[2] Gomezelj, D. O.: A systematic review of research on innovation in hospitality and tourism. International Journal of Contemporary Hospitality Management, 28 (3), 516–558, 2016.

[3] Işık et al., 2019; Gomezelj, 2016.

[4] Forster, J.: The sociological consequences of tourism. International Journal of Comparative Sociology, 5, 217, 1964.

Koens, K.; Postma, A.; Papp, B.: *Is overtourism overused? Understanding the impact of tourism in a city context*. Journal of Sustainable Tourism, 10 (12), 4384, 2018.

Van der Borg, J.; Costa, P.; Gotti, G.: *Tourism in European heritage cities*. Annals of Tourism Research, 23 (2), 306–321, 1996.

[5] Glass, R.: London: Aspects of change. London, MacGibbon and Kee, 1964.

Bolzoni, M.; Semi, G.: *Adaptive urbanism in ordinary cities: Gentrification and temporalities in Turin (1993–2021)*. Cities, 134, 104144, 2023.

Semi, G.; Tonetta, M.: *Marginal hosts: Short-term rental suppliers in Turin, Italy*. Environment and Planning A: Economy and Space, 53 (7), 1630–1651, 2021.

[6] Semi and Tonetta, 2021.

[7] Moulaert, F.; Martinelli, F.; Swyngedouw, E.; Gonzalez, S.: Towards alternative model(s) of local innovation. Urban Studies, 42 (11), 1969–1990, 2005.

development. This perspective aligns with the principles of social innovation in tourism (SIT), which seeks to mobilize tangible and intangible cultural heritage to support sustainable local development.[8] By framing innovation within a sustainable approach, tourism can be understood as a collective commitment to the future of local communities, where cultural heritage becomes a resource for empowerment and regeneration rather than a commodity to be consumed. This reorientation is essential for destinations like Venice, the case under study in this paper, where local actors seek to counteract depopulation caused by overtourism.

A sustainable approach underscores that tourism innovation should not be viewed merely as technological advancement or market competitiveness. Instead, it should be conceptualized as a transformative process guided by values and community needs, which integrates environmental, social, and cultural objectives. This approach encourages policymakers, practitioners, and researchers to consider the long-term impacts of tourism innovation, and to prioritize strategies that foster community well-being, cultural preservation, and sustainable development. In doing so, it offers a more comprehensive and meaningful understanding of tourism innovation that is capable of addressing the complex challenges facing contemporary destinations.

### 13.1.2 The Evolution of Tourism Innovation Toward Social Innovation

Within this broader framework, recent scholarship has identified several interconnected trends shaping innovation in tourism destinations. Innovation increasingly emerges through experience co-creation processes, understood as collective actions involving diverse stakeholders and enabled by elements of social capital such as trust, openness, networking, and collaboration, often supported by technological tools.[9] At the same time, the diffusion of ICTs, e-tourism platforms, virtual communities, and gamification has significantly transformed destination models, reshaping how tourism experiences are designed, communicated, and consumed.[10]

In parallel, the concept of smart destinations has evolved from the broader smart city paradigm, emphasizing the role of interconnected digital infrastructures—including ICT systems, the Internet of Things, cloud computing, and immersive technologies such as augmented and virtual reality—in connecting destination stakeholders. These technological ecosystems enhance opportunities for

---

[8] Busacca, M.; Tzatzadaki, O.: Social innovation in tourism: a systematic literature review and the proposal of an analytical framework. Journal of Enterprising Communities: People and Places in the Global Economy, 2025. doi: 10.1108/JEC-02-2025-0041.

[9] Trunfio, M.; Campana, S.: Drivers and emerging innovations in knowledge-based destinations: Towards a research agenda. Journal of Destination Marketing & Management, 14, 100370, 2019.

[10] Buhalis, D. and Law, R., Progress in information technology and tourism management: 20 years on and 10 years after the Internet—The state of eTourism research. Tourism management 29 (4): 609–623, 2008.

communication, collaboration, and knowledge sharing, thereby supporting more integrated and innovative destination management approaches.[11]

Innovation in tourism is also closely linked to evolving governance models. The literature highlights a shift toward e-participative and hybrid forms of destination governance, in which stakeholder engagement plays a central role in fostering innovation. Traditional top-down governance structures have increasingly given way to more inclusive and network-based models, a transition accelerated by the diffusion of digital platforms and ICT-enabled participation mechanisms.[12]

Finally, SIT is embedded within the broader understanding of social innovation as a process aimed at developing and implementing new ideas, strategies, or initiatives that address social needs, enhance social well-being, and generate positive societal impact. Social innovation seeks to tackle complex challenges such as inequality, environmental sustainability, access to services, and community development through collaborative, cross-sectoral approaches involving public institutions, private actors, nonprofit organizations, and local communities. By fostering systemic change and long-term improvements, SIT positions tourism as a vehicle for sustainable and socially inclusive development.

Indeed, SIT has gained increasing scholarly attention as it responds to the structural shortcomings of conventional tourism models, particularly in relation to social equity, community inclusion, and cultural sustainability.[13] SIT is generally understood as a set of initiatives that mobilize local cultural heritage, knowledge, and capabilities to generate social value and enhance community resilience.[14] Unlike innovation approaches primarily driven by technological advancement, SIT places emphasis on social processes and place-based dynamics, highlighting the central role of local actors—often small-scale entities such as entrepreneurs, cooperatives, museums, and social enterprises—and their participation in collaborative networks that support the co-creation of tourism experiences in interaction with governance structures.[15]

[11] Trunfio & Campana, 2019.

[12] Rodriguez, I., Williams, A. M., & Hall, C. M., Tourism innovation policy: Implementation and outcomes. Annals of Tourism Research, 49, 76–93, 2014.

[13] Busacca & Tzatzadaki, 2025.

[14] Busacca & Tzatzadaki, 2025.
Trunfio & Campana, 2019.
Wirth, S.; Bandi Tanner, M.; Mayer, H.: *Social innovations in tourism: Analysing processes, actors, and tipping points*. Tourism Geographies, 25 (5), 1322–1340, 2023.

[15] Kofler, I.; Marcher, A.; Volgger, M.; Pechlaner, H.: The special characteristics of tourism innovation networks: The case of the Regional Innovation System in South Tyrol. Journal of Hospitality and Tourism Management, 37, 68–75, 2018.

## 13.2 SIT Through the *Three-Pillar Model* Lens

### 13.2.1 Sustainable Purpose in SIT

SIT can be understood as a process through which local communities generate new social practices and organizational models to address the challenges and vulnerabilities created or amplified by tourism. Within this process, the Three-Pillar Model[16] offers a meta-level framework that places local communities at the core of tourism innovation, especially in contexts where tourism's impacts generate significant social challenges.

Rather than focusing primarily on market competitiveness, SIT is driven by the need to maintain community well-being, cultural identity, and territorial resilience. In this perspective, *sustainable purpose*[17] is not simply a moral commitment; it is a strategic orientation that shapes what tourism is for and what it should achieve. Sustainability in SIT is multidimensional, integrating social, environmental, and economic goals to ensure that tourism supports long-term development rather than temporary growth.[18]

This holistic purpose is crucial because it positions tourism as a collective resource for community regeneration, rather than as an external force that exploits local assets.[19]

A sustainable purpose in SIT is strongly connected to cultural preservation. Cultural and artistic heritage tourism initiatives, for instance, are not only "tourist

---

[16] Wollmann, P.; Kühn, F.; Kempf, M.; Püringer, R. (Eds.): Organization and leadership in disruptive times: Design and implementation using the 3-P-Model. Cham: Springer Nature, 2021.

Wollmann, P.; Stricker, L.; Pemler, D. (Eds.): Reading an Unknown Organization—Leaders in New Roles Thriving in Challenging Environments with the 3-P-Model Mindset. Cham: Springer Nature, 2024.

Wollmann, P.; Pemler, D.; Ndrevataj, M. (Eds.): The Sustainable Organization: How Organizations Address the 17 UN SDGs Using the 3-P-Model. Cham: Springer Nature, 2025.

[17] Wollmann et al., 2021; 2024; 2025.

[18] Antošová, G.; Rojas, I. D. M.; Mejía, M. P.; Gómez, H. Y. A.: *Sustainable tourism planning in Bahía Solano, Colombia*. European Journal of Tourism, Hospitality and Recreation, 10 (2), 165–176, 2020.

Batle, J.; Orfila-Sintes, F.; Moon, C. J.: *Environmental management best practices: Towards social innovation*. International Journal of Hospitality Management, 69, 14–20, 2018.

Lapointe, D.; Guimont, D.; Guillemard, A.; Benjamin, C.: *People, Place, Values: Living Lab as Social Innovation Processes for Tourism Communities*. Anais Brasileiros De Estudos Turísticos, 2021.

Phi, G. T.; Clausen, H. B.: *Fostering innovation competencies in tourism higher education via design-based and value-based learning*. Journal of Hospitality, Leisure, Sport & Tourism Education, 29, 100298, 2021.

Belliggiano, A.; Bindi, L.; Ievoli, C.: *Walking along the sheeptrack... rural tourism, ecomuseums, and bio-cultural heritage*. Sustainability, 13 (16), 8870, 2021.

[19] Tzatzadaki, O.: *Between proximity-based social innovation and tourism: initiatives of micro-regeneration in the Po Delta e Polesine region*. In: Donelli, C.; Pangallo, F.; Tzatzadaki, O. (Eds.): *Tracing Waste, Envisioning Sustainability: Reflections on Tourism and Community Practices for Collective Futures*. Roma: Editoriale Scientifica—PuntoOrg, 2025.

products" but also community tools that sustain local memory, practices, and identity.[20] In this sense, cultural heritage becomes an active resource for social cohesion, where residents are not passive hosts but active participants in the creation and narration of their own cultural narratives. Likewise, environmental sustainability initiatives emphasize the protection of natural habitats and biodiversity, strengthening local ecosystems while promoting responsible visitor behavior.[21] Economic sustainability is integrated through models that ensure tourism benefits remain within the community, promoting local employment and reducing economic leakage.[22]

The sustainable purpose pillar therefore provides a long-term orientation that legitimizes tourism as a social project. It helps to define the values and goals of SIT initiatives and ensures that innovation is aligned with community priorities. In practice, this purpose can manifest through initiatives that safeguard local heritage, promote inclusive participation, and encourage responsible consumption. It is also instrumental in aligning tourism initiatives with broader sustainability agendas, such as the SDGs, by linking local actions to global frameworks. In conclusion, sustainable purpose is the guiding principle that makes SIT a form of tourism innovation grounded in social and territorial responsibility.

### 13.2.2 Travelling Organization: Adaptation, Learning, and Local Agency

The *travelling organization* pillar[23] explains how SIT initiatives function as dynamic systems that continuously adapt to changing conditions. In contrast to conventional tourism development models, which often rely on fixed strategies and linear planning, SIT initiatives operate in contexts characterized by uncertainty and social complexity. Local actors—including social entrepreneurs, cooperatives, museums, and small businesses—must navigate shifting markets, demographic change, and policy fluctuations, often with limited resources. The travelling organization

[20] Le Bel, P. M.: Literary heritage and place building for communities: the case of Allier, France. Almatourism—Journal of Tourism, Culture and Territorial Development, 8 (7), 58–74, 2017.

Karzen, M.; Demonja, D.: *Importance of Storytelling: how to create more resilient cultural heritage*. Nova Prisut, 3, 653–668, 2020.

Corbisiero, F.: *Urban social innovation practices: the case of Rione Sanità in Naples*. Territorio, 99, 73–78, 2021.

Moleiro, D.: *Territorial cooperation networks, based on social innovation processes as new models of hospitality: The case of the historic houses*. Journal of Tourism and Development, 36 (2), 229–244, 2021.

Splendiani, S.; Forlani, F.; Picciotti, A.; Presenza, A.: *Social Innovation Project and Tourism Lifestyle Entrepreneurship to Revitalize the Marginal Areas. The Case of the via Francigena Cultural Route*. Tourism Planning & Development, 20 (5), 938–954, 2023.

[21] Batle et al., 2018.

Lapointe et al., 2021.

[22] Antošová et al., 2020.

Phi & Clausen, 2021.

[23] Wollmann et al., 2021; 2024; 2025.

perspective highlights that innovation in tourism is not a one-off event but a continuous process of learning, experimentation, and adaptation. This mindset is essential for building resilience, as it allows communities to respond to unexpected challenges and to seize emerging opportunities.

The adaptive nature of SIT is closely linked to local agency.[24] Travelling organizations are not passive recipients of external tourism trends; they actively shape tourism practices through collective action and creative problem-solving. This can involve the development of new tourism formats, reinterpretation of heritage, or creation of new community-led services. The process is often iterative: initiatives evolve through trial-and-error, feedback, and collective reflection. In this way, SIT can be seen as a form of social learning where the community gradually builds capabilities and knowledge, strengthening its capacity to manage tourism in a way that supports local well-being.[25] This approach also reduces dependency on external actors, enabling communities to retain control over tourism development and its benefits.

Furthermore, the travelling organization pillar supports the emergence of inclusive forms of tourism that prioritize social needs. By fostering experimentation and local learning, SIT can address social challenges such as unemployment, marginalization, and demographic decline through locally designed solutions. This creates a more equitable tourism system in which residents are empowered to participate in decision-making and to develop sustainable livelihoods.[26] In other words, the travelling organization pillar underscores that SIT is a transformative journey that builds local capacity and agency through continuous adaptation and learning.

### 13.2.3 Connectivity: Collaborative Governance, Networks, and Shared Value

*Connectivity*[27] is the pillar that explains how SIT initiatives are enabled through networks of relationships, partnerships, and collaborative governance. The nature of SIT requires the participation of multiple actors, as social challenges cannot be addressed by a single stakeholder. Multi-stakeholder collaborations unite communities, NGOs, public institutions, and private sector actors, creating a shared platform for innovation and collective action.[28] Connectivity is therefore not only about

[24] Busacca & Tzatzadaki, 2025.

[25] Wirth et al., 2023.
Işık et al., 2019.

[26] Wirth et al., 2023.
Busacca & Tzatzadaki, 2025.

[27] Wollmann et al., 2021; 2024; 2025.

[28] Amore, A.; Hall, C. M.: *From governance to meta-governance in tourism? Re-incorporating politics, interests and values in the analysis of tourism governance*. Tourism Recreation Research, 41 (2), 109–122, 2016.
Elias, S.; Barbero, A. C.: *Social innovation in a tourist coastal city: a case study in Argentina*. Social Enterprise Journal, 17 (1), 44–62, 2021.

communication but about building shared responsibility and mutual trust. In community-based tourism, governance mechanisms often involve formalized local institutions that facilitate dialogue and conflict resolution, enabling participatory decision-making and ensuring that tourism development aligns with local priorities.[29] This form of collaborative governance is essential to sustain SIT initiatives over time and to avoid tensions between stakeholders.

Connectivity is also supported by innovation and technology. ICT tools, such as digital platforms, online marketing, and smart destination management systems, strengthen stakeholder coordination and enable participatory management processes.[30] These technologies facilitate knowledge exchange, enhance visibility of local products, and support the diffusion of innovative practices across networks. In addition, connectivity is crucial for inclusive economic development, as partnerships and collaborative governance enable local entrepreneurship and equitable distribution of benefits.[31] Education and awareness-raising activities also reinforce

---

Nurhasanah, I. S.; Van den Broeck, P.: *Towards a sustainable metamorphosis of a small island tourism: dynamizing capacity building, alternating governance arrangements, and emerging political bargaining power*. Sustainability, 14 (12), 6957, 2022.

Sarkki, S.; Jokinen, M.; Heikkinen, H. I.; Nijnik, M.; Melnykovych, M.; Kluvánková, T.: *"Going out to get in"—Roles of forest conflicts in bottom-linked environmental governance progressing toward socio-political innovations*. Environmental Policy and Governance, 32 (6), 478–491, 2022.

[29] Amore & Hall, 2016.

[30] Lim, C.; Park, J.: Digital Omotenashi project: A tourists' application design by a design thinking approach. PACIS 2016 Proceedings, 2016.

Polese, F.; Botti, A.; Grimaldi, M.; Monda, A.; Vesci, M.: Social innovation in smart tourism ecosystems: How technology and institutions shape sustainable value co-creation. Sustainability, 10 (1), 140, 2018.

Van, N. T. T.; Vrana, V.; Duy, N. T.; Minh, D. X. H.; Dzung, P. T.; Mondal, S. R.; Das, S.: *The role of human–machine interactive devices for post-COVID-19 innovative sustainable tourism in Ho Chi Minh City, Vietnam*. Sustainability, 12 (22), 9523, 2020.

Errichiello, L.; Micera, R.: *A process-based perspective of smart tourism destination governance*. European Journal of Tourism Research, 29, 2909, 2021.

Presenza, A.; Panniello, U.; Messeni Petruzzelli, A.: *Tourism multi-sided platforms and the social innovation trajectory: The case of Airbnb*. Creativity and Innovation Management, 30 (1), 47–62, 2021.

[31] Zebryte, I.; Jorquera, H.: Chilean tourism sector "B Corporations": evidence of social entrepreneurship and innovation. International Journal of Entrepreneurial Behavior & Research, 23 (6), 866–879, 2017.

Quandt, C.; Ferraresi, A.; Kudlawicz, C.; Martins, J.; Machado, A.: *Social innovation practices in the regional tourism industry: Case study of a cooperative in Brazil*. Social Enterprise Journal, 13 (1), 78–94, 2017.

Mahato, S. S.; Phi, G. T.; Prats, L.: *Design thinking for social innovation: Secrets to success for tourism social entrepreneurs*. Journal of Hospitality and Tourism Management, 49, 396–406, 2021.

Tresiana, N.; Duadji, N.: *Developing forest coffee cultural tourism and historical heritage megalithic sites in social innovation governance: How does it work in a sustainable way?* Journal of Environmental Management & Tourism, 13 (4), 1036–1046, 2022.

Suriyankietkaew, S.; Krittayaruangroj, K.; Iamsawan, N.: *Sustainable Leadership practices and competencies of SMEs for sustainability and resilience: A community-based social enterprise study*. Sustainability, 14 (10), 5762, 2022.

connectivity by engaging tourists and residents in sustainable practices and strengthening community participation.[32]

Ultimately, connectivity transforms SIT into a collective process that integrates local knowledge, institutional support, and technological innovation. Through networks and partnerships, communities can create shared value and strengthen their capacity to manage tourism sustainably. This pillar highlights that social innovation in tourism is not only about creative solutions but also about the relational and governance structures that make these solutions possible and scalable.

## 13.3 Case Study: Venice and Surroundings

### 13.3.1 Methodological Aspects

The research adopted a two-stage methodological design combining theoretical investigation with empirical inquiry. The first stage consisted of a conceptual review aimed at identifying the foundational principles of SIT and its theoretical evolution. This phase included an analysis of international SIT case studies and a desk-based review of secondary sources. The latter focused specifically on the Venice context, examining institutional documents, tourism marketing plans, and policy reports to reconstruct recent regional trends and strategic orientations. The outcomes of this theoretical and documentary analysis were used to refine the research framework and to guide the selection and design of the empirical phase.

The second stage, conducted in the second half of 2024, involved an empirical investigation based on ten stakeholders located within the research area. The sample comprised a heterogeneous set of local stakeholders, including museum professionals, small-scale entrepreneurs, nonprofit organizations, consortia, ecomuseums, agritourism operators, and local institutions. These actors were selected because of their direct involvement in cultural heritage management and tourism-related social innovation. Data were collected through semi-structured and in-depth interviews, which explored the development and implementation of SIT initiatives and elicited participants' life stories and personal trajectories, with particular attention to inter-actor interactions, institutional support mechanisms, personal motivations, and the conditions enabling long-term sustainability and potential scalability.

Interviews were audio-recorded and transcribed with the support of *TurboScribe* software. The resulting transcripts were then analyzed using *Atlas.ti*[33] through a thematic coding process. The analysis focused on a set of pre-defined codes, related

[32] López, A.; Ramos, D.: Innovation in services: The case of rural tourism in Argentina. Emerging Markets Finance and Trade, 51 (3), 635–646, 2015.

Chiodo, E.; Adriani, H. L.; Navarro, F. P.; Salvatore, R.: *Collaborative processes and collective impact in tourist rural villages—insights from a comparative analysis between Argentinian and Italian cases*. Sustainability, 11 (2), 432, 2019.

Belliggiano et al., 2021.

[33] Smit, B.: Atlas.ti for qualitative data analysis. Perspectives in Education, 20 (3), 65–75, 2002.

to the above aspects (networks, motivation, governance, etc.), enabling a systematic interpretation of stakeholder narratives and experiences. To ensure confidentiality, the names of the organizations involved in the case studies are not disclosed. Instead, each stakeholder is described through its institutional profile and assigned a numerical identifier used throughout the text:

1. *Industrial heritage and cultural exhibition space*—managing and promoting an industrial heritage site, while hosting also cultural events.
2. *Venetian mask artisan and educational workshops*—producing traditional Venetian masks and offering short lectures, practical courses, guided visits, and team-building activities.
3. *Civic Museums network (Education)*—responsible for educational programs and activities across a network of civic museums.
4. *Civic Museums network (Communication)*—overseeing public relations, communication, and promotion for the civic museums' network.
5. *Sustainable fashion atelier*—designing and producing clothing adapted to the city's climate, using fair-trade and sustainable textiles.
6. *Agritourism and farm enterprise*—combining agricultural production with hospitality services in a rural lagoon setting.
7. *Industrial tourism promoter*—focused on valorizing and promoting industrial heritage sites through tourism initiatives.
8. *Plastic-free and local craft retailer*—selling unpackaged products and locally made artisanal items.
9. *Social promotion cycling association*—organizing bike festivals, storytelling rides, and gravel tours to engage communities with local culture.
10. *Lagoon ecomuseum*—preserving and promoting the cultural and environmental heritage of Venice, its city, and surrounding islands.

All entities included in the study are engaged in tourism, culture, and social innovation, aiming to enhance cultural heritage in its diverse forms while addressing local social challenges such as depopulation and unemployment.

### 13.3.2 Tourism and Venice

Venice is a paradigmatic case of overtourism, where the concentration of visitors creates intense pressures that significantly reduce residents' quality of life. Over time, the historic city has become a highly commodified space, where everyday life is increasingly shaped by tourism demands. This process contributes to a growing sense of alienation among local inhabitants, who may feel like "strangers in their own home."[34] Such alienation is not merely symbolic but reflects a deeper erosion

[34] Glass, R.: *London: Aspects of change*. London, MacGibbon and Kee, 1964.

Tulumello, S.; Allegretti, G.: *Articulating urban change in Southern Europe: Gentrification, touristification and financialisation in Mouraria, Lisbon*. European Urban and Regional Studies, 28 (2), 111–132, 2021.

Bolzoni & Semi, 2023.

Moulaert et al., 2005.

of community identity and social cohesion, as local traditions and practices are gradually displaced by the needs of visitors.[35]

The impacts of overtourism in Venice are not only economic but also cultural and social. The progressive depopulation of the historic center intensifies vulnerability for the remaining residents, transforming their daily existence into a survival challenge. As the resident population declines, social structures weaken and the city's collective memory is at risk of fading. Additionally, overtourism places significant strain on the built environment and infrastructure, increasing the city's exposure to environmental hazards such as high tides, which further threaten the preservation of Venice's unique heritage.

According to local institutional reports,[36] the current tourism dynamics show that Venice's tourism system is characterized by high seasonal concentration and short average stays, reinforcing the city's dependency on mass tourism and exacerbating local vulnerabilities. A SWOT[37] analysis of Venice's tourism context highlights both significant strengths and structural weaknesses. Venice benefits from a strong global brand, excellent international connectivity, and extraordinary natural and cultural assets, including the lagoon, beaches, and internationally renowned cultural events. The city also possesses a diversified tourism offer and a strong industrial sector (Porto Marghera), supported by active cultural and business associations and excellence in academic institutions. However, the city faces major challenges, including a high concentration of unregistered second homes, depopulation of the historic center, seasonal overcrowding, short average tourist stays, and an overreliance on tourism as the primary economic driver. These weaknesses are contrasted by opportunities such as the attraction of high-spending markets, growing political recognition of tourism as a strategic sector, access to EU funds (POR FESR), external investments in hospitality, and the expansion of tourism on the mainland and among students. Nonetheless, Venice remains exposed to threats such as the risk of losing its identity, the spread of low-cost tourism, declining medium- and long-term stays, damage to cultural heritage and local traditions, and the vulnerability of tourism monoculture to global crises and environmental risks.

In response to these challenges, the regional tourism objectives[38] focus on promoting and enhancing Venice's cultural and artistic heritage. Key goals include improving access to and knowledge of cultural assets such as museums, historical sites, and monuments through targeted initiatives and collaboration with local and

[35] Semi & Tonetta, 2021.

[36] Comune di Venezia: *Strategia per il turismo sostenibile di Venezia e la sua Laguna*. Comune di Venezia, 2024a.

Comune di Venezia: *Campagna di sensibilizzazione #EnjoyRespectVenezia*. Comune di Venezia, 2024b.

Comune di Venezia: *Venezia e la sua laguna: gestione e valorizzazione dei flussi turistici (Deliberazione di Giunta Comunale n. 148/2022)*. Comune di Venezia, 2022/2023.

Comune di Venezia: *Progetto di governance territoriale del turismo a Venezia*. Comune di Venezia, 2025.

[37] Strength/Weaknesses/Opportunities/Threats.

[38] Regione del Veneto: *Piano strategico per il turismo/Tourism strategic plan*. Regione del Veneto, 2024.

national institutions. The strategy also aims to develop cultural tourism by positioning the region as a destination of historical and artistic significance, supported by events, exhibitions, and curated routes that highlight the area's cultural peculiarities. Furthermore, the valorization of local traditions is pursued through festivals and cultural activities that celebrate craftsmanship, music, dance, and gastronomy. Finally, education and awareness programs are promoted to engage both residents and visitors, fostering appreciation for the city's heritage through workshops, seminars, and guided thematic tours. These objectives are designed to integrate cultural preservation with sustainable economic development, ensuring that tourism contributes positively to the well-being of local communities while safeguarding Venice's identity.

## 13.4 The Three-Pillar Model Unfolded: Evidence from Venice

### 13.4.1 Sustainable Purpose

Social innovation initiatives in tourism often originate from the individual motivations of their promoters, who seek to satisfy a range of personal needs that go beyond mere economic survival. At an individual level, social innovation initiatives address various needs for their promoters, ranging from physiological and employment needs to belonging, self-esteem, and self-actualization. This is particularly evident in projects focused on cultural heritage preservation, where actors commonly describe their work as a mission. Such initiatives are not simply business ventures or cultural activities; they represent a deep commitment to narrating and passing on the historical memory of their territory. As interviewee n.8 stated, "When you decide to stay and dedicate yourself to the city, it becomes a kind of mission." Interviewee n.8 confirmed this sense of purpose: "For me, this work is truly a mission." The strong emotional attachment and sense of responsibility are further emphasized in the claim that "I feel a great responsibility, and I do it with passion because it's part of us—it's a life mission" (interviewee n.5). These statements show that the sustainable purpose of SIT initiatives is often rooted in personal identity and a sense of belonging to the community.

Nevertheless, the mission-driven nature of these initiatives does not automatically translate into broad community recognition. The local community does not always recognize the importance of safeguarding cultural heritage, and this lack of shared awareness can hinder the sustainability of heritage-based tourism projects. However, when communities are able to connect with the historical meaning of heritage, they tend to embrace such initiatives more actively. As interviewee n.7 observed, "This process benefits the community, especially by fostering self-awareness of its past and roots, helping it recognize its identity." This statement underlines how cultural heritage projects can strengthen community cohesion by enhancing collective memory and shared identity. In many cases, the creation of a sense of community becomes an explicit objective, even when the final goals are not always clearly defined: "There isn't always a precise goal, but for me, creating a sense of community is essential" (interviewee n.2). Such perspectives demonstrate that sustainable purpose in SIT is often based on intangible social values rather than measurable economic outcomes.

Moreover, the perceived vitality of heritage plays a key role in community acceptance. Local residents are more likely to support and participate in heritage initiatives when they feel that the cultural assets are still alive and relevant to contemporary life. "Local communities are more open to these experiences when heritage is perceived as still alive." Interviewee n.7 noted, highlighting the need for heritage projects to communicate meaning and relevance rather than simply preserving artefacts. In this sense, social innovation in tourism becomes a process of re-education, where communities are encouraged to rediscover and revalue their historical roots. Interviewee n.10 noted that, "Communities need to be re-educated to recognize and revalue this self-knowledge." This view suggests that sustainable purpose is not only a personal mission but also a collective task that requires ongoing dialogue, awareness, and participation.

## 13.4.2 Travelling Organization

The "Travelling Organization" pillar emphasizes the need for continuous adaptation and learning within a dynamic socio-economic environment. In the context of SIT, this concept is particularly relevant to understanding how local actors navigate complex community landscapes and pursue collective goals through experimentation, collaboration, and flexible governance. A core characteristic of successful SIT initiatives is the desire to build networks that can mobilize multiple stakeholders and generate broader community benefits. The involvement of more stakeholders tends to create more beneficiaries, thus strengthening the resilience and legitimacy of innovation processes. However, the existence of a network is not enough; its effectiveness depends on the quality of interaction, the capacity for coordination, and the presence of leadership with strong communication skills. As interviewee n.6 highlighted, "Rather than resistance, I noticed a lack of willingness to network" which suggests that the obstacle is often not opposition but the absence of proactive engagement. This challenge is compounded by the demographic decline and weakening of social structures, as noted by interviewee n.8 "There is demographic decline and a weakening of the social fabric."

In this context, the travelling organization requires not only a mission but also a collective ability to navigate uncertainty through relationships and shared learning. The goal of many heritage-based initiatives is explicitly linked to protecting local cultural resources. Interviewee n.2 stated that "The goal was precisely to protect Venice's cultural heritage and artisan network," indicating a strategic orientation toward preserving intangible social infrastructures as well as physical artefacts. This protective objective often becomes a foundation for broader network-building. For example, the presence of artisan shops and local residents creates a living ecosystem that needs to be preserved and connected. "Here, there are artisan shops and local residents, and we want to preserve this network" (interviewee n.2) emphasizes the recognition of local interdependencies that constitute the community's social capital. In many cases, these networks support multiple initiatives simultaneously, generating a form of mutual reinforcement. Interviewee n.10 noted that "These businesses indirectly support our associations and the ecomuseum project, creating a support system that goes beyond just providing goods and services," highlighting how economic activity can be linked to cultural and social objectives through collaborative structures.

Despite these motivations, the consolidation of cohesive networks remains a significant challenge. The lack of established connections and continuity can limit the capacity of initiatives to scale and sustain themselves over time. As interviewee n.1 noted, "There is still no truly cohesive network," while interviewee n.9 added that "The lack of continuity can hinder the development of a stronger, more unified network." These observations indicate that the travelling organization in this context is not yet fully formed; rather, it is an evolving system that requires ongoing effort to create durable linkages among actors. Densely interconnected communities appear to be more successful in initiating regeneration processes, because they can share knowledge, resources, and responsibilities more effectively. Therefore, the travelling organization pillar highlights that the sustainability of social innovation in tourism is not only a matter of individual motivation but also of collective capacity to build, maintain, and adapt networks over time.

### 13.4.3 Connectivity

The Connectivity pillar emphasizes the importance of interlinking diverse stakeholders, resources, and institutional actors to enable sustained innovation and collective impact. In the context of social innovation in tourism, connectivity is essential because local initiatives frequently depend on collaborative networks to access funding, skills, and legitimacy. However, the empirical evidence from Venice highlights persistent barriers that weaken the long-term success of socially innovative initiatives. One major limitation is the presence of limited and ineffective institutional support, which often prevents local projects from scaling or stabilizing over time. This barrier is reinforced by deficiencies in communication and outreach: the lack of a coherent and effective communication strategy from institutions undermines the visibility and legitimacy of local initiatives. As interviewee n.5 noted, "There is a complete lack of governance and planning processes, and this is a major issue." This quote reveals how the absence of structured governance not only reduces efficiency but also erodes trust between local actors and public institutions.

Connectivity in SIT also requires stable and transparent institutional frameworks. However, in Venice, the system appears to depend heavily on the political vision of the current mayor and administration. This instability is perceived as a structural weakness that undermines long-term planning and consistent support. As interviewee n.2 argued, "In my opinion, at the administrative level, measures could be taken to prevent vacant spaces from being left unused, or lower taxes could be imposed." This statement illustrates that local actors often feel that administrative decisions are not strategically aligned with community needs and opportunities.

Moreover, dialogue with public institutions is frequently described as complex and bureaucratic. "Dialogue with public administration is always complex, especially when dealing with bureaucracy," observed interviewee n.6, pointing to the procedural obstacles that hinder collaboration and slow down project development.

A further structural challenge is the high turnover of public administration staff, which disrupts continuity and makes it difficult for local initiatives to build long-term relationships with decision-makers. "The biggest issue, however, is the frequent turnover of public administration staff, with political rotations and supervisory changes happening every three years" (interviewee n.8). This instability creates uncertainty for social innovators who require consistent support and clear institutional pathways. The resulting frustration is evident in the blunt expression by interviewee n.5: "I'm going to be ruthless here!" This statement reflects the emotional burden experienced by stakeholders who repeatedly face administrative barriers. In general, the Connectivity pillar demonstrates that while local networks and community initiatives possess strong potential, their long-term success is often constrained by institutional fragmentation, lack of governance, and unstable political leadership. Therefore, strengthening connectivity requires not only community collaboration but also sustained, transparent, and strategic institutional support.

## 13.5 Turning the Tide: Conclusions

The analysis of SIT initiatives in Venice demonstrates that their sustainability and impact depend on the interaction of purpose, organizational adaptability, and connectivity. In terms of Sustainable Purpose, the study reveals that cultural heritage projects are strongly motivated by intrinsic values such as personal growth, passion for culture, and a sense of responsibility toward the local community. These personal motivations contribute to the formation of a shared mission oriented to preserving historical memory and local identity. However, the findings also highlight critical challenges: long-term economic sustainability remains a major issue, and there is a need for greater inclusivity to address the needs of marginalized or disadvantaged groups. Moreover, the lack of information, limited access, and financial difficulties experienced by certain segments of the population indicate that the sustainable purpose of SIT must extend beyond heritage preservation to include social equity and accessibility. Thus, the sustainable purpose pillar should be understood as integrating cultural value with community well-being, balancing mission-driven goals with viable economic models.

The Travelling Organization pillar shows that SIT initiatives require continuous adaptation, learning, and leadership capacity to navigate complex socio-economic contexts. However, the evidence indicates that fragmented networks and communication barriers hinder collaboration and slow down the development of shared vision among stakeholders. The involvement of multiple actors increases the number of beneficiaries, but it also increases the need for coordination and governance capacity. Additionally, the lack of coordination between government levels (local, regional, national) further complicates the capacity of organizations to respond to changing conditions and pursue long-term strategies. This organizational fragmentation reduces the scalability and resilience of initiatives, limiting their ability to generate long-term regeneration processes. Therefore, SIT requires a travelling organization mindset capable of building cohesive networks, maintaining continuity, and adapting to institutional constraints through collaborative learning.

The Connectivity pillar emerges as a central determinant of the long-term success of SIT. Political and institutional support is crucial for project continuity, yet bureaucratic and administrative complexities create operational difficulties that undermine collaboration. The absence of effective governance and planning processes, combined with weak communication strategies, reduces the visibility and legitimacy of local initiatives. Moreover, the system's dependence on the vision of the mayor or current administration creates instability, and the frequent turnover of public administration staff further weakens continuity and trust. These structural weaknesses highlight the importance of strengthening multi-level governance, improving communication strategies, and creating stable support mechanisms. In particular, enhancing connectivity requires better coordination between government levels, clearer institutional pathways, and a stronger commitment to inclusive participation, ensuring that benefits reach marginalized communities.

Beyond their local relevance, the findings also highlight the broader contribution of SIT to the objectives of the 2030 Agenda for Sustainable Development. By promoting inclusive participation, safeguarding cultural heritage, and strengthening community resilience, SIT initiatives directly support several UN Sustainable Development Goals,[39] particularly SDG 8 (Decent Work and Economic Growth), SDG 10 (Reduced Inequalities), SDG 11 (Sustainable Cities and Communities), and SDG 17 (Partnerships for the Goals). The Venetian case demonstrates that heritage-based tourism, when guided by a clear sustainable purpose, adaptive organizational practices, and strong connectivity, can function as a lever for socially inclusive and territorially balanced development. However, the challenges identified—such as governance fragmentation, limited accessibility, and uneven participation—also reveal the risk of misalignment between local practices and global sustainability commitments. In this sense, SIT offers not only a practical pathway for advancing the Agenda 2030 at the local level, but also a critical framework for assessing how sustainability principles are translated into concrete governance arrangements and everyday practices within tourism destinations.

In conclusion, these findings highlight significant gaps in the scientific literature on SIT. Despite growing research interest, there is still insufficient attention to institutional barriers, governance fragmentation, and inclusivity issues that affect the sustainability of social innovation initiatives. Future research should explore the interplay between purpose, organizational adaptability, and connectivity in different territorial contexts, and examine how multi-level governance and inclusive strategies can enhance the long-term viability of SIT. Addressing these gaps would contribute to a more comprehensive understanding of SIT and support the development of policy frameworks that promote sustainable, inclusive, and resilient cultural heritage-based tourism.

[39] See https://sdgs.un.org/goals and https://sdgs.un.org/2030agenda

**Olga Tzatzadaki** is a researcher (PhD in Urban Planning and Public Policies), whose work focuses on social innovation in tourism, processes of local and proximity-based micro-regeneration, and territorial narratives. She collaborates with University Iuav of Venice, where she carries out academic tutoring activities within the second-level Master ProPART in Participatory Planning. Her research adopts community-based approaches and participatory governance models and explores literary narratives as a methodology for socio-spatial analysis, investigating the role of nonconventional languages in the critical interpretation of urban and territorial transformations. She contributed to European-funded research projects under the Italian National Recovery and Resilience Plan (PNRR) through NextGenerationEU (Interconnected Nordest Innovation Ecosystem), focusing on Tourism, Culture, and Creative Industries.

# Horizontal Networks for the Implementation of "Next-Door" Welfare Facilities: Case Study from the Third Sector

14

Giovanna Muzzi

**Abstract**

This chapter proposes a reflection from the intertwining of welfare services and "next-door/proximity" spaces with a bottom-up and horizontal approach, subsequently involving the territorial dimension, social issues, and the role of volunteering, proposing an approach that straddles research and activism. Starting from a nationwide survey in Italy to investigate the recent phenomenon of "P*ortineria di Quartiere*" ("Neighborhood concierge"), the focus is on experiences of horizontal networks that have launched "next-door" welfare facilities, mostly in challenging neighborhoods, highlighting opportunities and critical issues they are facing.

## 14.1 Introduction

Over the last few years, the debate on the need to rethink and reshape traditional welfare has returned to the fore and has been amplified during and following the pandemic period. This stems from the fact that welfare issues and needs in the broadest sense have changed rapidly over time and have required urgent action and rapid, decisive intervention.[1]

[1] Nel.lo O.; Blanco I.; Gomà R.:"El apoyo mutuo en tiempos de crisis. La solidaridad ciudadana durante la pandemia Covid-19". Ciudad de Buenos Aires: CLACSO -Consejo Latinoamericano de Ciencias Sociales. 2022.

G. Muzzi (✉)
Dipartimento di Culture del Progetto, Università Iuav di Venezia, Venezia, Venezia, Italy
e-mail: gmuzzi@iuav.it





P. Wollmann et al. (eds.), *The Sustainable Organization – Volume 2*, Future of Business and Finance, https://doi.org/10.1007/978-3-032-23802-3_14

Innovative and pioneering methods, alongside traditional ones, have sought to respond to new needs and demands for services, but they require a reformulation of the frameworks of meaning and governance within which they operate. Some organizations see the hybridization of different categories and methods of service delivery as a way of offering solutions and proposals where public actors and traditional forms of welfare are unable or fail to reach, particularly in the most challenging neighborhoods.

This contribution focuses on horizontal networks and "next-door/proximity" welfare centers as places where groups of people come together to respond to needs and work jointly to build a "next-door" welfare system.[2] The prevailing point of view concerns actors who, in relation to "next-door" spaces and the provision of welfare services, complement public action, namely third sector entities,[3] both for-profit and nonprofit, commercial entities and other forms of association (committees, informal groups, general associations), and considers spaces that are the result of horizontal and bottom-up activation processes.

The research focused on a specific type of local welfare service ("*Portineria di quartiere*" or Neighborhood concierge"), investigating the phenomenon on a broader scale with a nationwide survey across Italy.

In this text, the term "*Portineria di Quartiere*" ("Neighborhood concierge") refers to a place that has both a physical and a virtual dimension, with the aim of meeting the needs,[4] requests and demands of the population in a given area, thus defining the concierge service as a local service.

They can be defined as "neighborhood concierge" for places that provide innovative local services to meet the needs of residents, particularly the most vulnerable groups such as the elderly and people with a migrant background, have one—or more—physical location in the area in which they operate, and digital dimension for information and communication (website, social media channels, newsletters, and so on).

They also carry out activities to improve life in the neighborhood and promote a sense of belonging among residents, particularly in challenging neighborhoods where various forms of marginalization and social fragility overlap. They can also

[2] See also: Moreno, Carlos.: "Droit de cité: de la «ville-monde» à la «ville du quart d'heure»". Paris: Editions de l'Observatoire.2020; Manzini, E.: "Abitare la prossimità: idee per la città dei 15 minuti". Milano: Egea. 2021.

[3] Recently, in Italy, the reform of the Third Sector Code (Legislative Decree No. 117/2017) was approved to systematize entities operating in the third sector. This significant change came about through the establishment of the RUNTS - "*Registro unico del terzo settore*" (Single register of the third sector.) Although registration is not mandatory for organizations, it is becoming the tool through which the public sector formally recognizes organizations and associations operating in the third sector and monitors their activities, especially from a fiscal point of view. This contribution considers all entities operating in the third sector, regardless of whether or not they are registered in the national register.

[4] Bernardi, M.: «Portinerie di quartiere: innovazione sociale tra digitale e locale». In Enciclopedia Sociologica dei Luoghi, 1: 335–51. Milano: Ledizioni. 2019. Available at https://www.enciclopediasociologicadeiluoghi.it/wp-content/uploads/2021/02/ESL1-PortinerieDiQuartiere-Bernardi-336-351.pdf

be defined as urban welfare mechanisms in which the creation and management of new integrated public services for housing involves joint action between public actors and intermediaries from the third sector and local organizations.

## 14.2 Methodology

The research conducted is based on two general research questions concerning the ways in which local social and territorial services can contribute to the development of integrated public policies in the context of welfare and bottom-up/horizontal urban regeneration, and the forms and methods of interaction that exist between neighborhood concierge services and local institutions.[5]

Furthermore, qualitative methodological approaches have allowed for deeper reflection on the research questions identified, in order to identify the phenomenon under examination through a nationwide survey. The most general question concerns the forms and ways in which grassroots neighborhood spaces—and the activities promoted therein—are able to respond to the needs and demands expressed by the inhabitants of fragile neighborhoods.

Secondly, further research questions have arisen as to whether the term "neighborhood concierge" defines a type of welfare service that is different from other types of grassroots welfare initiatives, with a common underlying model, specific characteristics and shared distinctive features. Or whether, on the contrary, it is a sort of "label" that encompasses different types of initiatives.

The question arises as to whether this phenomenon has long-term prospects and what can be identified as the fundamental elements for ensuring a solid and lasting structure, the motivations and objectives for launching such an initiative and ensuring its continuity. The aim is to investigate whether and to what extent these initiatives arose from private and spontaneous impulses, or whether they were launched (or subsequently incorporated) within public policies. Finally, the study examines the existence of different forms of direct or indirect support from public institutions.

A further area of interest is investigating the forms and ways in which grassroots neighborhood spaces—and the activities promoted therein—manage to respond to the needs and demands expressed by the inhabitants of fragile neighborhoods. Questions are raised about the ways in which local socio-territorial presences can contribute to the construction of integrated public policies in the context of welfare and urban regeneration from the bottom up/horizontally. Hence, we propose to investigate whether and how neighborhood concierge services can be incorporated into public policies, becoming an integral part of a broader and more structured welfare system, and what forms and modes of interaction may exist between them and local institutions. We therefore intend to explore how they can also act as

[5] The possibility of having carried out this research derives from the fact that the author has held both the role of PhD student at Università Iuav di Venezia and active member of the local unit of Venice of the ETICity association, partner of the "*Portineria di Quartiere Mestre via Piave*", located in Venice mainland area.

intermediaries between residents and institutions, analyzing their role in facilitating dialogue and cooperation.

The nationwide survey stems from the need to investigate the extent of the phenomenon of "neighborhood concierge" in Italy, firstly by attempting to map on a large scale the existing situations or spaces associated with this practice. Secondly, the aim is to try to understand what elements they have in common and what elements differ, in order to determine which situations can be defined as "neighborhood concierge" and on the basis of which criteria. Thirdly, the survey aims to understand their capacity for advocacy, the degree of interaction and involvement with/within public institutions and public policies. Finally, there is interest in the similarities and differences between the Parisian model of *Lulu dans ma rue* (or some of its aspects) and the different variations that have been adopted by some of the portinerie in Italy that have taken it as a reference.

A qualitative approach was adopted, involving data collection through questionnaires for the nationwide survey on neighborhood concierge services. The research investigating this phenomenon was structured through a nationwide desk-analysis that identified around 60 cases in Italy that could be classified as "neighborhood concierge" (or other similar definitions).

First, a desk analysis of press coverage (carried out in spring 2022) was conducted, focusing in particular on neighborhood concierge services and issues relating to local welfare services and the establishment of new welfare facilities in vulnerable neighborhoods, identifying information and communications published on the main social media channels (Facebook, Instagram) and online searches. This was supplemented by snowball sampling,[6] thanks to information provided by the concierge services themselves or by other local authorities (municipal administrations, volunteer service centers, public bodies) that shared the contact details of other organizations, thus expanding the pool of organizations to be involved in the research. Further information was subsequently collected through the use of social networks[7] in cases where it was not possible to trace the concierge services through traditional channels (telephone, e-mail).

Half of these were involved in completing a questionnaire to explain the various ways in which this type of welfare service is organized (single activator or network), how it is managed, and the types of services and activities offered.[8] In addition, one

[6] Parker, C.; Scott, S.; GeddesA.: «Snowball Sampling». In SAGE Research Methods Foundations". Thousand Oaks: SAGE Publications Ltd. 2021. See at http: //methods. sagepub.com/foundations/snowball-sampling

[7] Browne, K.: «Snowball sampling: using social networks to research non-heterosexual women». In International Journal of Social Research Methodology 8 (1): 47–60.2005. https://doi.org/10.1080/1364557032000081663.

Baltar, F.; Brunet I.: «Social research 2.0: virtual snowball sampling method using Facebook». Internet Research 22 (1): 57–74.2021. https://doi.org/10.1108/10662241211199960.

[8] The survey was conducted with the support of CAVV-CSV Venezia (Coordination of Volunteer Associations of the Metropolitan City - CSV Venice), a third sector organization that supports volunteer organizations operating in the Metropolitan City of Venice. The collaboration took place over more than a year, from early 2022 to spring 2023, and saw an initial sharing of the interim

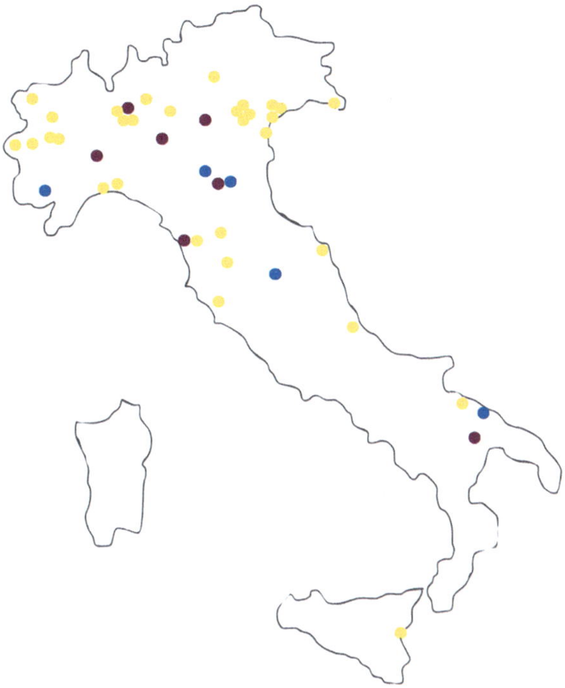

**Fig. 14.1** Map of "neighborhood concierge" identified in Italy (2002). Key: yellow: active, purple: closed, green: undefined status in, orange: opening soon. Source: Author's elaboration. Used with permission, all rights reserved

line of investigation concerned the urban environment in terms of where these facilities are located, with a focus on the urban and socio-spatial environment as well as the location of the facility itself. It should be noted that not all the entities participating in the survey identified themselves as "neighborhood concierge," or, in any case, the "next-door/proximity" welfare service is not one of the main activities of all the spaces identified. However, for the purposes of this survey, all the entities surveyed will be defined as "neighborhood concierge," specifying, where necessary, the differences and peculiarities that emerged (Fig. 14.1).

## 14.3 The Case Study

As this is a relatively emergent phenomenon, there are still few ad hoc reflections specifically concerning "neighborhood concierges," but an analysis of the objectives and activities of these centers reveals aspects related to issues such as local

progress of the research with the organization of a public conference in Mestre in October 2022, attended by representatives of some of the "neighborhood concierge" involved in the survey.

welfare spaces,[9] smart cities,[10] digital platforms,[11] the activation of networks within neighborhoods,[12] the sharing economy and the solidarity economy.[13] Since 2016, a number of practices have emerged in Italy that can be traced back to neighborhood concierge services,[14] with features reminiscent of the French experience of *Lulu dans ma rue*, from which many draw inspiration and assimilate some of the main characteristics.

The choice of the subject of the investigation—"neighborhood concierge"—is linked to the fact that it has been identified as a specific and emerging phenomenon that has expanded significantly since the recent pandemic, albeit on a fairly limited scale. There has been a proliferation of organizations[15] that have called themselves "neighborhood concierge" and have begun to provide welfare services to meet urgent needs, continuing to do so even after the emergency period has passed.

The survey has allowed us to highlight the territorial and social contexts in which concierges operate. As emerged from the state of the art, there is a decline in the welfare state and a contraction of public services, leaving some sociocultural and social-assistance[16] services uncovered, particularly for the most vulnerable populations. The issue of gentrification and expulsion of local populations in favor of wealthier social classes is also considered[17] as such processes in certain contexts deprive populations of services, even informal ones, welfare and local trade. A further area of analysis concerns the target populations, which are no longer limited to

[9] Munarin S.; Tosi M.C.: "Welfare space: on the role of welfare state policies in the construction of the contemporary city". LIST Lab Laboratorio Internazionale Editoriale. 2014.

Officina Welfare Space: "Spazi del welfare esperienze luoghi pratiche". Macerata: Quodlibet. 2011.

[10] Ménascé, D.; Vincent C.E.; Martin Moreau M.: «Smart Cities and new forms of employment». Field Actions Science Report, Special Issue 16: 16–21.2017.

[11] Flichy, P.: «Platforms at the Heart of New Form of Labour». DigiWorld Economic Journal, 108, 45–58.2017.

[12] Ranzini, A. L.: «Quartieri come reti. Le reti territoriali come dispositivi di inclusione tra competenza e rappresentanza». Doctoral Thesis. Venezia: Università Iuav di Venezia. 2020. https://air.iuav.it/handle/11578/286432.

[13] Laville, J.L.: «I servizi di prossimità: la costruzione sociale di un campo di attività economiche». In "Terzo settore, stato e mercato nella trasformazione delle politiche sociali in Europa", 115–55. Sociologia e politica sociale., Sez. 2.: Ricerche 7. Milano: Franco Angeli. 1997.

Cacciari, P.: "101 piccole rivoluzioni Storie di economia solidale e buone pratiche dal basso". Milano. Altreconomia. 2016.

[14] Taverna, E.: «Portinerie di quartiere in ogni città: tre buoni motivi per promuoverle». Labsus. 13/02/2018. https: //www.labsus.org/2018/02/portinerie-quartiere-citta-tre-buoni-motivi-promuoverle/

[15] Mainly, it refers to the so-called "Civil Society Organizations (CSOs)": Non State, not-for-profit, voluntary entities formed by people in the social sphere that are separate from the State and the market.

[16] León, M.; Pavolini E.: «'Social Investment' or Back to 'Familism': The Impact of the Economic Crisis on Family and Care Policies in Italy and Spain». South European Society and Politics 19 (3): 353–69.2014. https: //doi.org/10.1080/13608746.2014.948603

[17] Cremaschi, M.:"Tracce di quartieri: il legame sociale nella città che cambia". Milano: Franco Angeli. 2008

standard populations such as low-income families or those with a migrant background, but also include other categories with limited economic means and few supportive family and/or friendship ties, such as elderly and self-sufficient indigenous populations. As evidenced by the literature, and even more so following the pandemic, the spectrum of vulnerable or marginalized groups requiring welfare and community services has broadened.[18] The issues frequently considered are found in neighborhoods "in need,"[19] characterized by varying degrees of complexity and stratification of fragility, as also reported in the specific case study. In general, this model of "next-door" welfare facility directly offers and organizes local services and activities, as well as providing proper orientation to promote those existing in the neighborhood and the surrounding area. They offer their own activities and support the development of socialization initiatives in the neighborhood, providing meeting places where people can talk and be heard. For example, the most frequently promoted activities include group meetings, cultural activities, training courses, markets, and basic healthcare services.

The next area concerns the objectives, the needs they address, and the results achieved with respect to the own goals set by the various "neighborhood concierge." The importance of dialogue and listening to understand the needs and concerns of various populations is emphasized, as is the concierge service's ability to act as an intermediary body within civil society, becoming a spokesperson for collective needs. In fact, it is noted that the "neighborhood concierges" often respond to collective needs and interests and redirect to the services of the designated area, both institutional ones and those offered by other third sector entities, acting as an "antenna."

Furthermore, it emerged that the provision of local services is linked to the social and cultural offerings typical of "third places"[20] and hybrid spaces, as they respond simultaneously, and often even at the same time, to various needs, promoting the interaction between different needs and different responses. The specific services and activities offered are investigated, as it is increasingly found that new populations and new lifestyles imply different needs and therefore services other than the standard ones, with a view to promoting social relationships, whether the provision is free or paid.

The results of the research conducted through questionnaires attempt to provide an initial overview of the phenomenon. Many of the cases analyzed are located in neighborhoods or urban areas affected by various types of complexity linked to sociodemographic changes, socio-economic changes and issues related to urban transformation. In this context, it is important to emphasize the central importance

---

[18] Pavolini, E.; Sabatinelli, S.; Vesan P.: «I servizi di welfare in Italia alla prova della pandemia. Uno sguardo di insieme». Politiche sociali 8 (2): 211–32.2021. https: //dx. doi.org/10.7389/101678

[19] Ostanel, E.: "Spazi fuori dal Comune: Rigenerare, includere, innovare". Milano: Franco Angeli. 2018.

[20] Oldenburg, R.: The Great Good Place: Cafés, Coffee Shops, Community Centers, Beauty Parlors, General Stores, Bars, Hangouts, and How They Get You Through the Day. New York: Paragon House. 1989.

of a permanent facility, an open space that is continuously active and to which all people can freely turn. From a spatial point of view, these welfare facilities are located both in urban centers and in semi-peripheral or peripheral areas of cities, offering services aimed at the entire population, with particular attention to vulnerable groups, and tailor them to the territorial and socio-economic context in which they operate. They also have a virtual dimension that helps to promote networking through social media platforms and instant messaging applications. Furthermore, they usually reuse underutilized or unused spaces, contributing to the urban regeneration of the area in which they operate.

The research therefore explored whether/how spaces defined as "neighborhood concierge" can be tools for revitalizing urban contexts characterized by the presence and stratification of socio-economic fragility, through the promotion of which practices (related to welfare but also economic, cultural, and productive) and with the involvement of which actors. With regard to the latter, it was considered important to also investigate the methods of activation and involvement (in the case of multi-stakeholder practices), where decision-making and implementation power lies, and what strategies and tools are adopted to ensure economic and managerial sustainability.

Particular attention was paid to investigating whether—and through what means—the concierge services respond to the needs of the neighborhoods in which they operate and with what capacity for planning and outreach. Furthermore, given that this is an emerging phenomenon that is proposed as a solution to be promoted in fragile neighborhoods, the question was raised as to whether there was public recognition of the potential expressed from the bottom up through the concierge services or, if not, whether the services themselves could take action to build a dialogue with local administrations to affirm/demonstrate the value of their role in local welfare, also with a view to long-term sustainability.

Following analysis of the data collected regarding the management of the initiatives, various cases were identified driven by groups of people (institutional and/or noninstitutional) who often work with a medium-to-short time frame ("project-based") in complex and challenging contexts/neighborhoods to respond to needs.

Firstly, an analysis was made based on how a "neighborhood concierge" was conceived and launched, identifying those created on the initiative of individuals (e.g., a commercial business offering local services, a day center opening a concierge desk); "neighborhood concierge" created through the coordination of various networked entities, particularly those belonging to the third sector (cooperatives, associations, social enterprises); and few introduced by public bodies (local administrations, public housing authorities).

At the same time, these welfare facilities intersect with the methods used to obtain the main sources of funding, the methods of fundraising and the actors who make them available. Various cases have been identified, including projects funded through public or private tenders; projects resulting from co-design processes with third sector organizations; initiation and management within public policies; facilities managed with own funds and/or charitable donations.

Subsequently, starting from the data collected, an attempt was made to understand the various forms in which a "next-door" welfare facility is defined. To do so, a matrix has been developed taking into account the combination of space hybridization and offer of "next-door" services variables, considering three levels of gradient, from minimum (+) to maximum (+++). From this combination three types of "next-door" welfare facilities have been identified: desk—service—concierge.

The "concierge desk" offers a limited range of local services within a highly hybridized space (dedicated to various activities, even simultaneously). It involves the activation of certain local services, often within a dedicated time slot in a space dedicated to other purposes (it is one of many services offered), in which the concierge service is a (small) part of the activities carried out.

The "concierge service" has an intermediate level of both services and space hybridization. The provision of "next-door" services takes place continuously within a space mainly dedicated to other purposes/initiatives (e.g., bar, day care center, etc.), where the concierge service is one of several active elements.

The "neighborhood concierge" offers a wide range of local services within a relatively unhybridized space. It is primarily aimed at providing local services in a designated area, which is also made available for other activities, particularly socializing and gathering.

This type of classification could also be applied to other types of horizontal organizations, where the main activity has various gradients within the main space (Table 14.1).

With regard to the second level of horizontal network building that this type of organization engages in, it is clear that they are capable of interacting with other institutional and noninstitutional entities in the area in order to respond to a wide range of needs, requirements, and requests.

These are places that try to provide an immediate response and take action themselves, but above all, they redirect people to the right person/office fairly quickly, without the cumbersome and complicated procedures that access to services is increasingly subject to, making them less accessible (e.g., digital bookings/access to service and digital literacy of certain segments of the population, or poor/limited knowledge of the Italian language and bureaucratic mechanisms). They are easily accessible facilities with a primary and fundamental function of listening. In practice, a person enters, asks for information, and if possible, they are helped on the spot. Otherwise, they are redirected to local agencies/offices/services/organizations that are more specific and competent in relation to the request made.

**Table 14.1** Hybridization of space and offer of "next-door" services

| | | Hybridization of space | | |
|---|---|---|---|---|
| | | +++ | ++ | + |
| Offer of "next-door" services | +++ | | | Neighborhood concierge |
| | ++ | | Concierge service | |
| | + | Concierge desk | | |

Source: Author's elaboration. Used with permission, all rights reserved

This indicates the existence of a primary internal network and subsequently a much larger network that is connected to in order to try to respond to the needs and requests that are intercepted. Over time, in many cases, a substantial network of partners has been consolidated and expanded, collaborating in various capacities and with different roles. In this sense, the "*Portineria di quartiere*" can be considered the central hub of the funnel, which on the one hand acts as a catalyst for the needs and requests expressed by users and volunteers, while on the other hand collects information on the services offered in the area by public and private bodies and the third sector.

A significant aspect to highlight concerns the central role played by the space itself, which should ideally be open continuously. It also becomes significant for the construction of networks, connections, the formation of small self-help and exchange groups, and as a constant presence in a challenging neighborhood (Figs. 14.2, 14.3, 14.4, and 14.5).

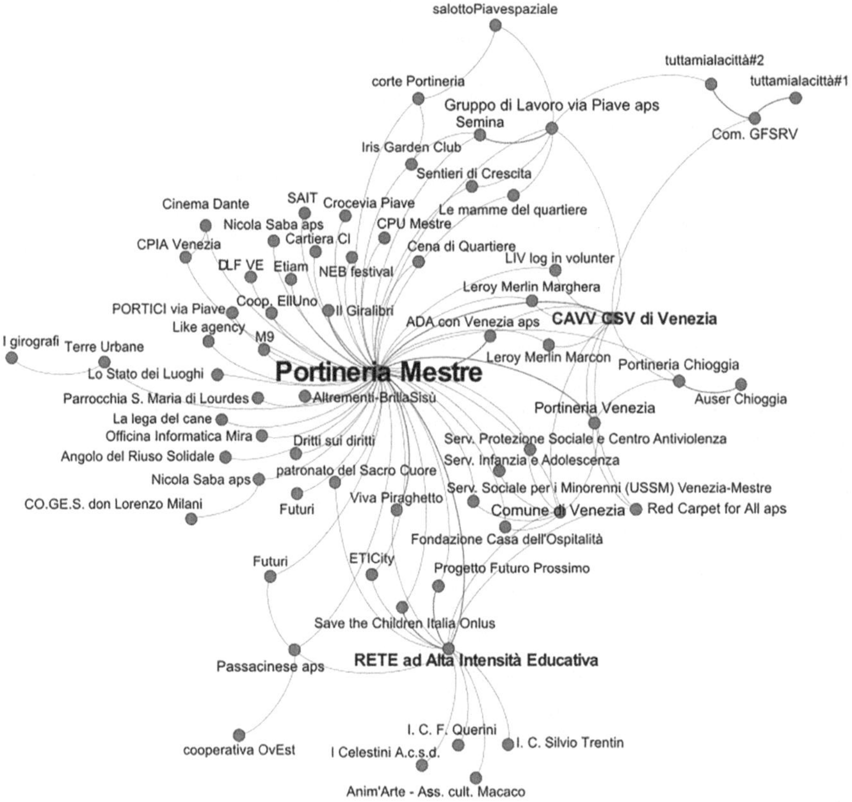


**Fig. 14.2** Local network activated by the "Portineria di Quartiere di Mestre via Piave," (Venice mainland area) that includes local government projects, third sector bodies, associations, local communities, informal groups and horizontal initiatives in the neighborhood. *Source*: Author's elaboration. Used with permission, all rights reserved

**Fig. 14.3** The entrance to the "Portineria di Quartiere Mestre via Piave." Photo credit Giovanna Muzzi, used with permission. All rights reserved

**Fig. 14.4** School support and homework help activities within the "Portineria di Quartiere Mestre via Piave." Photo credit Giovanna Muzzi, used with permission. All rights reserved

**Fig. 14.5** Concert of the "Perpetual Saxophone quartet" in the inner courtyard of the "Portineria di Quartiere Mestre via Piave." Photo credit Giovanna Muzzi, used with permission. 

On the other hand, some critical issues that have emerged concern economic sustainability in terms of the availability of ongoing funding, as these are often "project-based/call-based" initiatives or "ad hoc internal public funding" initiatives, as well as fragility in management, relying largely on volunteers and related issues regarding volunteer turnover.

## 14.4 The Relevance of the Case Study in Relation to the Agenda 2030 and the Three-Pillars Model

Regarding the 2030 Agenda for Sustainable Development,[21] the example of "next-door" welfare facilities analyzed here highlights the increasing cooperation between public and private sector organizations, specifically in the third sector, to achieve shared objectives in the broad sphere of welfare. Their actions can be interpreted as

[21] https://sdgs.un.org/2030agenda.

contributing to the objectives of the United Nations Sustainable Development Goals (SDGs),[22] particularly SDG 3 (Good Health and Well-Being), SDG 4 (Quality Education), SDG 10 (Reduced Inequalities), and SDG 11 (Sustainable Cities and Communities), which collectively emphasize social well-being, equitable access to opportunities, and the development of inclusive and sustainable communities.

Regarding the Three-Pillar Model,[23] the "neighborhood concierge" can be considered an example of a problem-solving mindset as they can be defined as small organizations with a high capacity for adaptation, tailored to the individual situation and context.

Regarding the Sustainable Purpose, these types of community welfare spaces have implemented different strategies to launch this type of initiative, drawing on different resources and implementing new ones. They have a well-defined objective of offering support and welfare services, in a broad sense, to residents, especially the most marginalized and vulnerable. This guiding objective provides guidance and certainty to those involved, often volunteers who, in order to dedicate their time and energy to a cause, must be clear about the reason, otherwise their motivation would wane. And it is precisely in a period of crisis and transformation like the recent pandemic that has seen people come together and organize themselves to respond to an emergency, giving rise to many "neighborhood concierge" throughout Italy, as an alternative and complementary place to public action, to guarantee services quickly and effectively. They can be seen as an alternative and complementary place, but not a substitute, for public action to guarantee services quickly and effectively, in particular for those sections of the population not always easily intercepted by the local administration.

Furthermore, in this case study it is noted that a critical issue emerges regarding the long-term continuation of activities and general sustainability. For this reason, too, "neighborhood concierge" they can be considered Travelling Organizations as they go through different types of flows and transformations, often demonstrating a capacity for learning and adaptability in relation to external requirements, which sometimes leads to internal transformation within the organization itself. This also derives from the fact that they are often located, not by chance, in challenging and complex urban contexts that require a continuous ability to listen and respond quickly to needs as well as to change the way of acting and offering services. Remaining rigid and immovable would make this type of "next-door" welfare facility less functional in a short time with respect to their objectives of responding promptly and accurately to the needs that emerge, given

---

[22] See https://sdgs.un.org/goals.

[23] Wollmann, P.; Kühn, F.; Kempf, M.; Püringer, R. (Eds.): Organization and leadership in disruptive times: Design and implementation using the 3-P-Model. Cham: Springer Nature, 2021.

Wollmann, P.; Stricker, L.; Pemler, D. (Eds.): Reading an Unknown Organization - Leaders in New Roles Thriving in Challenging Environments with the 3-P-Model Mindset. Cham: Springer Nature, 2024.

Wollmann, P.; Pemler, D.; Ndrevataj, M. (Eds.): The Sustainable Organization: How Organizations Address the 17 UN SDGs Using the 3-P-Model. Cham: Springer Nature, 2025.

that their strength, and fragility at the same time, lies in being adaptable and widespread.

The ability to create connections and networks is fundamental for "neighborhood concierge" as they are a place of exchange between requests and services. They often start from the very idea that they cannot function alone and that it is necessary to develop the ability to interconnect with other entities, whether public, third sector or private, as well as residents, to find various forms of necessary resources and strengthen local networks. Considering the third Pillar, the levels of connectivity that a "next-door" welfare facility is able to build over time also allow one to achieve greater stability and sustainability, interacting with the various types of stakeholders present in the territory.

## 14.5 Closing Remarks: Conclusions

In summary, a neighborhood concierge is a social and territorial facility that provides a variety of services in different fields, not only welfare services, and organizes activities. It is also a meeting place, often located in a complex and challenging neighborhood, with limited space but offering a kaleidoscope of initiatives and possibilities. It is dedicated to receiving requests and needs which, if it is not possible to resolve directly, it directs to the services available in the area. It is itself composed of a small network of partners that connects and integrates with larger systems, possibly collaborating with public and private institutions that already operate in the same context.

Possible guidelines and suggestions for similar "next-door" facilities include forming a small, consolidated initial group, identifying a specific, recognizable space that is open continuously and consistently, the ability to gradually structure possible connections and offshoots, and above all, a great deal of adaptability, flexibility, open-mindedness, and patience.

**Giovanna Muzzi** holds a master's degree in architecture and a PhD in Spatial Planning and Public Policy from the Università Iuav di Venezia, with a thesis on welfare-related spaces. Since 2018, she has held the position of teaching assistant in urban and regional planning and has obtained research grants at Università Iuav di Venezia and Università Ca′ Foscari di Venezia. She is currently vice-president of the ETICity association, project manager for ETICity and the "Portineria di Quartiere Mestre via Piave," and coordinator of urban regeneration and urban reactivation projects in collaboration with local organizations. She has a long experience with leading urban knowledge itineraries for training bodies (public offices, schools, international universities).

# Sustainable Development in World Heritage Policies and Management: The Italian Context

# 15

Remi Wacogne and Chiara Carolina Donelli

**Abstract**

Whilst UNESCO can be considered to have played a key role in the establishment of environmental conservation within the United Nations' policies, the World Heritage Convention of 1972 and the associated program's contribution to the sustainable development perspective has been widely emphasized. The adoption of Agenda 2030 and inclusion of target 11.4 'Strengthen efforts to protect and safeguard the world's cultural and natural heritage' under Sustainable Development Goal 11 'Make cities and human settlements inclusive, safe, resilient and sustainable', as well as the adoption by UNESCO of the Policy for the Integration of a Sustainable Development Perspective into the Processes of the World Heritage Convention, mark a significant progress in that sense. Yet, more than half a century after the adoption of the Convention many limits in its implementation have been highlighted, with special reference to protection and management processes at the local level. In addition to monitoring processes, we suggest that a closer look at policies adopted by Member States and lower-tier governments is needed. Focusing on the Italian context—the richest in World Heritage properties globally—in this contribution we propose a survey of WH-related provisions adopted at the national and regional levels, with reference also to local processes. The analysis shows that they include both explicitly dedicated and implicitly relevant policies, pertaining to three main fields: properties'

R. Wacogne (✉)
Department of Architecture and Arts, IUAV University of Venice, Venezia, Italy
e-mail: rwacogne@iuav.it

C. C. Donelli
Venice School of Management, Ca' Foscari University, Venice, Italy
e-mail: chiara.donelli@unive.it


P. Wollmann et al. (eds.), *The Sustainable Organization – Volume 2*, Future of Business and Finance, https://doi.org/10.1007/978-3-032-23802-3_15

management and governance, support and funding for their protection and management, and funding for their promotion, hereby including further applications to the WH List.

## 15.1 Introduction

The Convention concerning the Protection of the World Cultural and Natural Heritage, adopted in 1972 by the General Conference of UNESCO, has mobilized an articulate set of organizations operating at different levels, with reference to the document itself and to the related program, in and on places—more than 1200 by now—listed within this framework. Yet a certain asymmetry between the World Heritage (WH) program's stated universality and processes occurring within those places or 'properties', to use the Convention's term, has been increasingly emphasized.[1] Resuming the Three-Pillar Model on Sustainability and UN SDG Coverage[2] as for UNESCO itself and Advisory Bodies, State Parties' National governments, regional governments, and local authorities, dedicated agencies and stakeholders involved in the management of World Heritage properties, we suggest that this asymmetry can be understood in terms of: (1) varied purposes (other than implementing the WH Convention), (2) differentiated abilities to 'travel', and (3) uneven levels of connectivity (in general and more specifically between themselves).

The comprehensive protection of both cultural and natural heritage has been explicitly introduced in the global sustainability agenda only a decade ago through SDG 11, Target 11.4 which foresees to 'Strengthen efforts to protect and safeguard the world's cultural and natural heritage'. Among the previous Millennium Development Goals, n. 7 did aim to 'Ensure environmental sustainability' but was rather limited in ambition—to 'reverse' or 'reduce loss'—and scope—mainly 'environmental resources and biodiversity—and excluded the cultural dimension of

[1] Labadi, S. (2017), UNESCO, World Heritage, and Sustainable Development: International Discourses and Local Impacts, in: Gould, P./Pyburn, K. (2017, Eds), Collision or Collaboration. One World Archaeology, Springer, Cham, pp. 45–60; Donelli, C. C. (2025), Talking the Talk or Walking the Walk? Sustainability and SDGs in WHS Planning and Reporting, in Donelli, C. C. (2025), Talking the Talk of Culture and Sustainability. Exploring Sustainability Reporting in the Cultural Sector. FrancoAngeli, Milano, pp. 105–150.

[2] The so-called Three-Pillar Thinking Model for organization design and leadership consists of the pillars: **Sustainable Purpose** (the raison d'être of an organization, bringing orientation and certainty to the people for their joint endeavor and success), **Travelling Organization** (an organization accepting to be in a permanent state of flux and transformation), and **Connectivity** (the ability to interconnect on the journey with all needed). For details see: Wollmann, P.; Kühn, F.; Kempf, M. (Eds.): Three Pillars of Organization and Leadership in Disruptive Times—Navigating Your Company Successfully through the twenty-first Century Business World. Cham: Springer Nature, © 2020. See also Chap. 2 of this book or Chaps. 1 and 2 of the previous book: Peter Wollmann / Doris Pemler / Mersida Ndrevataj (eds.): The Sustainable Organization—How Organizations Address the 17 UN SDGs Using the 3-P-Model, Vol. 1, Cham: Springer Nature 2025.

heritage altogether'.[3] Instead, UNESCO can be considered to have played a key role in the establishment of environmental conservation within the United Nations' policies.[4] The World Heritage Convention of 1972 and the associated program's contribution to the sustainable development perspective, which associates nature conservation with the conservation of cultural properties and values, has been widely emphasized,[5] and formulated by UNESCO itself with reference to the 2030 Agenda through the *Policy for the Integration of a Sustainable Development Perspective into the Processes of the World Heritage Convention.*[6]

Yet, more than half a century after the adoption of the Convention many limits in its implementation have been highlighted, with special reference to protection and management processes at the local level.[7] At the same time, several threats are becoming increasingly critical, including mainly tourism development[8] and climate change,[9] in addition, in many cases, to armed conflicts. Based on available

[3] United Nations, The Millennium Development Goals Report (2015), https://www.un.org/millenniumgoals/2015_MDG_Report/pdf/MDG%202015%20rev%20(July%201).pdf (accessed the 15th December 2025).

[4] Maurel, C. (2013), L'Unesco, un pionnier de l'écologie? Une préoccupation globale pour l'environnement, 1945–1970, in: Monde(s), Vol 3, No. 1, pp. 171–192.

[5] Bandarin, F. (2015). Sustainability in the World Heritage Convention: The making of a policy framework. In M. Albert (Ed.), Perceptions of sustainability in Heritage Studies (pp. 35–46). De Gruyter; Wiktor-Mach, D. (2018), What role for culture in the age of sustainable development? UNESCO's advocacy in the 2030 Agenda negotiations, in: International Journal of Cultural Policy, Vol. 26, No. 3, pp. 312–327.

[6] Bille Larsen, P./Logan, W. (2018, Eds), Policy-making at the World Heritage-Sustainable development interface: Introductory remarks, in: Bille Larsen, P./Logan, W. (2018, Eds) World Heritage and Sustainable Development. New Directions in World Heritage Management, Routledge, pp. 3–20.

[7] Pforr, A. F./Pforr, C. (2024), 50 Years UNESCO World Heritage Convention: Curse or Cure?, in: Pforr, C./Pillmayer, M./Joppe, M./Scherle, N./Pechlaner, H. (2024, Eds), Tourism Policy-Making in the Context of Contested Wicked Problems: Politics, Paradigm Shifts and Transformation Processes, Emerald Publishing Limited, Leeds, pp. 37–47; Labadi, S. (2024), The World Heritage Convention at 50: management, credibility and sustainable development, in: Journal of Cultural Heritage Management and Sustainable Development, Vol. 14, No. 4, pp. 750–764.

[8] Bertacchini, E./Revelli, F./Zotti, R. (2024), The economic impact of UNESCO World Heritage: Evidence from Italy, in: Regional Science and Urban Economics, Vo. 105, n. 103996; Kutlu, D./Zanbak, M./Soycan, S./Kasalak, M.A./Aktaş Çimen, Z. (2024), The Influence of World Heritage Sites on Tourism Dynamics in the EU 27 Nations, in: Sustainability, Vol. 16, no. 9090; Škrabić Perić, B./Šimundić, B./Muštra, V./Vugdelija, M. (2021), The Role of UNESCO Cultural Heritage and Cultural Sector in Tourism Development: The Case of EU Countries, in: Sustainability, vol. 13, 5473.

[9] Cave, C. (2022), Climate Change and World Heritage: An Introduction, in: Albert, MT./Bernecker, R./Cave, C./Prodan, A.C./Ripp, M. (2022, Eds), 50 Years World Heritage Convention: Shared Responsibility—Conflict & Reconciliation. Heritage Studies, Springer, Cham, pp. 215–225; Daly, C. (2022), Climate Action and World Heritage: Conflict or Confluence? in: Albert, MT./Bernecker, R./Cave, C./Prodan, A.C./Ripp, M. (2022, Eds), 50 Years World Heritage Convention: Shared Responsibility—Conflict & Reconciliation. Heritage Studies. Springer, Cham, pp. 239–251; Reimann, L./Vafeidis, A.T./Brown, S./Hinkel, J./Tol, R. S. J. (2018), Mediterranean UNESCO World Heritage at risk from coastal flooding and erosion due to sea-level rise. Nat Commun 9,

documents, including the State-level Periodic Report issued by Italian authorities within the third cycle of Periodic Reporting, this chapter provides a survey of policies adopted at the national, regional and local levels with reference to World Heritage properties, in the country that still hosts the highest number of them globally.

## 15.2 Integrating Sustainable Development into World Heritage Processes: The Global Level

The *Policy for the Integration of a Sustainable Development Perspective into the Processes of the World Heritage Convention*, adopted by the General Assembly of the States Parties to the same Convention 2 months after the UN's 2030 Agenda, highlighted that both documents share a complementary if not common perspective, and the reciprocal benefits that their implementation should bring. The document refers especially, as 'dimensions of sustainable development', to 'Environmental Sustainability', 'Inclusive Social Development', 'Inclusive Economic Development', and 'Fostering Peace and Security'. It translated into a revision of the *Operational Guidelines for the Implementation of the World Heritage Convention*, with paragraph 119 now foreseeing that 'World Heritage properties may sustain biological and cultural diversity and provide ecosystem services and other benefits, which may contribute to environmental and cultural sustainability'. As for State Parties, paragraph 14bis within the same document encourages them 'to mainstream into their programs and activities related to the World Heritage Convention the principles of the relevant policies adopted by the World Heritage Committee, the General Assembly of States Parties to the Convention and the UNESCO Governing Bodies, [...] including the 2030 Agenda for Sustainable Development and international human rights standards.'[10]

The World Heritage Committee itself issued or endorsed other policy documents explicitly referring to sustainable development, including the Updated Policy *Document on climate action for World Heritage* (2021) and the *Sendai Framework for Action 2015–2030*. The *Strategic Action Plan for the Implementation of the Convention, 2012–2022* included, as Goal 3 (out of six), to ensure that 'heritage protection and conservation consider present and future environmental, societal and economic needs', with 'increased consideration of sustainable development through connecting conservation to communities' as the expected outcome. Ultimately, the *World Heritage Regional Action Plan for Europe and North America 2024–2031*

4161 (2018); Shirvani Dastgerdi, A./Sargolini, M./Broussard Allred, S./Chatrchyan, A./De Luca, G. (2020), Climate Change and Sustaining Heritage Resources: A Framework for Boosting Cultural and Natural Heritage Conservation in Central Italy, in: Climate, vol. 8(2), 26; UNESCO (2025), Climate change in Mediterranean World Heritage cities, https://unesdoc.unesco.org/ark:/48223/pf0000394251 (accessed the 15th December 2025).

[10] UNESCO World Heritage Centre (2025), The Operational Guidelines for the Implementation of the World Heritage Convention, available at https://whc.unesco.org/en/guidelines/ (accessed the 15th December 2025).

included as Strategic objective 4 to 'harness the full potential of World Heritage properties as drivers for sustainable development policies and as a means to achieving human well-being within planetary boundaries', with specific reference to Sustainable Development Goals.

But misalignment with the 2030 Agenda, along with erratic implementation of strategies adopted at various levels of World Heritage governance, have been widely pointed at,[11] suggesting the World Heritage Sustainable Development Policy's being reported as 'the most widely used policy across the Europe and North America region to set national policies or strategies for the protection of cultural and natural heritage'[12] to be questioned. For instance, among 1135 Voluntary Local Reviews mentioning 'Culture', 'World Heritage (or equivalent)' is cited only in 34, the same figure as 'conserve/conservation' and 'gastronomy/cuisine' and more than 'folk/folklore' (six occurrences) or 'performance' (31) for instance, but much less than 'museum(s)' (118) or 'recreation(al)' (231).[13]

Whilst, as observed by Daly,[14] all State Parties to the WH Convention are also signatories of the Paris Agreement, both the setting of strategies and goals that harness the mutual benefits of sustainable development and the world's heritage conservation, as well as their effective implementation, seem seldom satisfactory, if pursued at all.

---

[11] Donelli, C. C. (2025), Talking the Talk or Walking the Walk? Sustainability and SDGs in WHS Planning and Reporting, in Donelli, C. C. (2025), Talking the Talk of Culture and Sustainability. Exploring Sustainability Reporting in the Cultural Sector. FrancoAngeli, Milano, pp. 105–150; Fuhrmann, C. (2022), Sustainability—A Guiding Principle of the World Heritage Convention—What Has Been Achieved—What Is Missing—What Is the Future Perspective. In: Albert, MT./Bernecker, R./Cave, C./Prodan, A.C./Ripp, M. (2022, Eds), 50 Years World Heritage Convention: Shared Responsibility—Conflict & Reconciliation. Heritage Studies. Springer, Cham, pp. 445–457; Labadi, S. (2024), The World Heritage Convention at 50: management, credibility and sustainable development, in: Journal of Cultural Heritage Management and Sustainable Development, Vol. 14, No. 4, pp. 750–764; Pforr, A. F./Pforr, C. (2024), 50 Years UNESCO World Heritage Convention: Curse or Cure?, in: Pforr, C./Pillmayer, M./Joppe, M./Scherle, N./Pechlaner, H. (2024, Eds), Tourism Policy-Making in the Context of Contested Wicked Problems: Politics, Paradigm Shifts and Transformation Processes, Emerald Publishing Limited, Leeds, pp. 37–47.

[12] UNESCO World Heritage Centre (2024), Third Cycle Periodic Report for Europe and North America (Extract from document WHC/24/46.COM/10A.Rev., including Part I.), p. 33 (available at https://whc.unesco.org/uploads/pages/documents/document-153-73.pdf, accessed the 15th of December 2025).

[13] Culture 2030 Goal campaign (2021), Culture in the Localization of the 2030 Agenda: An Analysis of Voluntary Local Reviews, published in Barcelona, Paris, Abidjan, Montreal, The Hague and Brussels, in the frame of the 4th UCLG Culture Summit taking place on 9–11 September 2021, https://cultureactioneurope.org/wp-content/uploads/2022/01/af_culture2030goal_2021.pdf (accessed the 15th of December 2025).

[14] Daly, C. (2022), Climate Action and World Heritage: Conflict or Confluence? *op. cit.*

## 15.3 World Heritage-Related Policies in Italy: The National Level

Italy ratified the World Heritage Convention in 1977 by a dedicated act, but specific policies have been introduced only in the last few decades, and especially from the Law 20th February 2006, n. 77, entitled 'Special protection and enjoyment measures of Italian sites of cultural, landscape and environmental interest included in the "World Heritage List", placed under UNESCO's protection'.[15] In addition to a yearly-based funding scheme for World Heritage 'sites', the act incited collaboration between local authorities and stakeholders, especially as for the elaboration, adoption and implementation of dedicated Management Plans. Here again, this last phase appears often incomplete, which has been interpreted in terms of Management Plans' scarce bindingness on the one hand, and of complex management arrangements, scarce accountability conditions and management transparency and lack of participatory governance on the other.[16] To what extent the 'priority of intervention' foreseen by the act for World Heritage properties, 'should they be devolved funding according to current legislation', has been applied, is unclear; instead, as Sect. 15.5 will show in more detail, funding is aimed at 'a compatible management of sites and properties' but more often, at least supposedly, to 'a correct relationship between tourist flows and cultural services as provided'.

Within the State-level Periodic Report Third Cycle, the UNESCO Office (operating at the Ministry of Culture) reported as 'principal pieces of legislation for the protection, conservation and presentation of cultural and/or natural heritage' the Legislative Decree 22 January 2004, n. 42 or 'Cultural Heritage and Landscape Code', along with further amendments and provisions reorganizing the Ministry of Culture. Few references are made to World Heritage properties, which were introduced through the same amendments, and none to their effective management and governance.

But other provisions either specifically dedicated to, or at least referring to, World Heritage properties have been adopted, including Decree of the President of the Council of Ministers 10th May 2018, n. 76, which defined the characteristics of infrastructures to be submitted to public hearing (*Dibattito pubblico*). The scope of such procedures, consisting in public consultations over large public infrastructures' environmental impacts was thus broadened with specific reference to World Heritage properties, including their buffer zones, seven of which were to

[15] Tubertini, C. (2023), The Coordinated Protection of UNESCO Sites: Lights and Shadows of the Italian Experience. In: López Martín, A. G./Demarsin, B./Baroncini, E./Regueiro Dubra, R./Stoica, R.-I. (2023, Eds), Forever Young. Celebrating 50 Years of the World Heritage Convention, Stem Mucchi Editore, Modena, pp. 625–640.

[16] Badia, F. (2018), The management plan for the World Heritage Sites as a tool of performance measurement and sustainability reporting: Opportunities and limits in the Italian context, in S. Makuvaza (2018, Ed), Aspects of management planning for cultural World Heritage Sites, Springer, Cham, pp. 25–36.

be at least indirectly affected by the 17 infrastructures projects submitted between 2022 and 2023.[17]

Other national provisions have resumed Law n. 77/2006s layout, consisting in funding schemes. Following Covid-19 and with a view to 'bring back tourism attractiveness', a fund aimed at Municipalities which territories include World Heritage properties and/or (with a smaller endowment) are part of the Creative Cities Network was introduced in 2021 by the Ministry of Tourism and resumed in 2022 and 2023. Funded interventions consist in tourist websites, itineraries, marketing and events as well as works and/or staging for cultural venues and landscapes. A trend in expanding policies' focus to other UNESCO programs is evident in another scheme, the 'UNESCO Natural Sites Programme', instituted by the Ministry of the Environment in 2022. Providing funding for climate change adaptation to Municipalities in which territory World Heritage sites associated with 'natural' criteria are situated in its first year, it shifted its beneficiaries in favor of Man and the Biosphere reserves and Intangible Cultural Heritage elements in 2023. The program was not resumed since.

As suggested in this brief overview, national policies with reference to World Heritage properties and increasingly also to other UNESCO cultural programs have been characterized by some degree of experimentation, but also discontinuity in funding and incoherence, be it only in terms of Ministries overseeing dedicated programs—namely those in charge for Culture, the Environment and more recently Tourism (which was previously a matter administered together with Culture in a single Ministry). Whilst the National Statistics Institute includes the number of WH properties within its articulate set of 'fair and solidarity well-being' indicators,[18] the 2022 National Sustainable Development Strategy includes only two indicators referred more generally to cultural and natural heritage, respectively in terms of 'public expenditure per capita for biodiversity and landscapes' protection', and 'overnight stays in tourist accommodations'.[19] No comprehensive nor coherent strategy dedicated to the implementation of the WH Convention thus emerges in the Italian context, unlike, for instance, that of Ireland, despite the latter's shorter experience as a State Party and a more limited presence in the World Heritage List.[20] Whereas the varied but significant impacts of tourism,[21] as well as threats posed by

[17] Wacogne, R. (2024), When Heritage and Landscape Values Are Confronted by Planned Infrastructures: A Glance at 'Public Debate' ('Dibattito Pubblico') Procedures in Italy. Sustainability. 2024; 16(14):6218.

[18] ISTAT (2025), 9. Paesaggio e patrimonio culturale, in ISTAT, Rapporto Bes 2024: il benessere equo e sostenibile in Italia, available at https://www.istat.it/wp-content/uploads/2025/11/09-Paesaggio-e-patrimonio-culturale.pdf (accessed the 15th of December 2025).

[19] Ministero dell'Ambiente e della Sicurezza Energetica (2022), Strategia Nazionale per lo Sviluppo Sostenibile, available at https://www.mase.gov.it/portale/documents/d/guest/snsvs22_aggiornata_completa-pdf (accessed the 15th of December 2025).

[20] World Heritage in Ireland (2025), Strategy for World Heritage in Ireland 2025–2035, available at https://www.worldheritageireland.ie/strategy-for-world-heritage-in-ireland-2025-2035/ (accessed the 15th of December 2025).

[21] Bertacchini, E./Revelli, F./Zotti, R. (2024), The economic impact of UNESCO World Heritage: Evidence from Italy, in: Regional Science and Urban Economics, Vo. 105, n. 103996.

climate change on WH properties at the national level would require decisive action in that sense, in their response to the Periodic Reporting Third Cycle Italian authorities acknowledged that there is limited integration of the conservation and protection of cultural and natural heritage as a strategic element in national sustainable development policies and strategies in:

- Protecting biological and cultural diversity and providing ecosystem services and benefits
- Strengthening resilience to natural hazards and climate change
- Contributing to inclusion and equality
- Enhancing the quality of life and well-being
- Respecting, protecting, and promoting human rights
- Respecting, consulting, and involving indigenous peoples and local communities
- Achieving gender equality
- Promoting economic investment and quality tourism
- Strengthening capacity-building, innovation and local entrepreneurship (UNESCO World Heritage Centre, 2024b).

In the same document they declared to make no use of the *Policy Document on the Impacts of Climate Change on World Heritage Properties*, *Strategy for Reducing Risks from Disasters at World Heritage Properties*, *World Heritage Capacity Building Strategy*, nor of the *World Heritage Sustainable Development Policy* (The current WH situation in Italy is described in detail in Tables 15.1, 15.2, and 15.3).

**Table 15.1** UNESCO World Heritage properties in Italy per: typology and region (in listing chronological order) (authors elaboration on https://whc.unesco.org/en/list/ and ISTAT, 2025). Used with permission, all rights reserved

| Region | Cultural properties | Natural properties | Total |
|---|---|---|---|
| **Lombardia** | Rock Drawings in Valcamonica<br>Church and Dominican Convent of Santa Maria delle Grazie with "The Last Supper" by Leonardo da Vinci<br>Crespi d'Adda<br>Sacri Monti of Piedmont and Lombardy<br>Rhaetian Railway in the Albula/Bernina Landscapes<br>Mantua and Sabbioneta<br>Longobards in Italy. Places of the Power (568-774 A.D.)<br>Prehistoric Pile Dwellings around the Alps<br>Venetian Works of Defence between the 16th and 17th Centuries: Stato da Terra – Western Stato da Mar | Monte San Giorgio | 10 |

(continued)

**Table 15.1** (continued)

| Region | Cultural properties | Natural properties | Total |
|---|---|---|---|
| **Toscana** | Historic Centre of Florence<br>Piazza del Duomo, Pisa<br>Historic Centre of San Gimignano<br>Historic Centre of Siena<br>Historic Centre of the City of Pienza<br>Val d'Orcia<br>Medici Villas and Gardens in Tuscany<br>The Great Spa Towns of Europe | Ancient and Primeval Beech Forests of the Carpathians and Other Regions of Europe | 8 |
| **Veneto** | Venice and its Lagoon<br>City of Vicenza and the Palladian Villas of the Veneto<br>Botanical Garden (Orto Botanico), Padua<br>City of Verona<br>Prehistoric Pile Dwellings around the Alps<br>Venetian Works of Defence between the 16th and 17th Centuries: Stato da Terra – Western Stato da Mar<br>Le Colline del Prosecco di Conegliano e Valdobbiadene<br>Padua's fourteenth-century fresco cycles | The Dolomites | 8 |
| **Campania** | Historic Centre of Naples<br>18th-Century Royal Palace at Caserta with the Park, the Aqueduct of Vanvitelli, and the San Leucio Complex<br>Archaeological Areas of Pompei, Herculaneum and Torre Annunziata<br>Costiera Amalfitana<br>Cilento and Vallo di Diano National Park with the Archeological Sites of Paestum and Velia, and the Certosa di Padula<br>Longobards in Italy. Places of the Power (568-774 A.D.)<br>Via Appia. Regina Viarum | | 7 |
| **Sicilia** | Archaeological Area of Agrigento<br>Villa Romana del Casale<br>Late Baroque Towns of the Val di Noto (South-Eastern Sicily)<br>Syracuse and the Rocky Necropolis of Pantalica<br>Arab-Norman Palermo and the Cathedral Churches of Cefalú and Monreale | Isole Eolie (Aeolian Islands)<br>Mount Etna | 7 |

(continued)

**Table 15.1** (continued)

| Region | Cultural properties | Natural properties | Total |
|---|---|---|---|
| **Lazio** | Historic Centre of Rome, the Properties of the Holy See in that City Enjoying Extraterritorial Rights and San Paolo Fuori le Mura<br>Villa Adriana (Tivoli)<br>Villa d'Este, Tivoli<br>Etruscan Necropolises of Cerveteri and Tarquinia<br>Via Appia. Regina Viarum | Ancient and Primeval Beech Forests of the Carpathians and Other Regions of Europe | 6 |
| **Emilia-Romagna** | Ferrara, City of the Renaissance, and its Po Delta<br>Early Christian Monuments of Ravenna<br>Cathedral, Torre Civica and Piazza Grande, Modena<br>The Porticoes of Bologna | Ancient and Primeval Beech Forests of the Carpathians and Other Regions of Europe<br>Evaporitic Karst and Caves of Northern Apennines | 6 |
| **Piemonte** | Residences of the Royal House of Savoy<br>Sacri Monti of Piedmont and Lombardy<br>Prehistoric Pile Dwellings around the Alps<br>Vineyard Landscape of Piedmont: Langhe-Roero and Monferrato<br>Ivrea, industrial city of the 20th century | | 5 |
| **Puglia** | Castel del Monte<br>The Trulli of Alberobello<br>Longobards in Italy. Places of the Power (568-774 A.D.)<br>Via Appia. Regina Viarum | Ancient and Primeval Beech Forests of the Carpathians and Other Regions of Europe | 5 |
| **Friuli-Venezia Giulia** | Archaeological Area and the Patriarchal Basilica of Aquileia<br>Longobards in Italy. Places of the Power (568-774 A.D.)<br>Prehistoric Pile Dwellings around the Alps<br>Venetian Works of Defence between the 16th and 17th Centuries: Stato da Terra – Western Stato da Mar | The Dolomites | 5 |
| **Basilicata** | The Sassi and the Park of the Rupestrian Churches of Matera<br>Via Appia. Regina Viarum | Ancient and Primeval Beech Forests of the Carpathians and Other Regions of Europe | 3 |
| **Sardegna** | Su Nuraxi di Barumini<br>Funerary Tradition in the Prehistory of Sardinia – The domus de janas | | 2 |
| **Trento** | Prehistoric Pile Dwellings around the Alps | The Dolomites | 2 |
| **Liguria** | Portovenere, Cinque Terre, and the Islands (Palmaria, Tino and Tinetto)<br>Genoa: Le Strade Nuove and the system of the Palazzi dei Rolli | | 2 |

(continued)

**Table 15.1** (continued)

| Region | Cultural properties | Natural properties | Total |
|---|---|---|---|
| **Umbria** | Assisi, the Basilica of San Francesco and Other Franciscan Sites<br>Longobards in Italy. Places of the Power (568-774 A.D.) | | 2 |
| **Calabria** | | Ancient and Primeval Beech Forests of the Carpathians and Other Regions of Europe | 1 |
| **Abruzzo** | | Ancient and Primeval Beech Forests of the Carpathians and Other Regions of Europe | 1 |
| **Bolzano/ Bozen** | | The Dolomites | 1 |
| **Molise** | | Ancient and Primeval Beech Forests of the Carpathians and Other Regions of Europe | 1 |
| **Marche** | Historic Centre of Urbino | | 1 |
| **Valle d'Aosta/ Vallée d'Aoste** | | | 0 |

**Table 15.2** Provisions adopted by Italian Regional authorities with reference to World Heritage properties (authors' elaboration based on institutional websites). Used with permission, all rights reserved

| Region | Provision's title or main intervention | Provision's reference (in Italian) |
|---|---|---|
| **Campania** | Modification of a precedent annual budget to include WH properties among potential beneficiaries of funding originally intended for Intangible Cultural Heritage elements | Legge regionale 29 giugno 2021, n. 5—Misure per l'efficientamento dell'azione amministrativa e l'attuazione degli obiettivi fissati dal DEFR 2021—2023—Collegato alla stabilità regionale per il 2021, art. 65, comma 8 \|<br>L.R. 38/2017—Legge di stabilità 2018, art. 10 c. 2 bis |
| **Friuli** | 'Measures in support of regional heritage included in the UNESCO World Heritage List' | Legge regionale 25 luglio 2019, n. 11—TESTO VIGENTE dal 10/08/2024—Testo coord. alla Legge regionale 7 agosto 2024 n.7—Misure di sostegno a favore del patrimonio regionale inserito nella lista del patrimonio mondiale posto sotto la tutela dell'UNESCO, Art. 8 bis |
| | 'Regulation for the installation of renewable energy plants within the Regional territory' | Legge regionale 4 marzo 2025, n. 2 Norme per l'installazione di impianti a fonti rinnovabili sul territorio regionale. |

(continued)

**Table 15.2** (continued)

| Region | Provision's title or main intervention | Provision's reference (in Italian) |
|---|---|---|
| **Lazio** | 'Initiatives for the promotion of regional sites included in the World Heritage List' | L.R. 08 Novembre 2004, n. 14 Iniziative per la valorizzazione dei siti regionali iscritti nella lista del patrimonio mondiale dell'umanità |
| **Lombardia** | Specific article on 'Sites included in the UNESCO World Heritage List' within the updated regional policy for culture | Legge Regionale 7 ottobre 2016, n. 25 Politiche regionali in materia culturale—Riordino normativo, art. 18 (Siti inclusi nella lista del patrimonio mondiale dell'umanità dell'UNESCO) |
| **Piemonte** | 'Guidelines for local masterplans and building regulations' compliance with protection indications for the UNESCO site Vineyard Landscape of Piedmont: Langhe-Roero and Monferrato' | DGR n. 26-2131 del 21 settembre 2015, Linee guida per l'adeguamento dei piani regolatori e dei regolamenti edilizi alle indicazioni di tutela per il Sito Unesco: i paesaggi vitivinicoli del Piemonte Langhe-Roero e Monferrato |
| | Specific article on 'UNESCO Programmes' within the Regional policy for culture | Legge regionale 1 agosto 2018, n. 11. Disposizioni coordinate in materia di cultura, Art. 19. (Programmi UNESCO) |
| | Measure 'Promotion of sites included in the UNESCO World Heritage List' (within the European Regional Development-Funded Regional Operational Programme 2014-2020) | POR FESR 2014-2020, Asse V—Azione V.6c.7.1. Azione V.6c.7.1—Interventi per la tutela, la valorizzazione e la messa in rete del patrimonio culturale, materiale e immateriale, nelle aree di attrazione di rilevanza strategica tale da consolidare e promuovere processi di sviluppo—Misura Valorizzazione dei siti inseriti nella lista del patrimonio mondiale UNESCO |
| **Puglia** | 'Guidelines for funding aimed at the enhancement and promotion or UNESCO heritage in Puglia' | L. R. n. 35/2020, art.17—Approvazione delle Linee guida per la erogazione di finanziamenti finalizzati alla valorizzazione e promozione del patrimonio UNESCO in Puglia |
| **Sicilia** | Provision of funding for Municipalities in which WH properties and/or Geoparks are situated, within the Annual budget for 2024-2026 | Legge regionale 16 gennaio 2024, n. 1, Legge di stabilita' regionale 2024-2026, art. 2 comma 16 |
| **Toscana** | Stated Regional support for 'UNESCO sites' management programmes' as well as for local governments committed to application processes | Legge regionale 25 febbraio 2010, n. 21, Testo unico delle disposizioni in materia di beni, istituti e attività culturali. TITOLO III—Beni paesaggistici, CAPO I—Valorizzazione culturale dei beni paesaggistici, Art. 33 |

(continued)

**Table 15.2** (continued)

| Region | Provision's title or main intervention | Provision's reference (in Italian) |
|---|---|---|
| **Veneto** | 'Initiatives supporting the UNESCO application of the colline del Prosecco di Conegliano e Valdobbiadene' | Legge regionale 6 giugno 2019, n. 21—Iniziative a sostegno della candidatura UNESCO delle colline del Prosecco di Conegliano e Valdobbiadene |
| | Specific article on 'UNESCO Heritage' within the regional policy for culture | LR 6 maggio 2019, n. 17 Legge per la cultura, Art. 19 Patrimonio UNESCO. |
| | 'Interventions aimed at UNESCO sites', within the 'Specific Objective' consisting of 'Strengthening the role of culture and sustainable tourism for economic development, social inclusion and social innovation' (within the European Regional Development-Funded Regional Operational Programme 2021-2027) | PR Veneto FESR 2021-2027, Priorità 4—Un'Europa più sociale e inclusiva attraverso l'attuazione del pilastro europeo dei diritti sociali—Obiettivo Specifico 4.6—Rafforzare il ruolo della cultura e del turismo sostenibile nello sviluppo economico, nell'inclusione sociale e nell'innovazione sociale", Azione 4.6.1—Interventi a favore dei siti UNESCO. |

**Table 15.3** Projects funded in 2024 as part of Legge n. 77/2006 scheme (listed in funding amount ascending order, authors' elaboration). Used with permission, all rights reserved

| WH property (or group of) | Funded project's title | Applicant/beneficiary organization |
|---|---|---|
| Crespi d'Adda | The old wash house at Crespi d'Adda 'The recovery of water' | Municipality of Capriate San Gervasio |
| Historic Centre of Siena | Challenge for high schools in the province of Siena | Municipality of Siena |
| Italian World Heritage Properties' Association | Teaching without borders: a participatory and accessible platform for World Heritage | *Associazione Beni italiani Patrimonio Mondiale* |
| Historic Centre of Florence | Educating to World Heritage. Complementary activities | Municipality of Florence |
| The Trulli of Alberobello | Alberobello's Youth Management Plan | Municipality of Alberobello |
| Padua's fourteenth-century fresco cycles | World Heritage streets (temporary title) | Municipality of Padua |
| Botanical Garden (Orto Botanico), Padua | Accessible Botanical Garden's collections: toward an Orto 'for all' | Centro dl Ateneo Orto Botanico—Università degli Studi di Padova |
| Assisi, the Basilica of San Francesco and Other Franciscan Sites | Communicating Franciscan places: Francesco's herald students | Municipality of Assisi |
| The Great Spa Towns of Europe | A shared heritage | Municipality of Montecatini Terme |

(continued)

**Table 15.3** (continued)

| WH property (or group of) | Funded project's title | Applicant/beneficiary organization |
|---|---|---|
| Ivrea, industrial city of the 20th century | The UNESCO site Ivrea, industrial city of the 20th century: between conservation, vision and innovation | Municipality of Ivrea |
| Cathedral, Torre Civica and Piazza Grande, Modena | Research project to guarantee and share the site's integrity | Municipality of Modena—Museo Civico d'Arte |
| Cathedral, Torre Civica and Piazza Grande, Modena | Project Management Plan 2025-2030 | Municipality of Modena—Museo Civico d'Arte |
| Archaeological Area and the Patriarchal Basilica of Aquileia | Aquileia between science and culture | Fondazione Aquileia |
| Historic Centre of Florence | AI and World Heritage. An integrated information system for the Historic Centre of Florence | Municipality of Florence |
| Genoa: Le Strade Nuove and the system of the Palazzi dei Rolli | Let's protect ourselves: community and prevention of vandalism in the historic centre | Municipality of Genoa |
| Genoa: Le Strade Nuove and the system of the Palazzi dei Rolli | Twenty years of the UNESCO site Genoa: Le Strade Nuove and the system of the Palazzi dei Rolli | Municipality of Genoa |
| The Porticoes of Bologna | Lines of shade: climate itineraries under the porticoes | Municipality of Bologna |
| Piazza del Duomo, Pisa | The monumental Camposanto: a journey through time | Municipality of Pisa |
| Early Christian Monuments of Ravenna | Welcoming experiences without barriers. Interventions for the accessibility and increased enjoyment of the site | Municipality of Ravenna |
| Early Christian Monuments of Ravenna | Heritage communities. Fostering awareness and accessibility of the sites' values | Municipality of Ravenna |
| Portovenere, Cinque Terre, and the Islands (Palmaria, Tino and Tinetto) | Updating the World Heritage site's Management Plan and Disaster Risk Management Man | Municipality of Porto Venere |
| Ferrara, City of the Renaissance, and its Po Delta | Ferrara and its Po Delta: new tools to communicate the site's universal value | Municipality of Ferrara |
| Properties in Lombardia | Heritage for all: inclusion and accessibility at Lombard UNESCO sites | Regione Lombardia |
| Medici Villas and Gardens in Tuscany | Medici Villas and Gardens in Tuscany: towards a participatory governance for the site | Regione Toscana |

(continued)

**Table 15.3** (continued)

| WH property (or group of) | Funded project's title | Applicant/beneficiary organization |
|---|---|---|
| Sacri Monti of Piedmont and Lombardy | Updating the Management Plan | Ente di gestione Sacri Monti (Regional Trust) |
| City of Vicenza and the Palladian Villas of the Veneto | Projects under the 'Conservation' objective within the Management Plan | Municipality of Vicenza |
| Colline del Prosecco di Conegliano e Valdobbiadene | Implementation and compliance of the Technical Regulation foreseen by the Management Plan | Associazione per il Patrimonio Mondiale Colline del Prosecco di Conegliano e Valdobbiadene |
| Venetian Works of Defence between the 16th and 17th Centuries: Stato da Terra – Western Stato da Mar | From one fortress to the next through sustainable tourism | Municipality of Bergamo |
| Vineyard Landscape of Piedmont: Langhe-Roero and Monferrato | Actions and best practice for sustainable tourism in the Vineyard Landscape of Piedmont | Associazione per il Patrimonio del Paesaggi vitivinicoli di Langhe-Roero e Monferrato |
| Longobards in Italy. Places of the Power (568-774 A.D.) | Longobards 4.0: from the past, Guides for the future | Associazione Italia Langobardorum |
| Late Baroque Towns of the Val di Noto (South-Eastern Sicily) | Accessible Vai di Noto | Distretto Turistico Sud EST |

## 15.4 World Heritage-Related Policies in Italy: The Regional Level

Italy is characterized by a certain degree of territorial decentralization, in which regional authorities, either with ordinary or special status, share legislative powers with the State in many policy fields; this complex multi-level governance framework, along with a sectoral approach, has significant implications for sustainable development policies and their implementation.[22] As Table 15.2 shows, based on a survey of pages dedicated to regional UNESCO sites on institutional websites, it has also enabled relevant measures with reference to WH properties at the regional level, mostly in Regions which host a significant number of them—all regions but Valle d'Aosta/Vallée d'Aoste count at least one WH property or element or area (of a wider serial or transregional or transnational property), half of them have five or more, Lombardia as many as ten (Table 15.1).

Just as for sustainable development policies, urban and regional planning appears as a key competence of regional authorities for WH properties' conservation and management. Regional landscape plans have thus been mandated to

[22] Bianchi, S./Richiedei, A. (2023), Territorial Governance for Sustainable Development: A Multi-Level Governance Analysis in the Italian Context, in: Sustainability, vol. 15, no. 3: 2526.

include prescriptions also for World Heritage Sites, which implementation would require a comprehensive assessment but provide at least a formal integration in planning processes.[23] Specific regulations have also been adopted for both historic vineyards' landscapes listed in Italy, respectively the Vineyard Landscape of Piedmont: Langhe-Roero and Monferrato in 2015 and Colline del Prosecco di Conegliano e Valdobbiadene in 2019, which local master plans are to comply with. Friuli-Venezia Giulia has instead defined WH properties' core and buffer zones, as well as MAB Reserves, as the main field of application of the 'Regulation for the installation of renewable energy plants within the Regional territory' it adopted in 2025.

Another field for regional initiative is that of shared—if only to a limited extent—governance. Following Veneto, which has instituted a board gathering WH properties situated at least partially in its territory in 2006,[24] other regional governments have introduced forms of consultation between properties' managing authorities and their own administrations, in some cases through dedicated offices and/or with the support of in-house agencies or external organizations (e.g., Universities or cultural foundations). Here again, an increasing interest for other UNESCO programs is evident in the inclusion of Creative Cities, MAB Reserves and/or Geoparks within those arrangements, also as part of 'UNESCO clusters' in the case of Piemonte, following a concept focused on the potential of cultural heritage for territorial development.[25]

In addition to the Legge n. 77/2006—or anticipating it (with a regional act adopted in 2004) in the case of Lazio—several regional governments have introduced funding schemes, either in association with one-off provisions like those evoked above, and/or in a medium-term perspective within their annual budgets or (EU-funded) strategic programs. In Veneto the same regional act which foresaw the planning regulation specifically intended for the Colline del Prosecco di Conegliano e Valdobbiadene also secured funding for the then-candidate WH property's application process, to which success only a few months later it may have contributed. But such funding's aims are not always coherent with strategies and policies adopted by WH governing bodies: those intended to support further applications are thus in contradiction with the Global Strategy for a representative, balanced and credible World Heritage List, which should exclude further applications from Italy, the country being already the most represented.

[23] Giudice, B./Voghera, A. (2024), Planning for landscape and heritage. A community perspective to overcome risks and vulnerabilities in the Italian case study, in: International Journal of Disaster Risk Reduction, Vol. 110, 104610.

[24] Galli, F./Wacogne, R. (2023), Verso una gestione coordinata dei siti Patrimonio Mondiale? L'esperienza del Veneto. In: Adobati, F./De Bonis, L./Marson, A. (2023, Eds), Agire sul patrimonio, Atti della XXIV Conferenza Nazionale SIU Dare valore ai valori in urbanistica, Brescia, 23–24 giugno 2022, vol. 08, Planum Publisher e Società Italiana degli Urbanisti, Roma-Milano, pp. 96–101.

[25] Bertacchini, E./Segre, G. (2021), Designazioni UNESCO e sviluppo locale: una prospettiva di distretto culturale, in: Economia della Cultura, Rivista dell'Associazione per l'Economia della Cultura, No. 4/2021, pp. 603–608.

If 'sectoral fragmentation of competence collides with the holistic layout of sustainability' (Bianch/Richiedei, 2023), this applies also to World Heritage, in regions where dedicated provisions have been adopted, that is almost half of them. Different provision typologies and aims have thus been adopted, which contribution to sustainable development as well as compliance to general World Heritage policies yet appear much variable. Similarly to the imbalance still evident within the WH List at the global level, the uneven territorial distribution of properties between Italian regions appears as a constitutive factor shaping both the possibilities and the limits of implementing the WH Convention and related sustainable development objectives.

## 15.5 World Heritage-Related Policies in Italy: The Local Level

Among the WH governance levels considered here, the 'local' stands out as the focus of them all: properties or 'sites' are (de-)listed or submitted as such, ordained as targets or beneficiaries of national and regional policies, and at least potentially dealt with by a wide array of institutions and stakeholders with often competing interests and practices.[26] In order to provide an overview of WH-related policies at such scale, in a country that counts almost 60 properties, this section focuses on projects funded as part of the Legge n. 77/2006 scheme for the year 2024, before drawing upon three cases where the integration between heritage conservation and sustainable development appears particularly critical.

Table 15.3 lists projects presented by 'referent' managing authorities for about a half of Italy's WH properties. Coherently with the variety of the latter's' typologies, different types of authorities are represented: mostly Municipalities (but through different sectors, among which three 'UNESCO offices'), but also three dedicated local associations, two Regions and one Regional trust, one foundation and one University.

Although a deeper exploration of such projects' objectives and implementation would be required to assess their coherence with WH as well as sustainable development strategies and policies, the very titles show no explicit reference to the latter, but for several mentioning 'sustainable tourism'. Climate change is tackled only in relation to the Porticoes of Bologna, through the design of 'climate itineraries' which seem limited to fostering awareness rather than adaptation. Six projects focus on Management Plans or tools, including Portovenere, Cinque Terre, and the Islands (Palmaria, Tino and Tinetto)'s Disaster Risk Management Man as well as an 'integrated information system' for the Historic Centre of Florence. Instead, six projects focus on accessibility, and several on 'communication', awareness building or similar purposes.

[26] Boland, P./Hastings, T./Kumar, M.S./McKay, S. (2023), World Heritage Sites and the question of scale in governance and politics: A study of Stonehenge. In: International Journal of Cultural Property, no 2023;30(2):155–176.

Of course, this list of funded projects is only partially indicative of experimentations at work in Italian WH sites. For instance, the WH platform Canopy includes four Italian 'case studies' in terms of 'innovative strategies and practices that integrate heritage conservation with sustainable development', respectively in Bologna, Florence, Mantua and Sabbioneta, and Urbino.[27] The National Alliance for Sustainable Development has introduced a series of similar contents, reviewing 'actions related to energy efficiency, energy communities and decarbonation of historic centers' in October 2025.[28]

But local experimentations too may be incoherent with coherent with strategies and policies adopted by WH governing bodies as well as with the 2030 Agenda. The case of Venice and its lagoon is both iconic and revealing. After several 'Advisory' of 'Reactive Monitoring' missions commissioned by the World Heritage Committee over the last decade, the same overturned the draft decision enacting its inclusion within the World Heritage in Danger List in 2023. This and previous decisions are referred to by national, regional and local governments within several provisions, including the ban on large ships from the Giudecca canal and San Marco basin stated in 2021, the Access Fee enforced in 2023[29] as well as a project aiming at the 'management and enhancement of tourist flows' funded by the Ministry of Tourism, a part of which was devolved to the organization of a pop concert and an amusement fair in Mestre (under 'Intervento 8: Design and realization of events and initiatives').[30]

The same Colline del Prosecco of Conegliano and Valdobbiadene, for which a special planning regulation was adopted by the regional government of Veneto, have been impacted since their listing by intensifying wine production and tourism branding, raising concern over related environmental pressures, including pesticide use, monoculture expansion and deforestation, soil erosion and biodiversity loss.[31]

A third significant case is that of a 44-acre photovoltaic plant project in Putifigari, Sardinia, where the Domus de janas was recently listed as World Heritage. After negative advice from the Ministry of Culture, the project was successfully submitted to Environmental Impact Assessment, and seemingly ultimately disapproved by the Ministry of Environment and Energy Security in

[27] UNESCO World Heritage Centre (2025), World Heritage Canopy Heritage Solutions for Sustainable Futures, https://whc.unesco.org/en/canopy/ (accessed the 15th of December 2025).

[28] Alleanza Italiana per lo Sviluppo Sostenibile (2025), Rubrica: #UnescoSostenibile, https://asvis.it/rubrica-patrimonio-in-transizione/ (accessed the 15th of December 2025).

[29] UNESCO World Heritage Centre (2025), State of Conservation: Venice and its lagoon, https://whc.unesco.org/en/soc/4723 (accessed the 15th of December 2025).

[30] Comune di Venezia (2025), Venezia e la sua Laguna: gestione e valorizzazione dei flussi turistici, https://www.comune.venezia.it/it/content/venezia-e-la-sua-laguna-gestione-e-valorizzazione-dei-flussi-turistici (accessed the 15th December 2025).

[31] Basso, M., Fregolent, L. (2021). Fighting Against Monocultures: Wine-Growing and Tourism in the Veneto Region. In: Fregolent, L., Nel·lo, O. (eds) Social Movements and Public Policies in Southern European Cities. Urban and Landscape Perspectives, vol 21. Springer, Cham.

November 2025.[32] In this process the Regional and local governments have expressed shared concern about the project's impacts, and a Regional assessor suggested in occasion of a dedicated Municipal council that the Domus de janas 'is a heritage to maintain not only for Putifigari or Sardinia, but above all for Italy, which has kept its world record of inscribed properties thanks to the site's inscription'.[33]

## 15.6 Concluding Remarks

Policies adopted, and projects developed at, National, Regional, and local levels in Italy with reference to WH properties show limited alignment with the 2030 Agenda for Sustainable Development, but also with orientations formulated by the program's governing bodies. This is coherent with the scarce implementation of WH policies and strategies highlighted at the global level but is especially significant in the context of the most represented country within the List.

A common focus on promotion or enhancement (*valorizzazione*) of WH properties is evident from the national to the local level, a priority that seems questionable with respect especially to tourism regulation and climate change adaptation. Provisions adopted at the same level and in some cases by the same organization are sometimes contradictory, e.g., the Ministry of Environment's contested assessment of the photovoltaic plant at Putifigari's environmental impact (which it seems to have later overturned) less than 3 years after the introduction of the 'UNESCO Natural Sites Programme'.

At least in the Italian context, it seems that only a part of organizations dealing with World Heritage share the program's (sustainable) purposes at least apparently, if at all; and that those who do pursue them sincerely may succeed only thanks to their ability to travel and connect. Further acknowledging the workings and implications of 'WH Politics' from the global to the local level, as attempted here, should lead us to navigate them and connect with the various positive experimentations carried out in many places and organizations toward harnessing the mutual benefits of WH conservation and sustainable development.

[32] ANSA (2025), Annullata autorizzazione impianto fotovoltaico a Putifigari, https://www.ansa.it/sardegna/notizie/2025/11/29/annullata-autorizzazione-impianto-fotovoltaico-a-putifigari_fd408dfb-446b-471b-873b-d4922127503b.html#google_vignette (accessed the 15th of December 2025).

[33] Regione Sardegna (2025), Putifigari: assemblea cittadina sull'impianto fotovoltaico. La Regione al fianco degli Enti Locali: 'Nessun progetto autorizzato' (press release), https://www.regione.sardegna.it/notizie/putifigari-assemblea-cittadina-sull-impianto-fotovoltaico-la-regione (accessed the 15th of December 2025).

**Remi Wacogne** holds a PhD in Regional planning and public policy and has conducted research and teaching activities at Iuav and Ca′ Foscari universities in Venice. His research on heritage and landscape management and policies has been completed with several collaborations with local governments, public authorities, and NGOs.

**Chiara Carolina Donelli** holds a PhD in Economics and Management of Innovation and Sustainability from the University of Parma and is currently Assistant Professor Tenure Track of Accounting and Business Administration at the University of Pavia. Her research interests include sustainability and sustainability reporting, collaborative governance, and alternative (and counter) forms of accounting, particularly in the nonprofit and cultural sectors.

# 16 The Presence of Absence: Conflicts in Collaborative Spaces: A Comparative Ethnographic Study in France, Italy, and Czechia

Daphné Reguiessé

**Abstract**

This multicountry ethnographic study examines the complex daily dynamics of conflict, cooperation, and institutional influences in nine collaborative spaces across France, Italy, and Czechia. Challenging idealized views, the findings show coworking environments as contested arenas where power imbalances, socioeconomic inequalities, and cultural norms meet. Drawing from extensive fieldwork, the analysis unpacks the intricate interplay between individual experiences, organizational structures, and broader socioeconomic forces shaping these shared workspaces. Collaborative Working Spaces (CWS) emerge as sites of negotiation, exclusion, and conflict, deeply embedded within national contexts. Grounded in fieldwork and observations, this research deconstructs dominant myths surrounding coworking related to spontaneous collaboration, open accessibility, and technology-driven networking. The comparative approach analyzes how national contexts like economic models and labor conditions influence coworking development and user experiences. Beneath the community-building rhetoric lies a complex mix of social, economic, and cultural forces shaping the dynamics within these shared workspaces. This study seeks to understand the conflicts, cooperative practices, and cohabitation strategies emerging in coworking environments.

By comparing across countries, this research offers nuanced insights into how economic conditions, labor markets, and cultural factors shape coworking realities, often undermining the pursuit of community and collaboration through the reproduction of hierarchies and commodification of resources. The study's findings contribute to understanding how organizations address multiple UN Sustainable Development Goals (SDGs, https://sdgs.un.org/goals) through the

D. Reguiessé (✉)
Scienze Politiche Giuridiche e Studi Internazionali, Univeristy of Padova, Padova, Italy
e-mail: daphne.reguiesse@unipd.it



P. Wollmann et al. (eds.), *The Sustainable Organization – Volume 2*, Future of Business and Finance, https://doi.org/10.1007/978-3-032-23802-3_16

lens of CWS, particularly SDG 8 (Decent Work and Economic Growth); SDG 10 (Reduced Inequalities). The findings have implications for SDG 11 (Sustainable Cities and Communities).

## 16.1 Introduction

The rise of Collaborative Working Spaces (CWS) has come with narratives celebrating their potential as collaborative hubs fostering innovation and networking.[1] However, this study across France, Italy, and Czechia challenges such idealized views. Rather than harmonious environments, CWS emerge as contested arenas where power dynamics, economic pressures, and institutional logics intersect, shaping the lived experiences within.

These dynamics directly relate to UN Sustainable Development Goal (SDG)[2] number 8 (Decent Work) and SDG 10 (Reduced Inequalities) (see Fig. 16.1). This research aims to generate new insights into CWS, focusing on the interplay between collaboration and conflict in these environments in four European cities—Paris (France) and Padova (Italy). This approach allows for an examination of how CWS

**Fig. 16.1** UN SDG (source: https://www.un.org/sustainabledevelopment/), used with friendly permission of UN. The content of this publication has not been approved by the United Nations and does not reflect the views

[1] Capdevila, I. (2015). CWS and the localized dynamics of innovation. *European Planning Studies.*; Waters-Lynch, J., & Duff, C. (2021). The affective commons of Coworking. *Human Relations, 74*(3), 383–404. https://doi.org/10.1177/0018726719894633

[2] See: https://sdgs.un.org/goals

contribute to SDG 11 (Sustainable Cities and Communities) across different urban contexts. I seek to understand how these conflicts shape the dynamics of CWS and their broader socioeconomic impact. Additionally, I aim to enhance managerial capacity and build connections with key stakeholders, including scholars, practitioners, and policymakers. This study also explores how CWS align with or challenge the principles of SDG 17 (Partnerships for the Goals) through their collaborative practices, and SDG 12 (Responsible Consumption and Production) in their operational models. To achieve these goals, I have formulated the following research questions:

1. How do conflicts between coworkers, managers, and inhabitants shape CWS, and what are the implications for achieving SDGs 8 and 10?
2. How do interactions between CWS and their surrounding contexts influence collaboration and conflict?

To address these research questions, I employ a comprehensive methodological approach integrating literature review and Participatory Action Research[3] (PAR) based upon Grounded Theory Methodology[4] (GTM) approach. I hypothesize that CWS foster unique conflict dynamics that differ from traditional workplaces[5] due to the heterogeneous interactions among professionals from diverse organizations. This diversity will likely lead to increased relational conflicts, arising from differences in values, work habits, and expectations.[6] Besides, I also anticipate that the community-oriented culture prevalent in CWS may mitigate some tensions, facilitating constructive problem-solving and enhancing communication.[7] By grounding my analysis in established conflict theories, my framework aims to illuminate the complexities of interpersonal interactions in CWS and provide actionable strategies for managing potential conflicts, ultimately contributing to a productive and harmonious work environment in the evolving landscape of work.[8] I designed this

[3] Fals-Borda, O. (2001). Participatory action research in social theory: A research agenda for the twenty-first century. In O. Fals-Borda & H. Rahman (Eds.), Action and knowledge: Breaking the monopoly with participatory action research (pp. 49–68). The Apex Press.

[4] Gilgun, J. F. (2014). Qualitative research and grounded theory. Social Work Research, 38(1), 1–8.

[5] Roscigno, V. J., Hodson, R., & Lopez, S. H. (2009). Workplace incivilities: the role of interest conflicts, social closure, and organizational chaos. Work, Employment and Society, 23(4), 747–773.

[6] Akhavan, N. (2021). The rise of coworking spaces: A new era of work. Cambridge University Press.; Bednář, P., Palek, M., Pospíšil, J., & Šebek, P. (2021). Coworking spaces as a hub for innovation: An empirical study. Journal of Business Research, 137, 114–123. Berbegal-Mirabent, J. (2021). Coworking spaces: A new paradigm for workplaces. Journal of Organizational Behavior, 42(3), 254–275.

[7] Berbegal-Mirabent, J. (2021). Coworking spaces: A new paradigm for workplaces. Journal of Organizational Behavior, 42(3), 254–275.; Gandini, A., & Cossu, G. (2021). Coworking spaces as innovation hotspots. Journal of Innovation Management, 9(2), 101–120.

[8] Lacárcel, F., González, M., & Morales, R. (2024). Remote work dynamics in coworking spaces. Journal of Digital Work, 7(1), 50–70.

integrated approach to ensure that my analysis of CWS is comprehensive and contextually grounded, providing valuable insights into the complex interplay of collaboration and conflict within these environments as per the research goals.

All in all, the primary objective of this research is to analyze conflict dynamics within CWS and compare them with traditional organizational settings. By examining the interplay between collaboration and conflict in various coworking environments, the study aims to understand how these conflicts shape the socioeconomic trajectories of these spaces and their broader impact on work organization. The research employs a mixed-methods approach, integrating qualitative interviews,[9] participant observations, and document analysis, all based upon the theoretical frameworks of Grounded Theory Methodology[10] (GTM) and Participatory Action Research (PAR). This approach facilitates an in-depth exploration and empirical observation of conflict dynamics, allowing for a comprehensive understanding of both the subjective experiences of CWS participants and the objective structural factors influencing conflict. The literature review is a foundation for identifying key conflict dynamics and theoretical constructs relevant to CWS. It provides a comprehensive understanding of existing knowledge, which will inform the design of data collection instruments and the focus of empirical research.

## 16.2 Theoretical Framing

The study integrates theoretical perspectives from organizational sociology, labor studies, and urban geography to analyze the data through a multidimensional framework approached through a syncretic perspective (1, 2, 3). In this pattern, organizational theory[11] (1) provides insights into governance structures, management styles, and organizational cultures influencing conflict resolution and community-building efforts. Conflict theory (2) in precarious work contexts[12] offers a lens to understand tensions arising from resource commodification, blurred personal-professional boundaries, and social hierarchy reproduction. Spatial theory (3), like Lefebvre's production of space,[13] illuminates how architecture and design shape social interactions, power dynamics, and boundary negotiations[14] within CWS. While addressing

[9] Patton, M. Q. (1990). *Qualitative evaluation and research methods*. Sage Publications.

[10] Charmaz, K. (1990). *Constructing grounded theory: A practical guide through qualitative analysis*. Sage Publications.

Charmaz, K. (2006). Constructing grounded theory. In C. Willig & W. Stainton-Rogers (Eds.), *The SAGE handbook of qualitative research in psychology* (pp. 45–60). Sage Publications.

[11] Blagoev, B., Costas, J., & Kärreman, D. (2019). 'I are all herd animals': Community and organizationality in CWS. *Organization, 26*(6), 894–916.

[12] de Peuter, G., Cohen, N. S., & Saraco, F. (2017). The ambivalence of coworking: On the politics of an emerging work practice. *European Journal of Cultural Studies, 20*(6), 687–706.

[13] Lefebvre, H. (1974). *La production de l'espace*.

[14] Merrifield, A. (1996). Public space: Integration and exclusion in urban life. *City, 1*(5–6), 57–72.

the topic of CWS, I tapped into the concepts of precarity and affective labor,[15] shedding light on the emotional dimensions of work, particularly relevant in freelance and gig economies prevalent in coworking.

This syncretic approach, weaving three theoretical frameworks, enabled me to provide a nuanced and multidimensional analysis of the conflicts, cooperative dynamics, and cohabitation practices that shape the lived experiences within CWS across different national contexts. Grounded in neo-institutionalist theory, my analytical framework critically examines the prevailing "myths" that shape both scholarly and popular narratives around CWS and evaluates their divergence from empirical realities. First, the *Serendipity Myth* posits that spontaneous networking and innovation emerge organically within open-plan environments; however, our observations and interviews demonstrate that unstructured proximity often yields only superficial encounters, and that deliberate curation of events and seating arrangements is necessary to foster meaningful exchanges. Second, the *Accessibility Myth* assumes that CWS are inherently inclusive "third places" open to anyone; in practice, access is frequently gated by membership fees, invitation-only memberships, and opaque social networks, resulting in selective participation and reinforcing existing socio-economic hierarchies. Third, *the Technology Myth* equates greater technological provision—high-speed Wi-Fi, digital platforms, smart desks—with enhanced collaboration; yet our findings reveal that technology alone cannot substitute for the managerial practices and community norms that underpin genuine knowledge sharing. By deconstructing these myths, I highlight critical design implications—such as integrating intentional social infrastructure alongside physical layouts—and policy recommendations, including subsidized access models and guidelines for community facilitation, that more accurately align coworking environments with their purported goals of democratized innovation and social inclusion.

## 16.3 Methodology

The methodological framework for this research was designed to provide a robust foundation for analyzing conflict dynamics within collaborative working spaces (CWS[16]). It is structured in two phases: desk research and online preliminary analysis (phase 1) and empirical research (phase 2).

Initially, desk research aims to establish a comprehensive understanding of conflict dynamics by reviewing existing literature, which will guide the development of data collection tools and refine the research focus. During phase 1, I systematically reviewed 31 peer-reviewed articles on CWS, selected from an initial pool of 42—which were including third spaces and Fab Lab. The analysis followed Saldaña's (2015) coding framework,[17] supported by manual color-coded coding in Excel to

[15] Standing, G. (2011). *The precariat: The new dangerous class*. Bloomsbury Academic.

[16] Includes all collaborative working spaces such as CWS, incubators, accelerators, except marker-spaces and cultural spaces.

[17] Saldana, J. (2015). *The coding manual for qualitative researchers* (3rd ed.). SAGE Publications.

identify thematic patterns. I focused on extracting and comparing quotations that revealed recurring themes, which helped shape the analytical dimensions of the study. The review was structured in two parts. The first part investigated conflict dynamics within CWS, emphasizing organizational culture and member interactions. Abstracts, keywords, and methodological sections were examined to identify key features such as flexible structures, diverse user profiles, and community-oriented practices.[18] Special attention was paid to research design, data collection methods, and the geographical scope of the studies, spanning publications from 2013 to 2024. Following this, the online preliminary analysis has enabled to gather preliminary data on CWS through various digital sources, informing the selection of case studies and further research directions, building reports on National data sets for the three targeted countries.

Building on these foundational steps, the empirical research (phase 2) utilized an ethnographic approach along with semi-structured interviews and participant observations to delve deeply into conflict dynamics, guided by established conflict typologies and frameworks. This approach, grounded in Grounded Theory Methodology (GTM), ensures that the study is rooted in empirical evidence while systematically exploring and categorizing conflicts within CWS. This led me to conduct a multi-sited ethnographic study across CWS in Paris, other French cities, Padova (Italy), and Prague (Czechia). Research sites were selected based on criteria like membership models, managerial structures, and socioeconomic accessibility levels. The selection of six case studies for each country/city—for a total of three CWS with three main focus—next involved a 1-week immersive period in each space in both countries, guided by specific inclusion criteria: (1) Type and size: spaces catering to individuals, startups, and larger organizations (see the table below with the taxonomy Orel & Bennis[19]); (2) Location: a mix of central urban settings between big and medium cities—all hosting universities facilities—to reflect varied socioeconomic contexts; (3) Member diversity: a range of member profiles, including freelancers, entrepreneurs, and corporate teams; (4) Management models: different management approaches to explore how organizational structure impacts conflict dynamics. These criteria also align with the framework established by Cnossen and Stephenson,[20] and facilitate comparisons with similar studies conducted in Europe (Table 16.1).

Eventually, I focused on three types of CWS, specifically those intended for individual, corporate, and startup purposes. Exclusion criteria included collaborative spaces with negative reputations and inaccessible facilities due to logistical or geographical challenges. Additionally, Fab Labs and maker spaces were excluded, as were analyses of coworking solely as a practice rather than as a physical space.

[18] Ivaldi, G., Sannino, A., & Scaratti, G. (2021). Coworking spaces and their impact on work practices. *Cultural Historical Activity Theory, 12*(4), 33–58.

[19] Orel, V., & Bennis, J. (2021). Real estate and coworking: A complex relationship. *Journal of Real Estate Research, 45*(2), 215–230.

[20] Cnossen, H., & Stephenson, H. (2022). *Collaborative workspaces: The impact of design on social interactions and productivity.*

**Table 16.1** Drafted by the author using the source: CWS Taxonomy by Orel & Bennis, 2021 (see footnote 18). All rights reserved, used with permission

| Category | Description |
|---|---|
| **Individual use** | Spaces oriented towards independent professionals or freelancers who use coworking as a personal office |
| **Group use** | Spaces designed to accommodate teams or workgroups, often equipped with meeting rooms or collaborative areas |
| **Startup use** | Spaces dedicated to new businesses, typically with incubation, acceleration services, and networking support |

**Table 16.2** Type of space, based on Table 16.1. Elaboration from the author. Used with permission, all rights reserved

| Country | Coworking space | Type of space |
|---|---|---|
| **Italy** | Covox | Group use; Individual use |
| | Cowo Padova | group of start-ups; Individual use |
| | Le Village by CA | Startup-Focused |
| **France** | Greenspace | Group use; Startup-Focused; Individual use |
| | Les Liens | Individual use |
| | Le Village by CA (Paris) | Startup-Focused |
| **Czechia** | Locus | Individual use |
| | Opero | Group use; Startup-Focused; Individual use |
| | Impact Hub | Group use; Startup-Focused; Individual use |

Out of the three key research geographical sites, I included nine spaces: Greenspace, Les Liens, and Le Village by CA in Paris; Covox and Le Village in Padova; and Locus, Opero, and Impact Hub in Prague (Table 16.2).

Data collection involved ethnographic observation, interviews with managers and users, informal conversations, and digital surveys collected between November 2024 and May 2025. Observations focused on work routines, spatial dynamics, interpersonal interactions, and access mechanisms. To ensure analytical rigor and enrich the interpretation of coworking space dynamics, I applied a data triangulation strategy. This involves systematically comparing and integrating data from three primary sources:

1. **Semi-structured interviews** with managers, and some coworkers, which reveal individual perceptions, narratives, and interpretations of conflict and cooperation.
2. **Participant observations**, which offer insights into actual behaviors, interactions, spatial dynamics, and unspoken tensions in everyday life within CWS.
3. **Document analysis**, including internal regulations, communication materials, websites, and fieldwork notes, which provide context on norms, values, and institutional framing.

This triangulated approach strengthens validity by allowing the convergence of findings across different data types. For example, if interviews expose a sense of personal frustration over limited access to meeting rooms, and observations reveal recurrent scheduling conflicts or improvised solutions, the combined evidence supports a robust interpretation of resource-based conflict. By synthesizing subjective narratives and observed behaviors, I can identify underlying institutional logics and tensions—such as autonomy vs. control or inclusion vs. gatekeeping—that may not emerge from a single data source alone. This method also enhances the project's comparative dimension by making it possible to detect patterns and divergences across the different national contexts under study (Italy, France, Czech Republic). The ultimate goal of this procedure is to move beyond isolated anecdotes toward a multivoiced and comprehensive understanding of conflict as both a social process and a lens into broader institutional dynamics. Our ethnographic approach allows for a nuanced understanding of how CWS contribute to SDG 8 (Decent Work) and SDG 10 (Reduced Inequalities) in practice.

## 16.4 Findings

First, I present in this section the empirical analysis by country; each country represents a case study and will be further discussed in the second part of this *findings* Sect. 16.4.4.

### 16.4.1 Case Study 1: Italy (Padova)

The coworking landscape in Padova, Italy, revealed a deep divide between freelancers and institutional actors, with disparities in power and voice shaping the experiences within these shared workspaces.

**At Cofare,** a shared space between two startups, hierarchical tensions and conflicts emerged from the authoritarian leadership style and the lack of structured support systems for precarious workers. Privacy concerns and the blurring of work-life boundaries further exacerbated the challenges faced by coworkers, highlighting the need for more comprehensive approaches to address the precarious nature of freelance work.

**Covox,** a converted industrial building, aimed to prioritize well-being and flexibility. However, structural limitations, such as noise pollution and inadequate infrastructure, hindered the realization of this inclusive vision. The informal governance structure and lack of clear conflict resolution mechanisms contributed to ambiguities and unresolved grievances among users.

**Le Village by CA Triveneto,** a corporate-driven accelerator, exemplified the instrumental nature of interactions within hierarchized CWS. The focus on structured networking events and corporate partnerships challenged the organic community-building ethos, raising questions about the potential for such spaces to reinforce existing power dynamics and prioritize economic capital over genuine collaboration.

These observations in Padova, coupled with Table 16.3 analytical variables, underscored the emotional labor involved in cultivating a sense of "belonging"

**Table 16.3** Synthetic fieldwork data from Italian Case study. Elaboration from the author. Used with permission, all rights reserved

| Country | Coworking space | Flexibility of working hours | Hierarchical structure | Community support | Privacy spaces | Internal regulations | Freedom of customization | Economic accessibility(/ month) |
|---|---|---|---|---|---|---|---|---|
| **Italy** | Covox | High | Horizontal/ Founder-led | Moderate | Limited | Flexible | High | Medium (€250-300) |
| | Cowo Padova | Low | Authoritarian/ Founder-led | Poor | Limited | Rigid | Low | High (€350-400) |
| | Le Village by CA | Medium | Hierarchical | Event-based, discontinuous | Present | Balanced | Medium | High (€400-500) |

within CWS that can be apprehended on a larger level for the Italian context, as well as the tensions that arose when instrumental participation and economic imperatives overshadowed the pursuit of genuine connection and personal growth.

### 16.4.2 Case Study 2: France (Paris)

The CWS in France, including Greenspace, Coworking Republic, Les Liens, and Le Village by Crédit Agricole, exhibited a range of conflicts and cooperative dynamics shaped by the institutional context and the state's involvement in "community design."

In spaces like **Greenspace,** resource competition and tensions between permanent and temporary members emerged as significant sources of conflict. The informal hierarchies and territorial behaviors of long-term users challenged the purported ethos of inclusivity and collaboration, revealing the complexities of negotiating individual ownership and collective sharing within coworking environments.

**Les Liens**, a carefully curated "wellness" space, paradoxically reinforced socioeconomic barriers through the monetization of wellness trends and the exclusionary nature of its pricing model. The enforced silence and individualistic ethos cultivated an environment where work was inherently solitary, raising questions about the true democratization of such spaces and their ability to foster genuine connection.

**Le Village by Crédit Agricole,** a corporate-driven accelerator, exemplified the tension between nationalized models of coworking and the organic, community-driven ethos that characterizes many grassroots spaces. The hierarchical management structure and instrumental networking opportunities challenged the ideals of spontaneous collaboration and peer-to-peer engagement.

Across these spaces analyzed for the French case-study, the role of the state and top-down "community design" efforts emerged as a significant factor shaping the conflicts and cooperative dynamics within coworking environments. The observations in France, resumed in Table 16.4, highlighted the reproduction of social hierarchies and inequalities through spatial management and the potential for CWS to reinforce existing power structures rather than fostering genuine inclusivity and accessibility.

### 16.4.3 Case Study 3: Czechia (Prague)

The CWS in Prague, including Locus, Opero, and Impact Hub, exhibited a range of conflicts and cooperative dynamics that reflected the broader socioeconomic and cultural context of Czechia.

**At Locus,** grassroots, user-driven space, conflicts emerged from the contradictory values and philosophies that underpinned the coworking ethos. While the space promoted an organic community-building approach, tensions arose from the potential for "inequality regimes" to emerge, where informal hierarchies and insider dynamics benefited certain groups over others. The role of architecture and spatial design in shaping interactions and fostering a sense of belonging was evident, with the flexible seating arrangements and shared spaces facilitating fluid movement and spontaneous collaboration.

**Table 16.4** Synthetic fieldwork data from French Case study. Elaboration from the author. Used with permission, all rights reserved

| Country | Coworking space | Flexibility of working hours | Hierarchical structure | Community support | Privacy spaces | Internal regulations | Freedom of customization | Economic accessibility (month) |
|---|---|---|---|---|---|---|---|---|
| **France** | Greenspace | High | Hierarchical/ Founder-led | Strong | Present | Balanced | High | High (€390-450) |
| | Les Liens | High | Founder-led | Limited | Present | Balanced | Low | Affordable (€150-200) |
| | Le Village by CA (Paris) | Medium | Corporate, highly hierarchical | Limited (networking but no informal community building) | Present | Structured | Medium | High (€400-500) |

**Opero,** a feminist-oriented coworking space, aimed to create an inclusive and convivial community through conscious design choices and an ethic of care. However, paradoxes emerged as the monetization of wellness trends and the reinforcement of socioeconomic barriers challenged the space's aspirations for accessibility and inclusivity. The flat organizational structure and emphasis on personal growth revealed the complexities inherent in crafting sustainable work environments that balance professional demands with personal well-being.

**Impact Hub Praha**, a professionalized coworking space, stood apart as a hierarchized ecosystem governed by a corporate ethos focused on efficient service provision and individual productivity. The minimalist aesthetic and hierarchical management structure challenged the organic community-building ethos, raising questions about the potential for such spaces to reinforce isolating paradigms or prioritize surface-level corporate diversity norms over deeper strategies for user inclusivity. Conflicts arose from the instrumental nature of interactions, designed for business development and strategic positioning rather than genuine peer-to-peer collaboration or personal connection.

Throughout these spaces in Czechia, informal mediation practices and management styles played a crucial role in shaping the resolution or exacerbation of conflicts. Table 16.5 summarizes key contents retrieved from my fieldwork in Prague highlighting well the tensions between corporate and creative values, as well as the challenges of balancing openness with commercialism within coworking environments.

### 16.4.4 Takeaways

The ethnographic observations and empirical analysis across Czechia, Italy, and France reveal both commonalities and distinct national characteristics in the types and intensities of conflict, governance structures, and the role of CWS in shaping labor dynamics and social hierarchies that are summarized in Table 16.6.

From Table 16.6, which collects the fieldwork findings across the variables formerly identified, I have identified the redundant conflicts occurring in CWS, partly contradicting the image of these spaces as purely collaborative environments.

The comparative perspective adopted for this research revealed significant national differences in how CWS operated and who used them. Starting with the Italian case study, I observed that compared to France and Czechia, coworking provided an affordable option for precarious freelancers and those in irregular work situations who needed flexibility. In France, coworking was an elite, expensive service mainly used by established professionals who could afford the high daily rates of €50–60. In Czechia, a mix of models caters to entrepreneurs, startups, and remote workers, with varying levels of community engagement. In many spaces in France and Czechia, there was competition and tension over access to desks/seats, creating a "first-come, first-served" dynamic that went against the sharing ethos, highlighting a **competition for resources (key takeaway 1),** mostly in these cities.

The coworking space offers a routine, a familiar place, sometimes even friendships, but with differences related to its interactions among coworkers.

**Table 16.5** Synthetic fieldwork data from Czechian Case study. Elaboration from the author. Used with permission, all rights reserved

| Country | Coworking space | Flexibility of working hours | Hierarchical structure | Community support | Privacy spaces | Internal regulations | Freedom of customization | Economic accessibility (month) |
|---|---|---|---|---|---|---|---|---|
| **Czechia** | Locus | Medium | Horizontal/ owner-led | Strong | Present | Balanced | Medium | Affordable (200€- 260€) |
| | Opero | High | Hierarchical/ CEO-led | Limited (networking but no informal community building) | Present | Balanced | Medium | Affordable (100€- 250€) |
| | Impact Hub | High | Hierarchical/ Corporate-led / | Limited | Present | Balanced | Medium | Affordable (40€- 200€) |

**Table 16.6** Comparative analysis: Czech Republic, Italy, France. Author's elaboration. Used with permission, all rights reserved

| Analytical level | Theme | Czechia (Prague: Locus, Opero, Impact Hub) | Italy (Padova: Le Village, Cofare, Covox) | France (Paris: Greenspace, Les Liens, Le Village CA) |
|---|---|---|---|---|
| **Individual** | **Noise, Privacy, Engagement** | Quiet, respectful environment; phone booths available but not ideal | Mixed; Le Village lacks soundproofing; COFARE/COVOX more pragmatic and informal | Greenspace: quiet, controlled; Les Liens: hybrid zones for rest and work, supports deeper emotional presence |
| | **Diversity, Inclusion, Social Tensions** | International crowd; onboarding gaps; informal inclusion | Gender imbalance and emotional labor often unacknowledged; class distinctions subtle | Les Liens excels with emotional safety and affective inclusion; Le Village Paris shows hierarchical tendencies |
| | **Temporary Social Ties and Competition** | Short- to mid-term stays; ties rely on "connectors" | Frequent fragmentation; reliance on events for cohesion | Greenspace encourages self-expression; Les Liens balances affect and productivity |
| | **Health, Wellbeing, Ergonomics** | Basic ergonomics: poor chairs, limited standing desks | Complaints about lighting, posture, and stress; functional but uninspiring | Les Liens offers diversified spaces for body/mind balance; France shows better spatial design for well-being |
| | **Gender Dynamics, Emotional Labor** | Emotional labor informal (Eva); gender parity absent in discourse | Women bear logistical and emotional loads; not structurally addressed | Les Liens embraces emotionality and gender topics; Le Village Paris resists affective engagement structurally |

(continued)

**Table 16.6** (continued)

| Analytical level | Theme | Czechia (Prague: Locus, Opero, Impact Hub) | Italy (Padova: Le Village, Cofare, Covox) | France (Paris: Greenspace, Les Liens, Le Village CA) |
|---|---|---|---|---|
| **Organizational** | **Governance, Leadership, Role Clarity** | Horizontal, relational; informal boundaries | Often centralized (Chiara at Le Village); COFARE more flat and DIY | Le Village Paris highly corporate; Les Liens embraces relational and fluid roles |
| | **Visions, Goals, Collaboration** | Member-led projects; soft facilitation | High on vision, low on implementation; collaboration often symbolic | Le Village Paris pushes output; Greenspace and Les Liens nurture reflective autonomy |
| | **Participation and Engagement** | Events curated but uneven in participation | Events are unevenly attended; some highly curated, others spontaneous | Les Liens integrates rituals into everyday; France more invested in socio-cultural bonding |
| | **Privacy, Security, Trust** | Strong trust culture; app-based access; few issues | Le Village has trust but lacks privacy infrastructure | Greenspace and Les Liens display intentional architecture of trust and discretion |
| **Institutional** | **Economic Instability & Precarity** | Affordable, sustainable; model fits mobile professionals | High costs at Le Village alienate precarious startups; COFARE more accessible | Parisian spaces stratify: Greenspace affordable; Le Village is exclusive and highly branded |
| | **Territoriality, Gentrification** | Central but accessible; gentrification creeping in | Coworking often tied to urban redevelopment; COFARE critiques displacement | Les Liens critiques gentrification subtly through local integration; Greenspace is adaptive |

(continued)

**Table 16.6** (continued)

| Analytical level | Theme | Czechia (Prague: Locus, Opero, Impact Hub) | Italy (Padova: Le Village, Cofare, Covox) | France (Paris: Greenspace, Les Liens, Le Village CA) |
|---|---|---|---|---|
| | **Work-Life Balance, Personal Life** | Valued for walkability, but comfort of home still preferred | Flexibility does not guarantee balance; burnout possible | France emphasizes experiential quality and aesthetic care, enhancing balance |
| | **Inequality, Access, Elitism** | Egalitarian ethos, but informal barriers persist | Le Village reinforces elitism via selective support; COFARE/COVOX more open | Les Liens reflects on power and inclusion critically; Le Village Paris mirrors corporate inequality dynamics |

In Italy, there was a strong focus on socializing—frequent coffee breaks, informal chats, and daily rituals[21] that created a sense of familiarity. In France, coworking was much more functional—people came primarily to work, not socialize, keeping work and personal life separate. And in the Czech Republic, CWS had a more minimalist culture, with less interaction between people, more individual work, and longer quiet periods. This diversity shows coworking is not a one-size-fits-all model—it is a flexible framework that takes on very local cultural shapes, based on different norms around socializing, collaborating, and work-life balance. And this **tension between openness and exclusivity (key takeaway 2)** exists everywhere. I hence assessed that the sense of community and inclusion is not always guaranteed as much as it is claimed or branded. Those who don't speak the local language well, who are overworked and lack time for socializing, or who don't have the cultural skills to understand the implicit norms, often remain on the outside. Long-term members often formed insider groups and marginalized newcomers, creating social hierarchies within these supposedly inclusive spaces. In other words, coworking promises an open community, but access to that community gets filtered, sometimes in subtle ways. Adding to **this insider/outsider dynamic**, **the silent reality (key takeaway 3)** was striking: despite being promoted as networking hubs, many CWS were very quiet, individual workspaces. Informal socializing was rare, with networking limited to structured events. Work-life balance was emphasized over community-building, dicing the space into an atmosphere that I conceptualized as "*the sound of silence*"—by and large co-performed by trained sound engineers (the coworking managers, see below, key takeaway 5).

[21] Goffman, E. (1967). Interaction ritual: Essays on face-to-face behavior. Aldine Publishing Company.

**Another common feature was a sense of impermanence (key takeaway 4)**—in Paris or Prague (highly mobile cities), I found CWS often function as vital community hubs for people without local social ties—expats, digital nomads, or those starting over somewhere new. On the other hand, I witnessed that coworkers are often "in between" phases—they may have just moved cities, changed careers, or left a corporate job. Even those with hosting or facilitating roles frequently have unstable jobs, juggling different tasks from logistics to emotional support. The only relatively stable role is the coworking manager—and this is who I want to focus on next.

**Coworking managers emerged as key figures (key takeaway 5)**—not just coordinators, but what I call "*technicians of sociability*." They are like *sound engineers* at a concert—mostly invisible, but crucial to making things run smoothly. Managers determine how to welcome new employees, which events to host, and how to address tensions or conflicts. And they actively adapt shared practices to fit the local culture—for example, offering a welcome coffee ritual in Prague, an evening drink in Paris or no formalities in Padova. The presence of active community managers helped foster interactions[22] and resolve conflicts. Spaces without dedicated managers often struggled with social dynamics as the underlying goal is always the same: to create a sustainable sense of community within the coworking space.

**These key takeaways** are relevant to several UN Sustainable Development Goals (SDGs) and can be analyzed through the lens of the Three-Pillar Model[23]:

1. *SDG 8 (Decent Work and Economic Growth)*: The diverse coworking models influence working conditions and economic opportunities. This aligns with the "Sustainable Purpose" pillar, as CWS strive to create environments that foster economic growth and decent work. However, the challenge lies in formulating a convincing sustainable purpose that can overcome the perception bias and skepticism often associated with sustainability initiatives, as highlighted by Wollmann.

2. *SDG 10 (Reduced Inequalities)*: The dynamics observed in the spaces reveal tensions related to inclusion and social inequalities. This connects to the "Connectivity" pillar, highlighting the challenges in creating truly inclusive environments. The fragmentation of stakeholder opinions and ethics, as noted in the Three-Pillar Model, is particularly relevant in the CWS context and impacts efforts to reduce inequalities.

3. *SDG 11 (Sustainable Cities and Communities)*: CWS emerge as key elements in the evolution of urban environments and work communities. This relates to both the "Travelling Organization" and "Connectivity" pillars, as CWS adapt to changing urban needs and facilitate community building. The nonlinear changes

[22] Wijngaarden, Y. (2023). 'I like the "buzz", but I also suffer from it': Mitigating interaction and distraction in collective workplaces. *Human Relations*, *76*(12), 1881–1903.

[23] Developed in Wollmann, P.; Kühn, F.; Kempf, M. (Eds.): Three Pillars of Organization and Leadership in Disruptive Times—Navigating Your Company Successfully through the twenty-first Century Business World. Cham: Springer Nature, © 2020

in external conditions, as emphasized by Wollmann, pose significant challenges for CWS in maintaining this adaptability within urban contexts.

4. *SDG 17 (Partnerships for the Goals)*: The collaboration and networking practices observed offer insights into new forms of partnership for sustainable development, directly linking to the "Connectivity" pillar. The capability to connect in difficult environments, as described in the Three-Pillar Model, is crucial for CWS to foster effective partnerships.

Indeed, the comparative analysis highlights significant differences between countries in terms of:

1. Noise, Privacy, and Engagement
2. Diversity, Inclusion, and Social Tensions
3. Temporary Social Ties and Competition

These differences underscore the importance of contextualized approaches in implementing the SDGs, recognizing the cultural and socioeconomic specificities of each national context. This aligns with Wollmann's emphasis on the need for "*thinking without banister*" and adapting to unknown territories, especially when it comes to apprehend how CWS navigate the balance between:

1. **Sustainable Purpose:** CWS must articulate a clear, convincing purpose that addresses both their organizational sustainability and their contribution to global sustainability goals.
2. **Travelling Organization:** The adaptability of CWS to rapid changes demonstrates their capacity as "travelling organizations." However, they face challenges in navigating the "unknown territories" of sustainable development, requiring new learning approaches and openness to uncertainty.
3. **Connectivity:** CWS serve as hubs of connectivity, but face challenges in managing diverse stakeholder relationships and overcoming fragmentation of opinions and ethics.

## 16.5 Interpretation of Findings and Implications for SDGs 8, 10, and 11

### 16.5.1 Individual-Level Conflicts: Subjectivity, Autonomy, and Affective Labor

Across all three national contexts, individuals navigate CWS as sites of simultaneous opportunity and ambivalence. Consistent with literature, the data indicate that CWS produce a "managed informality"[24] that appeals to the desire for autonomy

[24] Merkel, J. (2019). 'Freelance isn't free': Co-working as a critical urban practice to cope with informality in creative labour markets. *Urban Studies, 56*(3), 526–547.

while demanding self-regulation and emotional labor. In Padua's Covox, the relaxed atmosphere and horizontal relationships offer psychological safety, yet beneath this apparent harmony lie subtle tensions around professional identity and blurred boundaries between work and leisure. Similarly, Prague's Opero and Locus display surface-level conviviality, but observation reveals affective performances tied to professional survival in neoliberalized urban centers.[25] These dynamics have implications for SDG 8 (Decent Work and Economic Growth), as they highlight the complex relationship between flexible work arrangements and worker well-being. The Parisian context, particularly in spaces like Greenspace, exemplifies this duality: while some users enjoy freedom and flexible routines, others experience alienation through commodified community, echoing critiques of affective subjectivation.[26] Gendered expectations—such as invisible reproductive labor and care responsibilities—were evident across all contexts, confirming the reproduction of traditional roles[27] in supposedly progressive environments.Such contradictions align with Rahim's (2001) typology of conflict,[28] where intrapersonal and interpersonal tensions arise not from overt disputes but from internalized pressures of performativity, economic precarity, and the ongoing negotiation of authenticity within professional relationships. The tensions observed in CWS reflect broader challenges in achieving SDG 10 (Reduced Inequalities), as these environments can both mitigate and reproduce existing social and economic disparities.

### 16.5.2 Organizational-Level Conflicts: Governance, Collaboration, and Inequality Regimes

At the organizational level, the findings reinforce and nuance the concept of CWS as "social operating systems"[29] characterized by heterarchical ideals but often governed through covert hierarchies and exclusionary gatekeeping. In Prague's Impact Hub and Paris's Le Village by CA, collaboration was instrumentalized through

---

[25] Merkel, J. (2019). 'Freelance isn't free': Co-working as a critical urban practice to cope with informality in creative labour markets. *Urban Studies, 56*(3), 526–547. Merrifield, A. (1996). Public space: Integration and exclusion in urban life. *City, 1*(5–6), 57–72.

Micek, G. (2020). Studies of Proximity in Coworking Spaces: The Basic Conceptual Challenges. *European Spatial Research and Policy*, *27*(1), 9–35. https://doi.org/10.18778/1231-1952.27.1.01

[26] Bandinelli, G. (2020). *Generational diversity in coworking spaces*. Routledge.

[27] Rodríguez-Modroño, F. (2021). Generational differences in coworking spaces. *Journal of Organizational Behavior, 42*(5), 584–597. Papageorgiou, A. (2022). Exploring everyday experiences of entrepreneurial labour gender and work in collaborative workspaces of Athens. *Gender, Place & Culture*.

[28] Rahim, M. A. (2003). Toward a Theory of Managing Organizational Conflict. SSRN Electronic Journal.

[29] Di Marino, M., Tabrizi, H. A., Chavoshi, S. H., & Sinitsyna, A. (2023). Hybrid cities and new working spaces—The case of Oslo. *Progress in Planning*, *170*, 100712. https://doi.org/10.1016/j.progress.2022.100712

curated events and mentorship schemes that favored specific user profiles, typically startups with high growth potential. These dynamics resonate with Gandini's (2015) concept of "inequality regimes,"[30] where access to resources, visibility, and influence is stratified by organizational design and embedded norms. This selective approach raises questions about the role of CWS in promoting inclusive economic growth, a key aspect of SDG 8. Even in Padua's Covox—where a flatter, cooperative culture prevails—subtle power asymmetries emerged around decision-making, with managerial staff informally mediating conflict without transparent protocols. This supports Ivaldi et al.'s (2021) thesis[31] on the "double binds" inherent in the "co-" ethos of coworking: while collaboration is valorized discursively, cooperation is unevenly distributed, producing latent disillusionment. The notion of "*organizationality*"[32] is useful here: CWS perform organizational forms without always providing clear structures of accountability.[33] This ambivalence becomes particularly fraught when users bring diverse expectations regarding community, support, and governance—having implications for SDG 16 (Peace, Justice and Strong Institutions)—while perpetuating informal conflict resolution practices—surely flexible—can lead to exclusion, especially for newcomers or marginalized workers who lack embedded social capital. The observed inequality regimes and exclusionary practices in CWS challenge the notion of these environments as inherently conducive to SDG 11 (Sustainable Cities and Communities), highlighting the need for more intentional efforts to create inclusive urban spaces.

### 16.5.3 Institutional-Level Conflicts: Urban Integration, Precarity, and Neoliberal Rationalities

Examining institutional-level conflicts in CWS provides insights into the challenges of implementing SDG 17 (Partnerships for the Goals) at a micro-scale. As we know, CWS across all three contexts are shaped by broader urban, economic, and political processes that generate systemic tensions. Consistent with several authors,[34] my analysis reveals that CWS are not merely neutral infrastructures for innovation but

[30] Gandini, A. (2015). Coworking: A place for creativity and innovation. *Creativity and Innovation Management, 24*(1), 16–28.

[31] Ivaldi, G., Sannino, A., & Scaratti, G. (2021). Coworking spaces and their impact on work practices. *Cultural Historical Activity Theory, 12*(4), 33–58.

[32] Blagoev, B., Costas, J., & Kärreman, D. (2019). 'I are all herd animals': Community and organizationality in CWS. *Organization, 26*(6), 894–916.

[33] See: Sailer, K., Thomas, M., & Pachilova, R. (2022). The challenges of hybrid work: an architectural sociology perspective.

[34] de Peuter, G., Cohen, N. S., & Saraco, F. (2017). The ambivalence of coworking: On the politics of an emerging work practice. *European Journal of Cultural Studies, 20*(6), 687–706.; Fast, K. (2024). Who Has the Right to the Coworking Space? Reframing Platformed Workspaces as Elite Territory in the Geomedia City. *Space and Culture*, *27*(1), 48–62. https://doi.org/10.1177/12063312221090429; Waters-Lynch, J., & Duff, C. (2021). The affective commons of coworking. *Human Relations, 74*(3), 383–404.

are embedded in neoliberal urban imaginaries that commodify community, flexibility, and space. This commodification challenges the implementation of SDG 11 (Sustainable Cities and Communities), particularly in creating inclusive and sustainable urban spaces. For instance, in Prague and Paris, gentrification pressures and rising rents mean that CWS like Locus and Greenspace often cater to privileged users, marginalizing freelancers and precarious workers who cannot afford premium memberships. In Padua, although the spatial location near the train station at Covox provided accessibility, the physical and symbolic boundary between the coworking community and the surrounding neighborhood remained largely intact. Such disconnections echo Merrifield's (1996[35]) critique of the decontextualization of workspaces in global cities, where enclaves of innovation coexist uneasily with exclusion zones. This spatial inequality directly impacts SDG 10 (Reduced Inequalities) and highlights the tension between innovation hubs and inclusive urban development. Furthermore, across all sites, the COVID-19 pandemic surfaced new forms of structural conflict around health insecurity,[36] remote work adaptability,[37] and care responsibilities.[38] These emerging challenges intersect with SDG 3 (Good Health and Well-Being) and SDG 8 (Decent Work and Economic Growth), emphasizing the need for holistic approaches to urban work environments. Such structural shifts intensified existing inequalities, with some users leveraging flexibility while others—particularly caregivers and early-career workers—faced mounting stress and loss of income. The promise of work-life balance through coworking often collapsed under these pressures, revealing the fragility of coworking's emancipatory claims when faced with systemic crisis.

### 16.5.4 Assessment

Across all three countries, resource competition, tensions between permanent and temporary members, and the presence (or absence) of effective community managers emerged as significant sources of conflict within CWS. These dynamics have implications for SDG 17 (Partnerships for the Goals), as they affect the potential for collaborative partnerships within and beyond CWS. The negotiation of individual ownership and collective sharing, as well as the blurring of work-life boundaries,

---

[35] Merrifield, A. (1996). Public space: Integration and exclusion in urban life. *City, 1*(5–6), 57–72.

[36] Smékalová, L., Matošková, J., Belvončíková, E., Kálmán, J., & Crhová, Z. (2022). 17 Work-life balance services in coworking spaces and the impact of COVID-19. *The COVID-19 Pandemic and the Future of Working Spaces*, 242.

[37] Ciccarelli, F. C. (2023). Exploring the potential of coworking spaces for quality of working life and wellbeing: A systematic review of academic literature. *CIDADES, Comunidades e Territórios, 46.*; Robelski, S., Keller, H., Harth, V., & Mache, S. (2019). Coworking spaces: The better home office? A psychosocial and health-related perspective on an emerging work environment. *International Journal of Environmental Research and Public Health, 16*(13), 2379.

[38] Ciccarelli, F. C., & Mariotti, I. (2024). Remote work, coworking spaces, and wellbeing during the COVID-19 pandemic and beyond: Exploring peripheral and rural areas. In *Pandemic Recovery?* (pp. 209–220). Edward Elgar

posed challenges for coworkers and highlighted the need for transparent rules and participatory planning. Across all three countries, CWS emerged as sites of soft power and labor control, where the ideals of community and collaboration intersected with the realities of economic pressures and the reproduction of social hierarchies. The observations highlighted the paradoxes[39] inherent in these shared workspaces, where the pursuit of well-being and inclusivity was often undermined by the commodification of shared resources and the reinforcement of existing socioeconomic inequalities.

However, the institutional contexts and economic models within each country shaped the specific manifestations of these conflicts and the strategies employed to address them. In France, CWS catered to an elite clientele, reflecting the market-driven model and high costs associated with these collaborative workspaces. In contrast, Italy's CWS served as a refuge for precarious freelancers, driven by the necessity of finding affordable and flexible work arrangements in a fragmented labor market. Czechia exhibited a mix of entrepreneurial and remote worker models, with varying levels of accessibility and community engagement. The governance structures and management styles within CWS also played a crucial role in shaping the resolution or exacerbation of conflicts. Spaces with flat, community-driven governance models tended to foster more organic collaboration and informal mediation practices, while hierarchical, corporate-driven spaces often prioritized instrumental networking and structured events. These varied governance models demonstrate the complex relationship between CWS and SDG 16 (Peace, Justice and Strong Institutions), particularly in fostering inclusive decision-making at the micro-level. Eventually, this analysis demonstrates how CWS address social, environmental, and economic considerations in their operations and impact on sustainable organizational practices. It also highlights the need for CWS to develop meta-capabilities, as suggested by Wollmann, to navigate the complexities of sustainability and SDG implementation in diverse national contexts.

## 16.6 Discussion

The findings show these spaces are arenas of conflict, negotiation, and social hierarchies. In France, the high costs created an elite, exclusive user base that contradicted ideas of egalitarian coworking. The spaces risked reproducing social inequalities by pricing out those with less financial means. These dynamic challenges the achievement of SDG 10 (Reduced Inequalities) within urban work environments. On the other hand, Italy's coworking scene provided an alternative for precarious workers shut out of stable employment. But questions remain about whether these spaces just enable further exploitation of a disposable workforce. Set

[39] Spinuzzi, C. (2012). Working Alone Together: Coworking as Emergent Collaborative Activity. *Journal of Business and Technical Communication*, *26*(4), 399–441., A., Strebkov, D., & Davis, S. N. (2019). The Autonomy Paradox: How Night Work Undermines Subjective Well-Being of Internet-Based Freelancers. *ILR Review*, *72*(1), 75–100. https://doi.org/10.1177/0019793918767114

at the crossroad between the two former, Czechia exhibited a diversity of coworking models trying to balance entrepreneurial vibrancy with community-building. But varying accessibility levels meant some spaces still excluded certain socioeconomic groups. These exclusionary practices have implications for SDG 8 (Decent Work and Economic Growth) and SDG 11 (Sustainable Cities and Communities), highlighting the need for more inclusive coworking models.

On the whole, I can assess that across countries, CWS emerged as sites where ideals of community collided with economic pressures and social divides. The reality of competition over resources and social stratification often undermined well-intentioned pursuits of work-life balance. Management approaches were key—community-driven models with dedicated facilitators fostered more genuine collaboration compared to corporate, hierarchical spaces focused on networking events. However, the overall "silent" atmosphere of many CWS raised doubts about how naturally social interaction and networking really occurred, despite the hype. While enabling new ways of working, CWS also revealed themselves as microcosms reflecting broader societal tensions[40] around precarious labor, inequality, community-building, and work-life balance in the modern economy. These tensions directly relate to multiple SDGs, particularly SDG 8 and SDG 3 (Good Health and Well-Being), emphasizing the complex role of CWS in sustainable urban development. Acknowledging coworking as sites of negotiation and conflict can inform nuanced approaches promoting genuine collaboration while addressing concerns over socioeconomic inequality reinforcement.[41]

In essence, this study found CWS are complex environments full of contradictions between their collaborative rhetoric and the reality of social conflicts playing out. National contexts and economic models significantly shaped these contradictory dynamics. This variability underscores the importance of context-specific approaches to implementing SDG 11 and SDG 17 (Partnerships for the Goals) in different urban settings.

The findings have implications for stakeholders like municipalities, urban planners, coworking operators, and policymakers navigating complexities of fostering inclusive, equitable coworking models. Encouraging mixed governance, investing in community facilitators, balancing functional design with social cohesion, and ensuring urban integration and responsiveness to diverse user needs are crucial considerations.[42]

---

[40] Merkel, J. (2019). 'Freelance isn't free': Co-working as a critical urban practice to cope with informality in creative labour markets. *Urban Studies, 56*(3), 526–547; Swezey, C., & Vertesi, J. (2019). Working Apart, Together: The Challenges of Co-Work. *Proceedings of the ACM on Human-Computer Interaction, 3*(CSCW).

[41] de Peuter, G., Cohen, N. S., & Saraco, F. (2017a). The ambivalence of coworking: On the politics of an emerging work practice. *European Journal of Cultural Studies*, *20*(6), 687–706.; Zheng, Y. et al. (2022). CWS as places where economic diversity can be articulated: Towards a theory of syntopia. *Organization, 27*(1), 60–87.

[42] Merrifield, A. (1996). Public space: Integration and exclusion in urban life. *City, 1*(5–6), 57–72.

## 16.7 Conclusions

This multicountry study challenges the rosy narratives around coworking as purely collaborative, open, and accessible spaces. As the coworking movement continues to evolve, it will be crucial to consider how these spaces can foster genuine connection, economic accessibility, and a holistic approach to well-being, while preserving the autonomy and flexibility that initially drew users to them. This evolution presents opportunities to align coworking practices more closely with SDGs 3, 8, 10, and 11, potentially transforming these spaces into microcosms of sustainable urban development. The tensions I observed—between community and exclusivity, well-being and monetization, flexibility and hierarchy—are not merely localized phenomena but rather microcosms of broader societal negotiations. By understanding the nuances of these coworking ecosystems, I can gain valuable insights into the changing nature of work, the blurring boundaries between professional and personal spheres, and the ongoing quest for balance and fulfillment in an increasingly complex world. Therefore, I encourage policymakers to recognize CWS' diverse motivations and experiences, tailoring regulations to unique national contexts.

For coworking operators, transparent rules, participatory planning processes, and inclusion measures addressing precarious work realities[43] are vital for fostering equitable, supportive communities. Ultimately, this study highlights CWS' complex, multifaceted nature, challenging oversimplified narratives. Embracing a nuanced, critical perspective enables a better understanding of the intricate forces shaping lived experiences within these shared workspaces, paving the way for more equitable, inclusive, and supportive coworking models embodying community, collaboration, and human-centered work practices.

By applying the Three-Pillar Model to CWS, we gain insights into how these spaces can better align their practices with sustainability goals and adapt to the challenges of implementing SDGs in diverse national contexts. This approach provides a framework for CWS to navigate the "unknown territories" of sustainable development and contribute more effectively to SDGs 8, 10, 11, and 17. Furthermore, the model allows us to address the challenges identified in this study, emphasizing to what extent CWS have the potential to contribute significantly to several SDGs, particularly in fostering inclusive economic growth, reducing inequalities, and creating sustainable urban communities. Future research should expand cross-country comparisons, assess coworking's evolution in post-pandemic economies, examine remote work's impact on collaborative workspace demand, and develop policy recommendations for inclusive, accessible models, particularly for marginalized and precarious workers.[44]

[43] Rodríguez-Modroño, P. (2021). Non-standard work in unconventional workspaces: Self-employed women in home-based businesses and CWS. *Urban Studies, 58*(12), 2261–2277.

[44] Knappert, L., Cnossen, B., & Ortlieb, R. (2024). *Inequality Regimes in Coworking Spaces: How New Forms of Organising (Re)produce Inequalities.*

**Daphné Reguiessé**, PhD, is a social scientist with an interdisciplinary background in urban sociology and anthropology, shaped by international research experiences across Europe and India. Her work focuses on cities, migration, collaborative spaces, and intercultural dynamics.

# Part V

# Conclusions and Take-Aways

# Conclusions and Takeaways 17

Peter Wollmann, Doris Pemler, and Mersida Ndrevataj

**Abstract**

In this chapter, we attempt to summarize the many and varied messages, key findings, and most interesting recommendations from the individual chapters at a meta-level. In doing so, we cover, among other things, the necessary classification of events and activities in historical dimensions, an understanding of the breadth of the sustainability movement in terms of topics and content, the necessary readjustment of perceptions of progress and levels already reached, and the need for an action-oriented mindset even under volatile, unclear, complex, or ambiguous external conditions. In this context we present an integrated and commented overview of selected important conclusions from the various contributions in the book and interpret them within the book's overall framework. The aim here is to give the readers a set of overall messages and suggestions for further thinking and/or acting, and animations which chapter especially should be studied in detail to cover the individual reader's key demand. Compared to Volume 1, the conclusion covers many additional new topics and facets, but the general direction remains unchanged.

P. Wollmann (✉)
Bonn, Germany
e-mail: pw@peterwollmann.com

D. Pemler
Ciudad Condal GmbH, Zürich, Switzerland
e-mail: doris@ciudad-condal.ch

M. Ndrevataj
IUAV University of Venice, Venice, Italy
e-mail: mndrevataj@iuav.it



P. Wollmann et al. (eds.), *The Sustainable Organization – Volume 2*, Future of Business and Finance, https://doi.org/10.1007/978-3-032-23802-3_17

## 17.1 A Fundamental Foreword

*"The Truth is Rarely Pure and Never Simple"*
*Oscar Wilde, The Importance of Being Earnest*

This book begins with the above quote from Oscar Wilde in order to raise awareness of the urgent need to change the way we perceive the world in general and tasks, events, progress, etc. in terms of the Sustainability Movement. Our perception has to be challenged and recalibrated as our actions often follow the analysis of our spontaneous impulses (which might be a victim of thinking traps, perception biases, neurological patterns[1]). The reality of the VUCA[2] world is confusing, unclear, ambiguous and often an unknown territory which we have to explore and for which we have to tailor or even to reinvent[3] our thinking styles, mindsets, methods, and tools. And we must always question ourselves—not only in terms of our perceptions: ("*What if we were wrong*"—Barack Obama 2016.[4] This is a challenging question which is very useful to avoid excessive certainty about one's own worldview. The question also exactly refers to Oscar Wilde's quote "The Truth is Rarely Pure and Never Simple" and fits to the very limited human ability to fully and finally understand a complex situation and especially to predict the future or even scenarios for the future).

In these disruptive and turbulent times, attempts at long-term, differentiated thinking in analyzing the present and assessing the future are often overshadowed by hectic headlines and selective attention. In Chap. 1, we also described this circumstance as a risk for a book like this one: between the completion and submission of the manuscript and its publication, will a new situation have arisen that could potentially devalue parts of the book? We countered this fear with a fundamental, long-term perspective, including from a historical point of view, but waited until shortly before the manuscript was submitted to write this last chapter. And now, 1 week before the deadline, an interesting column by Ullrich Fichtner has been published in the magazine "*Der Spiegel*"[5] supporting the book's outcome and direction. Fichtner refers to a thought experiment to better understand the character of time and, in particular, the age of the universe. For this, the age of the universe is mapped to only one calendar year, which means that humanity would have existed for only 8 min—and 8 or 10 years at the helm of a country would be mapped to less than the blink of an eye. What does this have to do with our perception of time? We should

---

[1] See Chap. 1, Sect. 1.2.4.

[2] **V**olatile, **U**nclear, **C**omplex, **A**mbiguous World, today often also described as BANI World (**B**rittle, **A**nxious, **N**onlinear, and **I**ncomprehensible)

[3] Using the Three-Pillar Thinking Style and Mindset Model, for detailed description see Chap. 2

[4] See https://www.nytimes.com/2018/05/30/us/politics/obama-reaction-trump-election-benjamin-rhodes.html

[5] Der Spiegel on 18.02.2026, 14.14 Uhr, see: https://www.spiegel.de/politik/donald-trump-warum-der-us-praesident-nur-ein-staubkorn-im-universum-ist-a-04d1c084-ad19-4fb4-8005-f5831e3894e7

not overestimate the moment. And as Fichtner stresses: "Thinking about space and time can help us stay calm."[6]

Fichtner continues: "Our planet is facing an exorbitant emergency: climate change, resource depletion, economic upheaval, and the extinction of more and more species. Hope through action shows us how to face this crisis and respond to it with unexpected resilience and creativity. This hope does not necessarily require optimism; we can even use this process where we see no hope left. The guiding impetus is our intention. Instead of weighing up the chances of success first and only taking action when we feel hopeful, we focus on our intention and let it guide us."[7]

Fichtner mentions in this context the American Buddhist and ecologist Joanna Macy[8] and praised her making many valid ideas powerful and popular, for example with her spiritual classic "Hope Through Action."[9] Key ideas from it are consistent with the American political scientist and psychologist Sarah Stein Lubrano's research and her publication on the subject.[10] Her credo is: "Don't discuss. Act." Action is in (political or societal) communication always more powerful than pure talking and discussing. In relation to the subject matter of our books, Volumes 1 and 2 (and also the upcoming Volume 3), this means focusing on the acting. And here, a reference from Volume 3 (Chapter 6) is of high importance mentioning the ME310 courses of Stanford University, a blueprint for innovation learning, giving students "in a real-world, engineering design experience in the spirit of a Silicon Valley startup, managing the uncertainty inherent in entrepreneurial design"[11] which has to be more understood as a mentoring concept than a classic university course. This is training in starting to act even if not everything is clear.

It is important to know that this proceeding is explorative, starting without a detailed plan over the full project life cycle, agile proceeding, rapid prototyping and learning on the journey (a classic "Travelling Organization Approach," see details in Chap. 2). To repeat: students are taught that they have to start acting in partly unknown territories and how to best cope with this. Many parts of the Sustainability Movement Activity Landscape have exactly these preconditions. The question of what to do under these circumstances is also answered in this book in the sense of an "Ethics of Responsibility" and a clear call for action and requirements for future education. In this context a quote from Hannah Arendt from 1950s might be interesting: "*Education is the point at which we decide whether we love the world enough to assume responsibility for it, and by the same token save it from that ruin which except for renewal, except for the coming of the new and the young, would be*

[6] Ibid, translated into American English using DeepL

[7] Ibid, translated into American English using DeepL

[8] https://en.wikipedia.org/wiki/Joanna_Macy

[9] https://openlibrary.org/books/OL25139864M/Active_hope

[10] https://www.sarahsteinlubrano.com

[11] https://explorecourses.stanford.edu/search?q=ME+310A%3A+Global+Engineering+Design+Thinking%2C+Innovation%2C+and+Entrepreneurship&view=catalog&filter-coursestatus-Active=on&academicYear=20252026

*inevitable. And education, too, is where we decide whether we love our children enough not to expel them from our world and leave them to their own devices, nor to strike from their hands their chance of undertaking something new, something unforeseen by us, but to prepare them in advance for the task of renewing a common world.*"[12]

The good news is that there are overwhelmingly many similar (successful) initiatives on education—and on better communication—in place worldwide as the readers will see browsing the book. Contrary to all fears, neither multilateralism nor measures against climate change etc. have been stopped across the board or even come to the end of their tether. As already mentioned in Volume 1,[13] the human lack of capabilities to imagine and think the future and to design future aspirations has a massive impact on understanding change developments which are nearly never linear[14] and to neutrally evaluate the progress of (political and societal) transformations—a fact that on the other hand severely affects the prognosis skills of people, especially if the situation is a lot better than perceived.

These systemic and personal human incapabilities were impressively shown by Hans Rosling in his book "*Factfulness*"[15] where he—only based on available facts—demonstrates the development of the state of the world in multiple perspectives. The interesting point here is that people, when asked about their perception of the state of the world in the most important perspectives covered by Rosling, are nearly always wrong (and too pessimistic), by the way, the more educated they were, the more pessimistic they were. It therefore seems as if brains like dramatic narratives, as can also be seen when looking at the success of attention seeking content in social media. So, Barack Obama's question might be further developed "What if we were wrong and still are wrong?". Is my thinking and perception styles effective and efficient—or does it produce wrong convictions and perception?

Considering the facts and results from the chapters and their authors carefully and critically, the message from Chap. 1 still remains very valid:

- There is a really broad movement of activities and actors in place covering a large number of important and significant areas. We call it the Sustainability Movement Activity Landscape covering Sustainability and the UN Sustainable Development Goals (SDG)[16]
- There is a lot more progress already achieved in many of the activities than perceived in the public perception

[12] Hannah Arendt: The Crises in Education, 1954. https://thi.ucsc.edu/wp-content/uploads/2016/09/Arendt-Crisis_In_Education-1954.pdf

[13] Partly quoted from Volume 1

[14] See Chap. 1 in Wollmann, P.; Pemler, D.; Ndrevataj, M. (Eds.): The Sustainable Organization - How Organizations Address the 17 UN SDGs Using the 3-P-Model. Cham Springer Nature, © 2025; (https://link.springer.com/book/10.1007/978-3-031-89549-4)

[15] Rosling, H: Factfulness New York 2008, Flatiron Books.

[16] https://sdgs.un.org/goals

- In general, there is a perception bias in the judgment of large transformational movements in place as people tend to expect linear developments (also as a consequence of neurological imprints)
- The movement is to a very large degree well rooted in the middle of society and irreversible, also as it covers systemic logic
- The movement will force many organizations (and private persons) to journey through unknown territories which requires the mindset of a Travelling Organization. The respective societal transformation will need a historical time span
- The coordination and cooperation across activities, projects and transformations is a Herculean task and can only be handled in an explorative and agile way. The expectation of a waterfall model with a detailed master plan in place is totally unrealistic (all such transformations have to agile as the future cannot be predicted as we stressed above)

In this context it might make sense to cite from Chap. 3: "In the long run, the intrinsic logic of (global and national) systems striving for optimized effectiveness and efficiency demands from different perspectives respective suitable global knowledge-based institutions with high legitimacy, rule-based interaction, cooperation and procedures and a reasonable degree of public support for the concept to achieve Sustainability and UN SDG Coverage and the respective Movement, which demands adequate public communication. However, reaching Sustainability and UN SDG Coverage including the institutions' interactions is clearly a highly complex, explorative and incremental step by step creation and optimization task with many variables to be taken into consideration. Expectations of politicians, experts and the public have to be respectively calibrated. In addition, the (interim and final) results of the transformation journey will for sure change over time under changed conditions, this has to be taken into account – and it is not a sign for bald planning or failure but for the success of an explorative, flexible proceeding."[17]

It is again very encouraging that the Three-Pillar (Mindset and Thinking) Model[18] perfectly fits the challenging situation and was applied in nearly all chapters or use cases respectively.

1. The first Pillar: **Sustainable Purpose**
   Sustainable Purpose is the *raison d'être* of an organization. Having a clear guiding objective provides orientation and certainty to people, fostering a sense of shared effort and success. The ***special challenge with Sustainable Purpose*** linked to a general idea of sustainability and the coverage of selected UN SDGs

[17] See the full text and its framing in Chap. 3

[18] Described in detail in Chap. 2. The Three Pillar Model is based on three key building blocks (pillars), defining the Three-Pillar (Mindset or Thinking) Model for organization design and leadership for the VUCA world. The author community's starting point in 2019 was to establish guiding principles for a world seemingly devoid of them. The model has since been frequently applied—and its applicability and usefulness proven—in various global use cases (so far >80) in the private and public sectors. The goal of the Three Pillar model is to support understanding and navigating the complexities of today's corporate and institution territory in all sectors.

in detail lies in the perception bias, the gap between the public perception of status, success, reachability of sustainability targets including the judgment of concrete parts of the activity landscape on the one hand and the fact-based reality (of the comprehensive big picture) on the other hand. To fight doubt, discouragement, even despair (up to doomsday mood) it is important to act, not to talk and to develop and show positive use cases and what an organization can reach with the help of good leadership.

2. The second Pillar: **Travelling Organization**
   Travelling Organization refers to an organization accepting to be in a permanent state of flux and transformation while interacting with the external requirements of its journey by means of rapid learning and adaptivity. Ideally, this is the mindset of loving the Unknown and being well able to manage it.[19] The ***special challenge with a Travelling Organization Mindset and Capabilities*** is today on the one hand that the north star of Sustainable Purpose in the context of Sustainability and UN SDG Coverage is challenged by a pessimistic view and a public perception bias already described above which has an impact on motivation and momentum. Going on an explorative journey through the Unknown being confronted with change is always demanding and even more challenging under the current preconditions. But our use cases show that even though the challenges for leadership are obviously tremendous, the journeys are manageable and at the end a mind-changer.
3. The third Pillar: **Connectivity**
   Connectivity focuses on the ability to interconnect during the journey with all necessary parties and resources inside and outside the silos, thus creating high effectiveness and consistency and also building shared learning networks[20] during this process. The ***special challenge with Connectivity*** in the context of activities on Sustainability in general and the coverage of selected UN SDGs in detail is the increasingly strong fragmentation among stakeholders in terms of opinions, perceptions, and ethics (ethics of conviction versus ethics of responsibility), all of which are sometimes strongly linked to a certain emotionality. This makes it difficult to agree on a common starting point for cooperation or even neutral exchange. High, often unattainable expectations are met with defeatism, but also with total rejection. The situation often requires the utmost diplomatic skills and special integrative leadership qualities. But the principle: start and don't talk too much, also fits here very well and seems reasonable.

[19] One has an intuitive idea of a Travelling Organization from the metaphor of an expedition to Africa in former times. The participants of the Travelling Organization must be curious, open to learn, courageous, keen to experiment, and they deal well with uncertainty, stress, unforeseen incidents in a totally new world. This mindset of the Travelling Organization is the core of a Learning Organization. It is impossible in and for the Travelling Organization not to learn. Learning is a question of survival.

[20] This means that organization has to be thought of within a network perspective, including interfaces to adjacent organizations and across organizational borders. In many cases, it might be impossible to learn effectively without collaborating closely with people from outside your silo.

## 17.2 Proceeding

We selected from each chapter or part respectively those conclusions which are of overarching importance and sorted them. Some of them were already mentioned in the fundamental foreword of the chapter.

Out of this selection we chose the following top eight general conclusions overarching the whole book. They share the character of recommendations for thinking, judging, and acting.

For the detailed framing of these conclusions, we recommend reading the chapter they are taken from. We did not want to produce too much redundancy by adding framing text from the original chapters as we believe that the conclusions are self-explanatory and do not require any additional explanations.

## 17.3 Our Top Eight General Conclusions and Recommendations

1. **Be always aware of your intellectual, mental, neurological, and cultural limitations:**
   The key question is: What if we were wrong and are still wrong? We must be aware that our intellectual, mental, neurological, and cultural conditioning (including our usual thinking styles and analytical methods) limits us in our ability to holistically understand the world in general—especially in historical dimensions—and especially the complex, multidimensional and, in sum, unmanageable landscape of activities on sustainability and coverage of the UN SDGs and to analyze their development. We expect familiar patterns, linear developments and methodological clarity, but we are finding ourselves—at least in part—in terra incognita. It is necessary to accept this and to find ways to best cope with this, for example actively look for different perspectives and thinking styles to approach a holistic understanding—and never be too convinced to know now the full truth and all its facets, there could always be an additional surprise. And we should often take the "Gapminder test" of Roslings foundation.[21]
2. **Be aware of your perception bias is often coming from your appetite for drama:**
   In particular, we must accept that our ability to recognize progress in the development of the world is very poorly developed. It is not for nothing that particularly educated people judge many issues (see development of child mortality, literacy and school attendance, spread of child labor, mortality, etc.).[22] This conclusion was also shown in Volume 1—and it is more true now than ever.

[21] Go to https://www.gapminder.org and select an icon under the headline "You are probably wrong about"

[22] Quote of Hans Rosling: "We have to learn to get a grip on our drama consumption. Without such control, our appetite for drama goes too far and prevents us from seeing the world as it is, leading us terribly astray" (Rosling, H: "Factfulness" New York 2008, Flatiron Books, S. 27)

3. **Regard "Sustainability" together with "Coverage of UN SDGs" empirically as a (historically) powerful long-term mega-trend:**
Striving for "Sustainability" and "Coverage of UN SDGs" is obviously a long-term megatrend to be judged in historical time dimensions, not short-term. It is important to challenge the world's understanding and perception of time. We show in the book again (as in Volume 1) that there are well documented and significant efforts on a global scale (UN level) and both on the regional and national scale in place to push the connected topics "Realization of the UN SDGs" and "Sustainability." It is clear that the implementation of individual projects and measures does not proceed in a straight line but is subject to their own problems, deviations and many counter-movements. This follows a well-known pattern and has also been the case with all former major developments and transformations. It will therefore only be possible to measure the success of this whole transformation effort for future generations from a historical perspective.
4. **Be aware that "Sustainability" and "Coverage of UN SDGs" have first and foremost a strong long-term economic and systemic logic and therefore a strong long-term systemic and operative underpinning:**
The general targets of the so-called Agenda 2030[23] are simply common sense—only the means of achieving them are controversial and must, of course, be discussed. In the long run, a green and circular economy will be more effective and efficient—and in parallel the costs of not striving for sustainability and coverage of the UN SDGs are overwhelming, so that even economic rationality requires following the megatrend. Even political changes in relevant countries will not be able to change its direction mid- or long-term as the main driver is economic efficiency, not ethical conviction. And this logic drives the operative underpinning of efforts to cover UN SDGs and reach sustainability for the world as well on a global scale (UN level) as also on regional and national level—even when some countries preliminarily withdraw support and efforts. As we show in the book, there are many concrete measures in place with named responsible organizations and individuals (see exemplarily the wonderful use case from the UK in Chap. 9). All official activities are subject to regular controlling and reporting.
5. **Be aware that the Movement for "Sustainability" and "Coverage of UN SDGs" requires in the long-term strong international institutions following the logic of effective and efficient systems and processes:**
As shown in the discussion in Chap. 3 (and also in Chaps. 5, 6, and 7), the meta-level systemic logic driving progress of humanity in history, namely improving effectiveness and efficiency, strives for strong international institutions in the context of the Sustainability Movement. This is proven from different perspectives, firstly from a mathematical or project/project portfolio driven

[23] https://sdgs.un.org/2030agenda.

logic, secondly from Rule of Law as well from a country as also from a multilateral UN perspective and thirdly, from the Global Governance view of system theory, applied in a knowledge-driven world. All perspectives are mainly based on the progress driving principles of gaining effectiveness and efficiency as system logic. A detailed deep dive use case regarding WIPO[24] is shown in Chap. 7, it is a role model for other institutions with a strong logical argumentation. It is obvious that the broad-based Movement as we might regard the nearly infinite Activity and Project Landscape to reach Sustainability and UN Sustainable Development Goals Coverage with billions of entries, which was reintroduced in Chap. 1, needs some institutional and rule-based fundaments to reach the needed effectiveness and efficiency—but this also valid for many other aspects of the world's development (same principle applied to other contexts).

6. **Be aware that there are successful models to sustainably develop whole countries beyond the pattern of the Western world but follow long-term the same goals and principles:**
   As shown in Chap. 6, countries might choose another journey to Sustainability and UN SDG Coverage as simply to follow the example given by the Western world. It would be a perception and judgment bias to underestimate these alternative journeys or respectively their development options as not each country has the exact same preconditions. It is important to prevent superficial or even arrogant perspectives but to take a more ethnological approach and check ambitions and progress. At the end, there might be in each of the use cases parts to regard as very positive as well as parts which need significant improvement. And coming back to the historical dimensions: even the West needed significant time to reach its current level of maturity.
7. **Be aware that the Sustainable Development in urban and rural areas is a special cornerstone for the success of the "Movement for Sustainability and Coverage of UN SDGs":**
   The book shows some few but impressive use cases from different countries in Chaps. 4 and 10–16, covering different topics including: the challenges of sustainable management of the food value chain, given the heterogeneity of actors it involves (Chap. 10); the rural development and the institutional learning in the Schleiden-Eifel region in Germany, following the 2021 flood event (Chap. 11); the rural development in India through the case of Bunker Roy's Barefoot College (Chap. 12); social innovation in tourism in Venice, Italy through the local cultural heritage initiatives (Chap. 13); proximity welfare centers and the horizontal networks activated through and for such initiatives (Chap. 14); the World Heritage policies and management for sustainable development at national, regional and local level in Italy (Chap. 15); coworking spaces and their contribution to Sustainability, examined through a comparative study across

[24] World Intellectual Property Organization

France, Italy, and Czechia (Chap. 16)—while Chap. 4 provides the conceptual foundation for all of these contributions, situating them within a common argument: that urban and rural areas form a system of interdependence, and that sustainable development depends on governance frameworks capable of managing it as such.

8. **Be aware that the right and significantly improved communication is a special cornerstone for the success of the "Movement for Sustainability and Coverage of UN SDGs":**
   One of the most important points of analysis and findings in Volume 1 was that communication is absolutely essential, not only within the Movement but especially with the public, and that it must be greatly improved. This is particularly true because of the perception and thinking traps mentioned above. The topic of communication is therefore addressed in detail at various points in this book: in a separate Chap. 10, showing animating use cases in different contexts, implicitly also in Chap. 1, including in the interview with Ovais, former deputy head of UNFCCC and responsible for several COP events, further implicitly in Chaps. 3, 7–9, and also Chapter 15 of the upcoming Volume 3 (with a context to the urgent need of investment in transformational education and consultancy). It remains a hot topic that will require a great deal of attention in the coming years and demands a different perspective: it is not only important that experts and stakeholders within the system understand each other, but also that the system is understood and accepted by people outside it. There are already various initiatives around the world for a new, more advanced form of journalism (in times of seemingly overpowering social media).

## 17.4 Some Selected Additional Specific Individual Conclusions from the Chapters[25]

- **The use case "Insights from the Unique Model of Qatar"** (Chap. 5):
  The analysis presented in Chap. 5 offers an interesting new perspective beyond usual Europe- or Western-centric thinking which is highly important to understand different options to reach sustainability and to identify valuable cooperation and learning opportunities. Qatar's developmental trajectory demonstrates that leadership, rather than governance structures in isolation, can be a decisive force in advancing sustainable development. The Qatari case complicates conventional assumptions that democratic institutions are a necessary precondition for progress toward the Sustainable Development Goals (SDGs). Instead, it highlights how the particular qualities of leadership discussed in this chapter—visionary, transformational, adaptive, modern, and sustainable—can be mobilized to generate development outcomes that are measurable and durable. To reiterate,

[25] Mostly taken as a 1:1 excerpt from the conclusion section of the chapter

this is not to suggest that governance systems are irrelevant, but rather that the role of leadership as a driver of direction, coherence, and long-term commitment deserves greater analytical attention. The reality of impressive progress and growth in Qatar, particularly its economic success, illustrates that impactful leadership is an essential and effective component of development. The qualities of Qatar's leadership identified here constitute a model that has enabled Qatar to transition from an oil-dependent, peripheral economy into a diversified and globally engaged state capable of promoting economic growth, encouraging social stability, and maintaining cultural continuity. While no single one of these qualities is necessarily unique or distinctive by itself, the manner in which they have been brought together to reinforce each other by Qatar's leadership is what has made them so effective. Visionary planning provides direction, transformational initiatives reshape (but do not disrupt) economic and social structures, adaptive measures ensure resilience to changing conditions, modern strategies support innovation without erasing tradition, and sustainability anchors growth in long-term security.

- **The use case "Multilateral State-to-State Dispute Resolution Mechanism within the WIPO Framework"** (Chap. 7):
  The analysis presented in Chap. 7 has demonstrated that the absence of a centralized, specialized[26] mechanism for the resolution of state-to-state intellectual property disputes poses a significant challenge to the stability, coherence, and legitimacy of the global IP system. As states increasingly rely on knowledge-based economic growth, and as intellectual property rights play a pivotal role in shaping patterns of trade, investment, and technology, the need for predictable, authoritative, and fair dispute settlement becomes crucial.
  The paralysis of the WTO Appellate Body since 2019 has fundamentally altered the state-to-state dispute settlement landscape, specifically in terms of IP. Once considered a model of rules-based adjudication, the WTO dispute settlement system now suffers from a loss of credibility and enforceability, with the "appeal into the void" phenomenon undermining legal certainty and creating opportunities for strategic noncompliance. In its wake, alternative venues such as IP chapters in FTAs have proliferated. While these may offer ad hoc solutions, their fragmented nature, overlapping jurisdictions, and varying obligations have exacerbated forum shopping and procedural inconsistencies. The resulting institutional vacuum erodes the normative coherence of the multilateral IP regime and threatens core rule-of-law principles such as predictability, stability, and equality of treatment.
  In this context, the creation of a permanent, WIPO-led dispute settlement body emerges as a structural necessity for the future of international IP governance. WIPO, as the sole specialized UN agency devoted exclusively to IP, possesses the institutional expertise, technical capacity, and convening power to fill this

[26] World Intellectual Property Organization

gap. A WIPO-based dispute resolution mechanism would serve as a coherent, authoritative forum capable of reconciling the technical complexity of IP with the imperatives of international economic governance. Crucially, it would contribute to the realization of Sustainable Development Goal 16, which calls for effective, accountable, and inclusive institutions at all levels. Access to justice in the IP context contributes to enabling equitable participation in the global knowledge economy and the protection of the rule of law in cross-border innovation systems.

However, the legitimacy and effectiveness of such a body will depend heavily on its institutional design. As discussed, the selection of adjudicators must embody both inclusivity and competence. Geographic, gender, and legal system diversity are essential for ensuring representativeness and legitimacy, but these must be balanced with the need for deep technical expertise in IP law and related fields. Binding rules on diversity, coupled with transparent, merit-based selection procedures, should form the backbone of the appointment process. Lessons from the International Criminal Court's gender quotas and multi-round elections can provide a valuable template, ensuring that diversity is operationalized rather than left to rhetorical commitment.

Furthermore, judicial capacity building should be institutionalized through a permanent WIPO Judicial Academy for IP Law. This body would provide continuous training, peer exchange, and resources for judges, fostering high-quality adjudication and contributing to sustainable judicial capacity building.

- **A use case from UK "Translating Climate Negotiation into Local and Regional Action"** (Chap. 8):

  The shown case study of the Yorkshire and Humber Climate Commission (YHCC) in the UK illustrates how place-based institutions can act as "connective tissue" between global goals and local realities through independent governance, service orientation, political neutrality, evidence-led framing of co-benefits, and inclusive co-production.

  International negotiations can articulate a shared purpose and provide long-range direction, but they cannot determine how policy lands in different national contexts. National governments can legislate, regulate and finance, yet their decisions only acquire meaning when translated into the daily realities of people, companies and communities. Local and regional actors, meanwhile, sit closest to where climate impacts are felt and where change is lived, but they often carry responsibilities that exceed their formal powers or resources and are frequently overwhelmed by the enormity of competing challenges.

  What often appears as a failure of one layer is, in reality, a failure of alignment between them. The international system presumes domestic machinery ready to operationalize global ambition. National governments presume that local actors can deliver complex transitions with limited fiscal space and uneven capacity. Local communities presume that national or global frameworks will eventually adjust to their needs, pressures and opportunities. These assumptions leave large gaps through which progress can easily stall, even when the underlying intent is widely shared. The result is a system in which each level is partially right about where the bottleneck lies, yet none alone can unblock it.

The experience of the Yorkshire and Humber Climate Commission (YHCC) illustrates what becomes possible when these governance layers are intentionally bridged. By creating a space where public, private and third-sector actors can work collectively—grounded in evidence, aligned around shared regional priorities, and supported by practical tools—place-based institutions can act as translators and accelerators of climate action. The YHCC does not replace international agreement or national policy; rather, it supplies what neither can provide: a coherent, grounded, region-wide framework of legitimacy, evidence, networks and practical delivery, which is the connective tissue through which both can be meaningfully enacted. Its independent governance model strengthens trust; its service orientation enables local authorities to act on common priorities; its evidence base reframes climate action as economic and social opportunity; and its inclusive practice embeds the transition in the lived concerns and values of communities. In doing so, it shows that when governance is rooted in local context, trusted by a diversity of actors, and capable of convening across boundaries, implementation gathers a momentum that formal authority alone cannot produce, linking the global sense of mission through national legislation to the local realities that determine whether that mission succeeds.

The lesson is not that every region should replicate Yorkshire and Humber's model exactly, but that place-based governance is indispensable to the implementation challenge facing the world. International agreements can define purpose, but they cannot create legitimacy on the ground. National governments can set policy direction, but they cannot build the networks of trust required for societal transformation. Local communities can innovate, but they cannot do so at scale without aligned frameworks, investment and enabling conditions.

- **A use case from the "Food Supply Chain"** (Chap. 10):

  The analysis shows that in the future; sustainability management systems will become the standard for companies in this industry and the value chains in which they operate. For the management of the companies involved, this means that tasks will become more complex, as they will relate not only to traditional targets but also to environmental and social ones.

  Not only will there be a greater number of control variables to manage, but conflicts between the target variables will also have to be resolved.

  Furthermore, traditional supplier/customer relationships are changed, and new partnerships and organizational models are emerging along the value chain, which must be incorporated into strategy and organization.

  Data management will become increasingly important. The definition of standards and interfaces will be a key success factor. Overall, the development of sustainability management will accelerate the digitalization of industries.

  However, sustainability is not just a corporate goal. Sustainability also means transferring behavior from the "private" to the "professional" sphere and vice versa. This makes it even more important to align the culture of companies and supply chains with this goal.

The food supply chain is certainly the most complex and extensive value chain. But right there the 8 billion consumers are confronted with sustainability daily. The food supply chain brings business processes into society and must therefore also take social requirements into account in a special way. As a pioneer for many other supply chains and a development laboratory for methods and behavior, it has a role that can shape and accelerate the comprehensive introduction of sustainability management systems worldwide.

However, the food supply chain does not only have a special role as a pioneer in the introduction and implementation of sustainability management systems. It combines well with the Three-Pillar Model in companies, as the dynamics with which the food supply chain is changing must also be reflected in companies. The pillars and the commitment of organizations to lasting change are a decisive competitive advantage.

Integrating sustainable benefits into the company's mindset is a basic prerequisite for thinking and acting across organizational and cultural boundaries within the supply chain.

When you consider the challenges facing companies, companies must embark on a journey, as both internal strategic understanding and an outward-looking organizational self-concept must be developed for employees, stakeholders, and the organization. The discussions about nutrition, the introduction and weighting of sustainability are diverse, and there are no standard solutions as yet. Openness to experience and learning from others are key success factors in generating corporate success from these approaches.

The third pillar of the "Three-Pillar Model," Connectivity, is a crucial one. Both effectiveness and efficiency are based on connectivity in a food supply chain system. This means Connectivity between individuals, between people, companies, products, and processes. Without Connectivity, the complex food supply chain in particular cannot be managed effectively.

For optimized management of the food supply chain, it is necessary to develop innovative management systems and continue to adapt them during implementation. This is one of the success factors for the future.

- **A use case on "Social Innovation in Tourism in Venice: A Three-Pillar Model Analysis of Purpose, Adaptability and Connectivity"** (Chap. 13):
  The analysis of Social Innovation in Tourism (SIT) initiatives in Venice demonstrates that their sustainability and impact depend on the interaction of purpose, organizational adaptability, and connectivity. In terms of Sustainable Purpose, the study reveals that cultural heritage projects are strongly motivated by intrinsic values such as personal growth, passion for culture, and a sense of responsibility toward the local community. These personal motivations contribute to the formation of a shared mission oriented to preserving historical memory and local identity. However, the findings also highlight critical challenges: long-term eco-

nomic sustainability remains a major issue, and there is a need for greater inclusivity to address the needs of marginalized or disadvantaged groups. Moreover, the lack of information, limited access, and financial difficulties experienced by certain segments of the population indicate that the sustainable purpose of SIT must extend beyond heritage preservation to include social equity and accessibility. Thus, the sustainable purpose pillar should be understood as integrating cultural value with community wellbeing, balancing mission-driven goals with viable economic models.

The Travelling Organization pillar shows that SIT initiatives require continuous adaptation, learning, and leadership capacity to navigate complex socio-economic contexts. However, the evidence indicates that fragmented networks and communication barriers hinder collaboration and slow down the development of shared vision among stakeholders. The involvement of multiple actors increases the number of beneficiaries, but it also increases the need for coordination and governance capacity. Additionally, the lack of coordination between government levels (local, regional, national) further complicates the capacity of organizations to respond to changing conditions and pursue long-term strategies. This organizational fragmentation reduces the scalability and resilience of initiatives, limiting their ability to generate long-term regeneration processes. Therefore, SIT requires a travelling organization mindset capable of building cohesive networks, maintaining continuity, and adapting to institutional constraints through collaborative learning.

The Connectivity pillar emerges as a central determinant of the long-term success of SIT. Political and institutional support is crucial for project continuity, yet bureaucratic and administrative complexities create operational difficulties that undermine collaboration. The absence of effective governance and planning processes, combined with weak communication strategies, reduces the visibility and legitimacy of local initiatives. Moreover, the system's dependence on the vision of the mayor or current administration creates instability, and the frequent turnover of public administration staff further weakens continuity and trust. These structural weaknesses highlight the importance of strengthening multi-level governance, improving communication strategies, and creating stable support mechanisms. In particular, enhancing connectivity requires better coordination between government levels, clearer institutional pathways, and a stronger commitment to inclusive participation, ensuring that benefits reach marginalized communities.

Beyond their local relevance, the findings also highlight the broader contribution of SIT to the objectives of the 2030 Agenda for Sustainable Development. By promoting inclusive participation, safeguarding cultural heritage, and strengthening community resilience, SIT initiatives directly support several UN Sustainable Development Goals, particularly SDG 8 (Decent Work and Economic

Growth), SDG 10 (Reduced Inequalities), SDG 11 (Sustainable Cities and Communities), and SDG 17 (Partnerships for the Goals). The Venetian case demonstrates that heritage-based tourism, when guided by a clear sustainable purpose, adaptive organizational practices, and strong connectivity, can function as a lever for socially inclusive and territorially balanced development. However, the challenges identified—such as governance fragmentation, limited accessibility, and uneven participation—also reveal the risk of misalignment between local practices and global sustainability commitments. In this sense, SIT offers not only a practical pathway for advancing the Agenda 2030 at the local level, but also a critical framework for assessing how sustainability principles are translated into concrete governance arrangements and everyday practices within tourism destinations.

In conclusion, these findings highlight significant gaps in the scientific literature on SIT. Despite growing research interest, there is still insufficient attention to institutional barriers, governance fragmentation, and inclusivity issues that affect the sustainability of social innovation initiatives. Future research should explore the interplay between purpose, organizational adaptability, and connectivity in different territorial contexts, and examine how multi-level governance and inclusive strategies can enhance the long-term viability of SIT. Addressing these gaps would contribute to a more comprehensive understanding of SIT and support the development of policy frameworks that promote sustainable, inclusive, and resilient cultural heritage-based tourism.

shows that in the future, sustainability management systems will become the standard for companies in this industry and for the value chains in which they operate. For the management of the companies involved, this means that tasks will become more complex, as they will relate not only to traditional targets but also to environmental and social ones.

- **A use case on "The Presence of Absence: Conflicts in Collaborative Spaces: A Comparative Ethnographic Study in France, Italy, and Czechia"** (Chap. 16):

  This use case bases on a multicountry study which challenges the rosy narratives around coworking as purely collaborative, open, and accessible spaces. As the coworking movement continues to evolve, it will be crucial to consider how these spaces can foster genuine connection, economic accessibility, and a holistic approach to well-being, while preserving the autonomy and flexibility that initially drew users to them. This evolution presents opportunities to align coworking practices more closely with SDGs 3, 8, 10, and 11, potentially transforming these spaces into microcosms of sustainable urban development. The tensions I observed—between community and exclusivity, well-being and monetization, flexibility and hierarchy—are not merely localized phenomena but rather microcosms of broader societal negotiations. By understanding the nuances of these coworking ecosystems, I can gain valuable insights into the changing nature of work, the blurring boundaries between professional and personal spheres, and the ongoing quest for balance and fulfillment in an increasingly complex world. Therefore, I encourage policymakers to recognize CWS' diverse motivations and experiences, tailoring regulations to unique national contexts.

For coworking operators, transparent rules, participatory planning processes, and inclusion measures addressing precarious work realities are vital for fostering equitable, supportive communities. Ultimately, this study highlights CWS' complex, multifaceted nature, challenging oversimplified narratives. Embracing a nuanced, critical perspective enables a better understanding of the intricate forces shaping lived experiences within these shared workspaces, paving the way for more equitable, inclusive, and supportive coworking models embodying community, collaboration, and human-centered work practices.

By applying the Three-Pillar Model to CWS, we gain insights into how these spaces can better align their practices with sustainability goals and adapt to the challenges of implementing SDGs in diverse national contexts. This approach provides a framework for CWS to navigate the "unknown territories" of sustainable development and contribute more effectively to SDGs 8, 10, 11, and 17. Furthermore, the model allows us to address the challenges identified in this study, emphasizing to what extent CWS have the potential to contribute significantly to several SDGs, particularly in fostering inclusive economic growth, reducing inequalities, and creating sustainable urban communities. Future research should expand cross-country comparisons, assess coworking's evolution in post-pandemic economies, examine remote work's impact on collaborative workspace demand, and develop policy recommendations for inclusive, accessible models, particularly for marginalized and precarious workers.

## 17.5 Some Closing Thoughts

To start with some fundamental statements:

- Since Sustainability is increasingly held hostage in the process of worldwide political polarization, we have to stress that in a mindset of equality of all people, it is in our shared interest to ensure sustainable solutions for the future, regardless of political views.
- We can hope and trust that action and our shared endeavors will expose the politics adverse to a sustainable future for all as a fight analogous to the one of Don Quixote, who is fighting a lost cause when the times had already changed.

The conclusions above cover a broad range of different focus topics and show different facets of the potential of intellectual capital in the global transformation endeavor to cover the UN SDGs and reach a sustainable development of the world. It is evident that even though some impressive concepts and methods are already in place, there exists no overarching recipe to solve the individual problems and challenges. There can be no "one-size-fits-all"—also because we are breaking completely new ground in many areas. It turned out that the Three-Pillar Model as a meta-level thinking style and mental setting is in all cases helpful as it does not provide recipes but rather an attitude on how to approach challenges regarding the ever-unknown territories of change and the future.

**Peter Wollmann** has been a responsible manager, initiator, mentor and facilitator in large, predominantly global transformations and strategic developments for almost 40 years in large organizations and since 2017 as independent consultant. He is also a member of an advisory board of a R&D oriented company. Peter holds a degree in mathematics and physics from the University of Bonn. He is the author and publisher of numerous books and articles on strategy, leadership, organization design and project and project portfolio management.

**Doris Pemler** is an entrepreneur, artist and researcher based in Zurich. She received her M.A. in Egyptology, Sudan Archaeology and Archaeology of the Ancient Near East from Humboldt University zu Berlin and FU Berlin. Her research focuses on iconography, gender and intercultural communication in Ancient Egypt and Sudan. As a photographer she contributed to several exhibitions in Germany and Switzerland. As CEO of Ciudad Condal GmbH she leads an aspiring company focused on import and distribution of high-end sustainable products in Switzerland.

**Mersida Ndrevataj** is an urban scholar with a background in architecture, urban planning, and environmental psychology. She holds a PhD in Urban Planning and Public Policy from IUAV University of Venice, where her research focused on the implementation of public policies in cultural heritage management. She is currently a lecturer at IUAV University of Venice and collaborates with several COST Action research networks focused on urban studies. Her work brings together academic research and field-based inquiry by adopting audiovisual and sensory approaches to investigate urban spaces, everyday socio-spatial practices, and lived experiences.

The manufacturer's authorised representative in the EU is Springer Nature Customer Service Centre GmbH, Europaplatz 3, 69115 Heidelberg, Germany. If you have any concerns regarding our products, please contact ProductSafety@springernature.com

Printed and bound by CPI Group (UK) Ltd, Croydon, CR0 4YY
07/10/2026
02229357-0009